D1091446

Universitext

Editors

J. Ewing
P.R. Halmos
F.W. Gehring

Universitext

Editors: J. Ewing, F.W. Gehring, and P.R. Halmos

Booss/Bleecker: Topology and Analysis
Charlap: Bieberbach Groups and Flat Manifolds
Chern: Complex Manifolds Without Potential Theory
Chorin/Marsden: A Mathematical Introduction to Fluid Mechanics
Cohn: A Classical Invitation to Algebraic Numbers and Class Fields
Curtis: Matrix Groups, 2nd ed.
van Dalen: Logic and Structure
Devlin: Fundamentals of Contemporary Set Theory
Edwards: A Formal Background to Mathematics I a/b
Edwards: A Formal Background to Mathematics II a/b
Endler: Valuation Theory
Frauenthal: Mathematical Modeling in Epidemiology
Gardiner: A First Course in Group Theory
Godbillon: Dynamical Systems on Surfaces
Greub: Multilinear Algebra
Hermes: Introduction to Mathematical Logic
Humi/Miller: Second Course in Ordinary Differential Equations
for Scientists and Engineers
Hurwitz/Kritikos: Lectures on Number Theory
Kelly/Matthews: The Non-Euclidean, The Hyperbolic Plane
Kostrikin: Introduction to Algebra
Luecking/Rubel: Complex Analysis: A Functional Analysis Approach
Lu: Singularity Theory and an Introduction to Catastrophe Theory
Marcus: Number Fields
McCarthy: Introduction to Arithmetical Functions
Mines/Richman/Ruitenburg: A Course in Constructive Algebra
Meyer: Essential Mathematics for Applied Fields
Moise: Introductory Problem Course in Analysis and Topology
Øksendal: Stochastic Differential Equations
Rees: Notes on Geometry
Reisel: Elementary Theory of Metric Spaces
Rey: Introduction to Robust and Quasi-Robust Statistical Methods
Rickart: Natural Function Algebras
Smith: Power Series From a Computational Point of View
Smoryński: Self-Reference and Modal Logic
Stanisić: The Mathematical Theory of Turbulence
Stroock: An Introduction to the Theory of Large Deviations
Sunder: An Invitation to von Neumann Algebras
Tolle: Optimization Methods

Mayer Humi
William Miller

Second Course in Ordinary Differential Equations for Scientists and Engineers

Springer-Verlag
New York Berlin Heidelberg London Paris Tokyo

Mayer Humi
Department of Mathematical Sciences
Worcester Polytechnic Institute
Worcester, MA 01609
USA

William Miller
Department of Mathematical Sciences
Worcester Polytechnic Institute
Worcester, MA 01609
USA

Library of Congress Cataloging-in-Publication Data
Humi, Mayer.
Second course in ordinary differential equations for scientists and engineers.
(Universitext)
1. Differential equations. I. Miller, William,
II. Title.
QA372.H875 1988 515.3′52 87-32732

© 1988 by Springer-Verlag New York Inc.
All rights reserved. This work may not be translated or copied in whole or in part without the written permission of the publisher (Springer-Verlag, 175 Fifth Avenue, New York, NY 10010, USA), except for brief excerpts in connection with reviews or scholarly analysis. Use in connection with any form of information storage and retrieval, electronic adaptation, computer software, or by similar or dissimilar methodology now known or hereafter developed is forbidden.
The use of general descriptive names, trade names, trademarks, etc. in this publication, even if the former are not especially identified, is not to be taken as a sign that such names, as understood by the Trade Marks and Merchandise Marks Act, may accordingly be used freely by anyone.

Camera-ready copy provided by the authors.
Printed and bound by R.R. Donnelley & Sons, Harrisonburg, Virginia.
Printed in the United States of America.

9 8 7 6 5 4 3 2 1

ISBN 0-387-96676-5 Springer-Verlag New York Berlin Heidelberg
ISBN 3-540-96676-5 Springer-Verlag Berlin Heidelberg New York

515.352
H882

89-1695

To Shulamit

To Gerry

10/6/89 _ MW

PREFACE

The world abounds with introductory texts on ordinary differential equations and rightly so in view of the large number of students taking a course in this subject. However, for some time now there is a growing need for a junior-senior level book on the more advanced topics of differential equations. In fact the number of engineering and science students requiring a second course in these topics has been increasing. This book is an outgrowth of such courses taught by us in the last ten years at Worcester Polytechnic Institute.

The book attempts to blend mathematical theory with nontrivial applications from various disciplines. It does not contain lengthy proofs of mathematical theorems as this would be inappropriate for its intended audience. Nevertheless, in each case we motivated these theorems and their practical use through examples and in some cases an "intuitive proof" is included. In view of this approach the book could be used also by aspiring mathematicians who wish to obtain an overview of the more advanced aspects of differential equations and an insight into some of its applications. We have included a wide range of topics in order to afford the instructor the flexibility in designing such a course according to the needs of the students. Therefore, this book contains more than enough material for a one semester course.

The text begins with a review chapter (Chapter 0) which may be omitted if the students have just recently completed an introductory course. Chapter 1 sets the stage for the study of Boundary Value Problems, which are not normally included in elementary texts. Examples of important boundary value problems are covered in Chapter 2. Although systems of ordinary differential equations are contained in beginning texts, a more detailed approach to such problems is discussed in Chapter 3. Chapters 4 through 10 examine specific applications of differential equations such as Perturbation Theory, Stability, Bifurcations. (See Table of Contents.)

A word about the numbering system used in this book. The sections of each chapter are numbered 1,2,3... Occasionally there are sub-sections numbered 1.1, 1.2,... Definitions, Theorems, Lemmas, Corollaries, Examples start at 1 in each section. Formulas are tied to sections e.g. (2.1), (2.2) are in section 2. Exercises start at 1 in each chapter.

Finally, special thanks are due to Mrs. C. M. Lewis who typed the manuscript. Her efforts went well beyond the call of duty and she spared no time or effort to produce an accurate and esthetically appealing "camera ready" copy of the book. Thanks are also due to the staff at Springer Verlag for their encouragement and guidance.

CONTENTS

CHAPTER 0. REVIEW

1. SOLUTION OF SECOND ORDER ORDINARY DIFFERENTIAL EQUATIONS BY SERIES

1.1 INTRODUCTION.

Probably one of the least understood topics in a beginning ordinary differential equations course is finding a series solution of a given equation. Unfortunately this lack of knowledge hinders a student's understanding of such important functions as Bessel's function, Legendre polynomials and other such functions which arise in modern engineering problems.

In this chapter we shall undertake a general review of the ideas behind solving ordinary differential equations by the use of infinite series. We shall first examine Taylor series solutions which can be expanded about ordinary points. If the point about which we wish to find the expansion is singular (i.e. not ordinary), we go on to investigate the types of singular points and finally discuss how, under certain conditions, a Frobenius series applies to them. Although our discussion will be limited to second order differenital equations, these methods can be applied to higher order equations.

1.2 ORDINARY AND SINGULAR POINTS.

We recall that a second order linear ordinary differential equation can be written in the form

$$a_0(x)y'' + a_1(x)y' + a_2(x)y = f(x)$$

where $a_0(x)$, $a_1(x)$, $a_2(x)$, $f(x)$ are assumed to be continuous real-valued functions on an interval I.

If $f(x) = 0$, the equation is said to be homogeneous and it is to the solution of such types of equations we shall devote our attention, that is, in what follows we shall look at methods used to solve the equation

$$a_0(x)y'' + a_1(x)y' + a_2(x)y = 0 \quad .$$

The first step in solving an equation by use of series is to determine whether the point $x = x_0$ about which we wish to expand the series is ordinary or singular. This is done by dividing both sides of the equation by $a_0(x)$ so that our equation looks like

$$y'' + \frac{a_1(x)}{a_0(x)} y' + \frac{a_2(x)}{a_0(x)} y = 0$$

If $\dfrac{a_1(x)}{a_0(x)}$ and $\dfrac{a_2(x)}{a_0(x)}$ can be expanded in a Taylor series about $x = x_0$, then $x = x_0$ called an *ordinary point*. If this is not the case then $x = x_0$ is called a *singular* point.

In many situations $\dfrac{a_1(x)}{a_0(x)}$ and $\dfrac{a_2(x)}{a_0(x)}$ are rational functions and it is known that such a function has a Taylor series expansion about all points $x \in I$ except where the denominator vanishes.

Example 1: If every coefficient of the differential equation is a constant, then every value of x is an ordinary point.

Example 2: What are the ordinary points of

$$x^2(x^2-1)y'' + (x+1)y' + xy = 0.$$

First we rewrite the equation in normal form

$$y'' + \frac{1}{x^2(x-1)}\,y' + \frac{1}{x(x^2-1)}\,y = 0.$$

Since the denominators of coefficients vanish at $x = 0, +1, -1$ these are singular points. All other values of x are ordinary points.

Example 3: Examine

$$xy'' + \sin xy' + x^2y = 0$$

for singular and ordinary points.

Rewriting the equation we have

$$y'' + \frac{\sin x}{x}\,y' + xy = 0.$$

The Taylor series expansion of the y'-coefficient is

$$\frac{\sin x}{x} = 1 - \frac{x^2}{3!} + \frac{x^4}{5!} - \ldots = \sum_{n=0}^{\infty} (-1)^n \frac{x^{2n}}{(2n+1)!}.$$

We can use the ratio test to easily show $\frac{\sin x}{x}$ converges for all x. Therefore, every point is an ordinary point.

More will be said concerning singular points in Section 2.

EXERCISE 1

1. List all ordinary and singular points

   (a) $y'' + 2y' + y = 0$

   (b) $y'' - xy = 0$

   (c) $xy'' - y = 0$

   (d) $x(x-1)y'' + (x+1)y' + (x+2)y = 0$

   (e) $(x^2+x-6)y'' + (x^2+x+1)y' + (x-1)y = 0$

2. List all ordinary and singular points

   (a) $y'' + \sin xy' + e^x y = 0$

   (b) $3\sqrt{x+1}\;y'' - xy' + 3\sqrt{x-1}\;y = 0$

(c) $\sin xy'' + \cos xy' + xy = 0,\ x \in [-2\pi, 2\pi]$

(d) $y'' + \sum_{n=0}^{\infty} \frac{x^n}{n!} y' + \sum_{n=0}^{\infty} \frac{x^{2n}}{(2n)!} y = 0$

1.3 TAYLOR SERIES SOLUTIONS.

Once we have ascertained that $x = x_0$ is an ordinary point then we know we can find a solution in a Taylor series expanded about $x = x_0$, that is there is a solution of the form

$$y = \sum_{n=0}^{\infty} c_n (x-x_0)^n . \tag{1.1}$$

It is our job to find the values of the coefficients c_n which give us a solution. This is done by substituting the series (1.1) into the given differential equation. The method can be best explained by example.

Example 4: Find the Taylor series solution of

$$y'' + xy' + 2y = 0 \tag{1.2}$$

about $x = 0$.

Step 1: Obviously $x = 0$ is an ordinary point so we assume a solution of the form

$$y = \sum_{n=0}^{\infty} c_n x^n . \tag{1.3}$$

Differentiating twice with respect to x we find

$$y' = \sum_{n=0}^{\infty} n c_n x^{n-1} \tag{1.4}$$

and

$$y'' = \sum_{n=0}^{\infty} n(n-1) c_n x^{n-2} . \tag{1.5}$$

Notice that the first term in the series y' and the first two terms in the series y'' are zero, therefore we can rewrite (1.4) and (1.5) in the form

$$y' = \sum_{n=1}^{\infty} nc_n x^{n-1} \tag{1.6}$$

$$y'' = \sum_{n=2}^{\infty} n(n-1)c_n x^{n-2} \quad . \tag{1.7}$$

Substituting (1.3), (1.6), and (1.7) into equation (1.2) we have

$$\sum_{n=2}^{\infty} n(n-1)c_n x^{n-2} + x \sum_{n=1}^{\infty} nc_n x^{n-1} + 2 \sum_{n=0}^{\infty} c_n x^n = 0 \tag{1.8}$$

or

$$\sum_{n=2}^{\infty} n(n-1)c_n x^{n-2} + \sum_{n=1}^{\infty} nc_n x^n + \sum_{n=0}^{\infty} 2c_n x^n = 0.$$

Step 2: Reindex individual series so that all powers of x are the same. Since two series are already of the form x^n we shall change the first series on the lefthand side. Let n be replaced by $n + 2$, i.e.

$$\sum_{n=2}^{\infty} n(n-1)c_n x^{n-2} = \sum_{n+2=2}^{\infty} (n+2)(n+1)c_{n+2}x^n$$

$$= \sum_{n=0}^{\infty} (n+2)(n+1)c_{n+2}x^n \quad .$$

Using this result in (1.8) we obtain the equation

$$\sum_{n=0}^{\infty} (n+2)(n+1)c_{n+2}x^n + \sum_{n=1}^{\infty} nc_n x^n + \sum_{n=0}^{\infty} 2c_n x^n = 0. \tag{1.9}$$

Step 3: We notice now that although each series contains the term x^n, they do not start at the same time. The series start at $n = 0, 1, 0$ repectively. Our goal is to combine all three series into one series but since some series contain terms not present in others we must make the following adjustment. Bring out from under the Σ sign those terms in a series which do not appear in all of them. That is, we write (1.9) in the form

$$2c_2 + \sum_{n=1}^{\infty} (n+2)(n+1)c_{n+2}x^n$$
$$+ \sum_{n=1}^{\infty} nc_n x^n + 2c_0 + \sum_{n=1}^{\infty} 2c_n x^n = 0. \tag{1.10}$$

Now since each series starts at the same point (i.e. $n = 1$) and has the same power (i.e. x^n), we can combine them into one series. Therefore, we can write (1.10) as

$$2c_2 + 2c_0 + \sum_{n=1}^{\infty} \{(n+2)(n+1)c_{n+2} + (n+2)c_n\}x^n = 0.$$

Step 4: Now if a series converges over an interval to the value zero, it is necessary that every coefficient of x^n be zero. Applying this requirement to our problem we have

$$2c_2 + 2c_0 = 0 \quad \text{or} \quad c_2 = -c_0 \tag{1.11}$$

and

$$(n+2)(n+1)c_{n+2} + (n+2)c_n = 0, \; n \geq 1 .$$

It is convenient to solve the second equation above for the higher coefficient in terms of the lower one. Thus

$$c_{n+2} = -\frac{c_n}{n+1}, \; n \geq 1 .$$

This equation is known as the *recurrence relation* and it is by using this formula repeatedly that we get the desired coefficients. (Notice that the *recurrence relation* yields equation (1.11) for $n = 0$ but this is not always true!).

Step 5: To find the general solution we must find two linearly independent solutions from expression (1.3). Usually this is done by stating that the lowest arbitrary coefficient is not zero while the next larger arbitrary coefficient is zero. A second linearly independent solution can be found by interchanging the above conditions. With this in mind let $c_0 \neq 0$, $c_1 = 0$, then

$$c_2 = -c_0$$

$$c_3 = -\frac{c_1}{2} = 0$$

$$c_4 = -\frac{c_2}{3} = \frac{c_0}{3}$$

$$c_5 = 0$$

$$c_6 = -\frac{c_4}{5} = -\frac{c_0}{3\cdot 5}$$

which gives us one solution

$$\begin{aligned} y_1(x) &= \sum_{n=0}^{\infty} c_n x^n = c_0 + c_1 x + c_2 x^2 + \ldots \\ &= c_0 - c_0 x^2 + \frac{c_0}{3} x^4 - \frac{c_0}{3\cdot 5} x^6 \\ &= c_0\left(1 - x^2 + \frac{x^4}{3} - \frac{x^6}{3\cdot 5} + \ldots\right) \quad . \end{aligned}$$

To find the second solution let $c_0 = 0$, $c_1 \neq 0$. Then

$$c_0 = 0$$

$$c_1 \neq 0$$

$$c_2 = -c_0 = 0$$

$$c_3 = -\frac{c_1}{2}$$

$$c_4 = 0$$

$$c_5 = -\frac{c_3}{4} = \frac{c_1}{2\cdot 4}$$

$$c_6 = 0$$

$$c_7 = \frac{c_5}{6} = -\frac{c_1}{2\cdot 4\cdot 6} \quad ,$$

which yields a second linearly independent solution

$$\begin{aligned} y_2(x) &= c_1 x - \frac{c_1}{2} x^3 + \frac{c_1}{2\cdot 4} x^5 \\ &= c_1\left(x - \frac{x^3}{2} + \frac{x^5}{2\cdot 4} - \frac{x^7}{2\cdot 4\cdot 6} + \ldots\right) \quad . \end{aligned}$$

The general solution of equation (1.2) is

$$\begin{aligned} y(x) = {} & c_0\left(1 - x^2 + \frac{x^4}{3} - \frac{x^6}{3\cdot 5} + \ldots\right) \\ & + c_1\left(x - \frac{x^3}{2} + \frac{x^5}{2\cdot 4} - \frac{x^7}{2\cdot 4\cdot 6} + \ldots\right) \quad . \end{aligned} \tag{1.12}$$

Comments: 1. One can easily see that the two solutions are linearly independent because there is no constant k such that

$$y_2(x) = ky_1(x) \quad .$$

2. Although a solution of the form (1.12) is adequate for some purposes, it is often more desirable to find a formula which generates the individual terms. Unfortunately this is an art which can only be learned by practicing on many examples. However, the example above is not too difficult and we can write the answer as follows:

$$y(x) = c_0 \sum_{n=0}^{\infty} (-1)^n \frac{2^n n!}{(2n)!} x^{2n} + c_1 \sum_{n=0}^{\infty} \frac{(-1)^n}{2^n n!} x^{2n+1} .$$

3. To find the interval over which solutions are valid we use the ratio test. Let us look at the series $y_1(x)$. Applying the test we have

$$\lim_{n\to\infty} \left| \frac{(-1)^{n+1} \dfrac{2^{n+1}(n+1)!}{(2(n+1))!} x^{2(n+1)}}{(-1)^n \dfrac{2^n n!}{(2n)!} x^{2n}} \right|$$

$$= \lim_{n\to\infty} \frac{2(n+1)}{(2n+1)(2n+2)} |x|^2 = \lim_{n\to\infty} \frac{1}{2n+1} |x|^2$$

$$= 0|x|^2 < 1.$$

Therefore, series converges for all x. In a similar way we can show $y_2(x)$ converges for all x. Thus, the solution $y(x)$ is valid everywhere.

Example 5: As an example with a slightly different twist let us solve Airy's equation

$$y'' + xy = 0 \quad .$$

Assuming the solution is of the form

$$y = \sum_{n=0}^{\infty} c_n x^n$$

we find using equation (1.3) and (1.7) that

$$\sum_{n=2}^{\infty} (n-1)nc_n x^{n-2} + x \sum_{n=0}^{\infty} c_n x^n = 0 \quad .$$

Reindexing and rearranging we are lead to the equation

$$2c_2 + \sum_{n=0}^{\infty} \left\{(n+2)(n+3)c_{n+3} + c_n\right\}x^{n+1} = 0$$

from which we see that $c_2 = 0$ over which we have no control and the *recurrence relation* is given by

$$c_{n+3} = - \frac{c_n}{(n+2)(n+3)} \quad .$$

By selecting $c_0 \neq 0$, $c_1 = 0$ and $c_0 = 0$, $c_1 \neq 0$ and recalling $c_2 = 0$ we are led to the general solution

$$y(x) = c_0(1 - \frac{x^3}{2\cdot 3} + \frac{x^6}{2\cdot 3\cdot 5\cdot 6} - \ldots)$$
$$+ c_1(x - \frac{x^4}{3\cdot 4} + \frac{x^7}{3\cdot 4\cdot 6\cdot 7} - \ldots) \quad .$$

Notice the terms $x^2, x^5, \ldots, x^{2+3n}$ are all missing because $c_2 = 0$.

EXERCISE 2

1. Find the series general solutions in the form

$$y = \sum_{n=0}^{\infty} c_n x^n \quad .$$

(a) $y'' + 4y = 0$

(b) $y'' + 2y' + 4y = 0$

(c) $2y'' + 3y' = 0$

(d) $y'' + xy = 0$

(e) $y'' + xy' + 2y = 0$

(f) $y''' + xy = 0$

(g) $y''' + x^2y' = 0$

2. Find the series general solutions in the form

$$\sum_{n=0}^{\infty} c_n (x-a)^n$$

(a) $y'' + (x-1)^2y = 0$, $a = 1$

(b) $y'' + xy' + y = 0, \quad a = 2$

Hint: Make sure coefficients are in powers of $(x-2)$.

2. REGULAR SINGULAR POINTS.

In paragraph 1.2 we discussed the distinction between ordinary points and singular points. When we attempt to solve a differential equation about a singular point it is necessary to distinguish between two types of singular points - regular and irregular.

Suppose $x = x_0$ is a singular point for the differential equation

$$y'' + \frac{a_1(x)}{a_0(x)}\, y' + \frac{a_2(x)}{a_0(x)}\, y = 0 \quad .$$

If

$$\lim_{x \to x_0} (x-x_0)\, \frac{a_1(x)}{a_0(x)}$$

and

$$\lim_{x \to x_0} (x-x_0)^2\, \frac{a_2(x)}{a_0(x)}$$

exist, then $x = x_0$ is said to be a *regular singular* point. If a singular point is *not* regular then it is *irregular*.

Example 1: Discuss the singular points of the differential equation

$$x(x-1)^3\, y'' + 2(x-1)^2\, y' + x^3 y = 0 \quad .$$

First rewrite equation in form

$$y'' + \frac{2}{x(x-1)}\, y' + \frac{x^2}{(x-1)^3}\, y = 0 \quad .$$

We see immediately that $x = 0, x = 1$ are singular points.

Now for $x = 0$ we write

$$\lim_{x\to 0} x \frac{2}{x(x-1)} = -2$$

$$\lim_{x\to 0} x^2 \frac{x^2}{(x-1)^3} = 0 \quad .$$

Since both limits exist $x = 0$ is a regular singular point. For $x = 1$

$$\lim_{x\to 1} (x-1) \frac{2}{x(x-1)} = 2$$

$$\lim_{x\to 1} (x-1)^2 \frac{x^3}{(x-1)^3} = \pm \infty \quad .$$

Since only one limit exists, $x = 1$ is an irregular singular point.

EXERCISE 3

1. Find all singular and ordinary points for differential equations listed below. Are singular points regular or irregular?

   (a) $(x-2)(x+1)y'' + (x-2)xy' + 2y = 0$

   (b) $x^3(x-1)y'' + xy' + (x-1)y = 0$

   (c) $x(x-2)(x+4)y'' + (x-2)^2y' + x(x-2)y = 0$

   (d) $x^2y'' + xy' + y = 0$

   (e) $(x+5)^4(x+2)^3y'' + (x+5)(x+2)y' + y = 0$

   (f) $x^3(x-1)^3y''' + x(x-1)y'' + xy' + (x-1)y = 0$

   Hint: Extend ideas in Section 2 naturally.

3. SERIES SOLUTIONS NEAR A REGULAR SINGULAR POINT

If we are searching for a series solution about a regular singular point $x = x_0$ a Taylor series generally will not work and it is necessary to use a more generalized series known as a Frobenius series.

This series solution takes on the form

$$y = (x-x_0)^r \sum_{n=0}^{\infty} c_n(x-x_0)^n$$

where r is a number real or complex which will be determined from the given differential equation.

We shall demonstrate Frobenius method by an example.

Example 1: Find the general solution of

$$3x^2y'' + xy' + (x-1)y = 0 \quad . \tag{3.1}$$

about $x = 0$.

Step 1: Show $x = 0$ is a regular singular point

Step 2: Assume solution of the form

$$y = x^r \sum_{n=0}^{\infty} c_n x^n = \sum_{n=0}^{\infty} c_n x^{n+r} \quad .$$

Computing the first and second derivatives of this series we find

$$y' = \sum_{n=0}^{\infty} (n+r)c_n x^{n+r-1}$$

$$y'' = \sum_{n=0}^{\infty} (n+r)(n+r-1)c_n x^{n+r-2} \quad .$$

Step 3: Substitute the series in Step 2 into (3.1) which yields

$$3x^2 \sum_{n=0}^{\infty} (n+r)(n+r-1)c_n x^{n+r-2} + x \sum_{n=0}^{\infty} (n+r)c_n x^{n+r-1} + (x-1) \sum_{n=0}^{\infty} c_n x^{n+r} = 0$$

or

$$\sum_{n=0}^{\infty} 3(n+r)(n+r-1)c_n x^{n+r} + \sum_{n=0}^{\infty} (n+r)c_n x^{n+r} + \sum_{n=0}^{\infty} c_n x^{n+r+1} - \sum_{n=0}^{\infty} c_n x^{n+r} = 0 \quad . \tag{3.2}$$

Step 4: Reindex the third series on lefthand side so that power of x matches other series, that is

$$\sum_{n=0}^{\infty} c_n x^{n+r+1} = \sum_{n=1}^{\infty} c_{n-1} x^{n+r} \quad .$$

Step 5: Now (3.2) becomes

$$\sum_{n=0}^{\infty} 3(n+r)(n+r-1)c_n x^{n+r} + \sum_{n=0}^{\infty} (n+r)c_n x^{n+r}$$

$$+\sum_{n=1}^{\infty} c_{n-1}x^{n+r} - \sum_{n=0}^{\infty} c_n x^{n+r} = 0 \quad . \qquad (3.3)$$

Notice that although the powers of x in each series are now the same, the series no longer start at the same value of n. The first, second, and fourth series start at $n = 0$ while the third series begins at $n = 1$. Since our goal is to have the powers of x all the same in each series and to have them start at the same number n so that they can be combined under one Σ sign, we proceed as follows.

Step 6: In the series which starts with $n = 0$ remove the term associated with $n = 0$, and rewrite (3.3) as follows:

$$3r(r-1)c_0x^r + \sum_{n=1}^{\infty} 3(n+r)(n+r-1)c_nx^{n+r} + rc_0x^r$$
$$+ \sum_{n=1}^{\infty} (n+r)c_nx^{n+r} + \sum_{n=1}^{\infty} c_{n-1}x^{n+r} - c_0x^r$$
$$- \sum_{n=1}^{\infty} c_nx^{n+r} = 0$$

or

$$(3r(r-1)+r-1)c_0x^r + \sum_{n=1}^{\infty} \{[3(n+r)(n+r-1)$$
$$+ (n+r) - 1]c_n + c_{n-1}\}x^{n+r} = 0 \ . \qquad (3.4)$$

Step 7: If the series (3.4) is to hold for all x in some interval, then each coefficient of $x^{n+r}(n = 0,1,\ldots)$ must equal zero.

First we see that

$$(3r(r-1) + r-1)c_0 = 0 \quad .$$

Normally we choose $c_0 \neq 0$, and, therefore,

$$(3r(r-1) + r-1) = (3r^2-2r-1) = (3r+1)(r-1) = 0 \ .$$

This equation is called the *indicial equation* and the roots $r = 1$, $r = -\frac{1}{3}$ are called the *indices*.

From the theory of Frobenius series if the sum of the roots does not equal zero or an integer, each root will yield a solution and the solutions will be linear independent.

If the sum of the roots equals zero or an integer, the larger root will yield a solution directly whereas the other linearly independent solution can only be found by a rather laborious technique which can be found in a detailed study of Frobenius series.

Step 8: The coefficient under the Σ sign must also be zero which tells us that

$$[3(n+r)(n+r-1) + (n+r)-1]c_n + c_{n-1} = 0$$

or

$$(3(n+r)+1)(n+r-1)c_n + c_{n-1} = 0$$

or

$$c_n = -\frac{c_{n-1}}{(n+r-1)(3n+3r+1)} .$$

This equation is called the *recurrence relation* and is used to find values of c_n.

Step 9: Let $r = 1$ then

$$c_n = -\frac{c_{n-1}}{n(3n+4)} .$$

Therefore,

$$c_1 = -\frac{c_0}{1\cdot 7}$$

$$c_2 = -\frac{c_1}{2\cdot 10} = \frac{c_0}{1\cdot 2\cdot 7\cdot 10}$$

$$c_3 = -\frac{c_2}{3\cdot 13} = \frac{c_0}{1\cdot 2\cdot 3\cdot 7\cdot 10\cdot 13}$$

and one solution is given by

$$y_1(x) = c_0 x\left(1 - \frac{x}{1\cdot 7} + \frac{x^2}{1\cdot 2\cdot 7\cdot 10} - \frac{x^3}{1\cdot 2\cdot 3\cdot 7\cdot 10\cdot 13} + \ldots\right).$$

The second linearly independent solution is found by taking $r = -\frac{1}{3}$ from which

$$c_n = -\frac{c_{n-1}}{(3n-4)n} .$$

Thus, letting $c_0 = \bar{c}_0$ to differentiate from previous solution we find

$$c_1 = -\frac{\bar{c}_0}{-1\cdot 1} = \bar{c}_0$$

$$c_2 = -\frac{c_1}{2\cdot 2} = \frac{\bar{c}_0}{2\cdot 2}$$

$$c_3 = -\frac{c_2}{5\cdot 3} = -\frac{\bar{c}_0}{2\cdot 5\cdot 2\cdot 3} .$$

The linearly independent solution is

$$y_2(x) = \bar{c}_0 x^{-\frac{1}{3}}\left(1-x + \frac{x^2}{2\cdot 2} - \frac{x^3}{2\cdot 5\cdot 2\cdot 3} + \ldots\right) .$$

Combining solutions y_1 and y_2 the general solution is given by

$$y(x) = c_0 x\left(1 - \frac{x}{1\cdot 7} + \frac{x^2}{1\cdot 2\cdot 7\cdot 10} - \frac{x^3}{1\cdot 2\cdot 3\cdot 7\cdot 10\cdot 13} + \ldots\right)$$
$$+ \bar{c}_0\, x^{-\frac{1}{3}}\left(1-x + \frac{x^2}{2\cdot 2} - \frac{x^3}{2\cdot 5\cdot 2\cdot 3} + \ldots\right)$$

EXERCISE 4

1. Use Frobenius method to find the general solution of the given differential equations in the form $x^r \sum_{n=0}^{\infty} c_n x^n$.

   (a) $x^2y'' + 3xy' - 4y = 0$

   (b) $x^2y'' + xy' + x^2y = 0$ find one solution only.

   (c) $x^2y'' + xy' + (x^2 - \frac{1}{9})y = 0$

   (d) $2x^2y'' + 3xy' + xy = 0$

   (e) $6x^2y'' + (x+1)y' + y = 0$

   (f) $x^2y'' + y' - \frac{5}{16}\, y = 0$

2. Starting with the Frobenius series

   $$y(x) = (1-x)^r \sum_{n=0}^{\infty} c_n(1-x)^n$$

and substituting in Legendre's equation
$(1-x^2)y'' - 2xy' + 2y = 0$

(a) Show that indicial equation is $r^2 = 0$.

(b) Since the indices are equal, we can find only one solution of the Legendre equation using the method discussed in this section. Write out the *recurrence relation*.

(c) Show that one solution is $y = 1-(1-x)$ or $y = x$.

3. Bessel's modified equation of order α is given by
$x^2y'' + xy' - (x^2 + \alpha^2)y = 0$.

(a) If $\alpha = 1$, find the solution of the equation for the higher index.

(b) If $\alpha = \frac{1}{2}$, find the general solution.

4. Laguerre's differential equation is given by
$xy'' + (1-x)y' + ny = 0$.

(a) Find the general solution of the differential equation.

(b) Show that if n is an interger, one solution is given by $L_n = \sum_{k=0}^{n} \frac{(-1)^k n! x^k}{(k!)^2 (n-k)!}$
which is known as a Laguerre polynomial.

CHAPTER 1: BOUNDARY VALUE PROBLEMS

1. INTRODUCTION.

When a person begins the study of ordinary differential equations, he is usually confronted first by initial value problems, i.e. a differential equation plus conditions which the solution must satisfy at a given point $x = x_0$.

Example 1: Given the third order differential equation

$$y''' + 2y'' - y' - 2y = 0$$

with conditions

$$y(0) = 1$$
$$y'(0) = -3$$
$$y''(0) = 2 \quad .$$

Such a system is known as an *initial value problem.*

There is another problem which arises very often in solving problems coming from the physical world. Consider the nth order linear differential equation

$$L(y) = a_0(x)y^{(n)} + a_1(x)y^{(n-1)} + \dots + a_n(x)y = g(x) \qquad (1.1)$$

and the $i = m$ boundary conditions where $0 \leq m \leq 2n$,

$$\begin{aligned} U_i(y) &= \alpha_i y(a) + \alpha_i' y'(a) \\ &+ \dots + \alpha_i^{(n-1)} y^{(n-1)}(a) \\ &+ \beta_i y(b) + \beta_i' y'(b) \\ &+ \dots + \beta_i^{(n-1)} y^{(n-1)}(b) = \gamma_i \end{aligned} \qquad (1.2)$$

where $\alpha_i^{(k)}$, $\beta_i^{(k)}$, γ_i $(k = 1,\ldots,n-1)$ are constants. The conditions U_i must also be linearly independent. Since there are 2n unknowns $y(a),\ldots,y(b),\ldots,$ we know from linear algebra that in order to find a nontrivial solution, the maximum number of equations possible is 2n. The differential equation along with boundary conditions is known as an *nth order boundary value problem*.

Example 2: Given the fourth order linear differential equation

$$y^{(4)} - 16y = x$$

along with the conditions

$$U_1(y) = y(0) - y(L) = 1$$

$$U_2(y) = y'(0) = 3 \qquad m = 3$$

$$U_3(y) = y(L) + y'(L) = 0 \quad .$$

This system is a fourth order boundary value problem.

Associated with the nonhomogeneous boundary value problems is the *reduced homogeneous* system which is written

$$L(y) = 0$$

$$U(y) = 0 \qquad i = m \quad .$$

As we prepare to solve this system the following questions arise:

1. Are there any solutions at all to the related system? If the answer is no, then the system is said to be *incompatible*.
2. If there are solutions to the related system, how many linearly independent solutions are there? We know that an nth order differential equation can have at most n linearly independent solutions. Suppose there are k linearly independent solutions

where $k \leq n$, then the system is said to be *k-ply compatible*. We call

$$y(x) = c_1y_1(x) + \ldots + c_ky_k(x)$$

the general solution of the system and k is called the *index of compatibility*.

These terms can be applied to the nonhomogeneous system as well. If $y_p(x)$ is any solution of the nonhomogeneous system (1.1) and (1.2), and if the related system is k-ply compatible

$$y(x) = y_p(x) + c_1y_1(x) + \ldots + c_ky_k(x)$$

is called the *general solution* of the nonhomogeneous system.

If we have found the general[1] solution to the related homogeneous equation, we know that it can be written in the form

$$y(x) = c_1y_1(x) + \ldots + c_ny_n(x).$$

To see if the system is compatible, we substitute the solution $y(x)$ into the m boundary conditions $U_i(y) = 0$. Upon doing this we arrive at the m equations

$$\begin{aligned}
c_1U_1(y_1) + c_2U_1(y_2) + \ldots + c_nU_1(y_n) &= 0 \\
c_1U_2(y_1) + c_2U_2(y_2) + \ldots + c_nU_2(y_n) &= 0 \qquad (1.3) \\
&\vdots \\
c_1U_m(y_1) + c_2U_m(y_2) + \ldots + c_nU_m(y_n) &= 0 \quad .
\end{aligned}$$

[1]The words "general solution" have different meanings in different situations. As used here, we mean a linear combination of n linearly independent solutions of $L(y) = 0$.

$$\text{Let} \quad \mathbf{U} = \begin{bmatrix} U_1(y_1)U_1(y_2) & \cdots & U_1(y_n) \\ U_2(y_1)U_2(y_2) & \cdots & U_2(y_n) \\ U_m(y_1)U_m(y_2) & \cdots & U_m(y_n) \end{bmatrix}$$

which is an $m\times n$ matrix.

We know from linear algebra that if the rank if **U** is r, then the maximum number of linearly independent sets of c_i's is $n - r$.

We can summarize these ideas in the following theorem.

Theorem 1: Consider the homogeneous system

$$L(y) = 0$$

$$U_i(y) = 0 \qquad 0 \leq i = m \leq 2n \quad .$$

If the rank of the matrix **U** is $n - k$, then the system is k-ply compatible. There is a similar theorem concerning a nonhomogeneous system.

Consider the nonhomogeneous system,

$$L(y) = g(x)$$

$$U_i(y) = \gamma_i \qquad .$$

If

$$y = y_p + c_1y_1 + \ldots + c_ny_n$$

is the general solution of the nonhomogeneous differential equation $L(y) = g(x)$ then the system will be compatible if the system of equations

$$\begin{array}{l} c_1U_1(y_1) + c_2U_1(y_2) + \ldots + c_nU_1(y_n) = \gamma_1 - U_1(y_p) \\ c_1U_2(y_1) + c_2U_2(y_2) + \ldots + c_nU_2(y_n) = \gamma_2 - U_2(y_p) \\ \quad \vdots \\ c_1U_m(y_1) + c_2U_m(y_2) + \ldots + c_nU_m(y_n) = \gamma_m - U_m(y_p) \end{array} \qquad (1.4)$$

has a solution.

Let $\mathbf{U}_\gamma$ represent the $m \times (n+1)$ matrix

$$\mathbf{U}_\gamma = \begin{bmatrix} U_1(y_1)U_1(y_2) & \cdots & U_1(y_n)\gamma_1 - U_1(y_p) \\ U_2(y_1)U_2(y_2) & \cdots & U_2(y_n)\gamma_2 - U_2(y_p) \\ \cdot & \cdot & \cdot \\ \cdot & \cdot & \cdot \\ \cdot & \cdot & \cdot \\ U_m(y_1)U_m(y_2) & \cdots & U_m(y_n)\gamma_m - U_m(y_p) \end{bmatrix} \qquad (1.5)$$

.

Using certain theorems from matrix theory, we can state the following theorem:

Theorem 2: Consider the nonhomogeneous system

$$L(y) = g(x)$$
$$U_i(y) = \gamma_i \qquad 0 \leq i = m \leq 2n .$$

If the rank of the matrix $\mathbf{U}$ and the augmented matrix $\mathbf{U}_\gamma$ have the same rank $n - k$, then there are solutions of the differential system and the system is k-ply compatible.

Example 3: Find the index of compatibility and solutions of the homogeneous system

$$y^{(iv)} + y'' = 0$$
$$U_1(y) = y(0) + y(\pi) = 0$$
$$U_2(y) = y'(\pi) = 0 .$$

The general solution of the differential equation is

$$y(x) = c_1 + c_2x + c_3\sin x + c_4\cos x.$$

If we let $y_1 = 1$, $y_2 = x$, $y_3 = \sin x$, $y_4 = \cos x$, we see that

$$U_1(y_1) = 2,\ U_1(y_2) = \pi,\ U_1(y_2) = 0,\ U_1(y_4) = 0$$
$$U_2(y_1) = 0,\ U_2(y_2) = 1,\ U_2(y_3) = -1,\ U_2(y_4) = 0 .$$

Substituting these values into the matrix $\mathbf{U}$ we find

$$\mathbf{U} = \begin{bmatrix} 2 & \pi & 0 & 0 \\ 0 & 1 & -1 & 0 \end{bmatrix} .$$

Using Gauss' reduction one can easily show that the rank of **U** is 2. Applying these results to Theorem 1 we see that $n = 4$, $n - k = 2$ and, therefore, that the system is 2-ply compatible.

In order to find two linearly independent solutions we return to the equations given by (1.3) and find that

$$\begin{aligned} 2c_1 + \pi c_2 \quad &= 0 \\ c_2 - c_3 &= 0 \end{aligned} .$$

Upon solving this system, we discover

$$\begin{aligned} c_1 &= -\frac{\pi}{2} c_3 \\ c_2 &= c_3 \end{aligned} .$$

To find one solution let $c_3 = 1$, $c_4 = 0$. Using these numbers we arrive at the solution

$$y_A(x) = -\frac{\pi}{2} + x + \sin x .$$

Another linearly independent solution can be found by allowing $c_3 = 0$ and $c_4 = 1$ from which it follows that

$$y_B(x) = \cos x .$$

The solutions y_A and y_B form a linearly independent family of solutions. Any other solution of the system will be a linear combination of y_A and y_B.

Example 4: (Note this example is connected to Example 3.)

Find if there are any solutions to the nonhomogeneous system

$$y^{(iv)} + y'' = 2e^x$$
$$U_1(y) = y(0) + y(\pi) = 1$$
$$U_2(y) = y'(\pi) \qquad = -1 \quad .$$

If there are solutions, compute the index of compatibility and write out a linearly independent family of solutions.

Using the method of undetermined coefficients it is easy to show that $y_p = e^x$. The general solution to the nonhomogeneous differential equation is

$$y(x) = e^x + c_1 + c_2 x + c_3 \sin x + c_4 \cos x.$$

Using information from the previous examples we find that (1.5) becomes

$$\mathbf{U}_\gamma = \begin{bmatrix} 2 & \pi & 0 & 0 & -e^\pi \\ \\ 0 & 1 & -1 & 0 & -1-e^\pi \end{bmatrix} .$$

Notice $-e^\pi = \gamma_1 - U_1(y_p)$ and $-1-e^\pi = \gamma_2 - U_2(y_p)$.

Once again using Gauss reduction we find that the augmented matrix $\mathbf{U}_\gamma$ has a rank of 2. Since the rank of $\mathbf{U}$ (Example 3) equals the rank of $\mathbf{U}_\gamma$, we know from Theorem 10.2 that there are solutions to the system. Since the ranks of both matrices are 2, we see that the system is 2-ply compatible.

If we return to the equation (1.4) we arrive at the system of equations

$$2c_1 + \pi c_2 \qquad = -e^\pi$$
$$c_2 - c_3 \qquad = -1-e^\pi \quad .$$

Solving this system we have

$$c_1 = (\pi-1)\frac{e^\pi}{2} + \frac{\pi}{2} - \frac{\pi}{2} c_3$$
$$c_2 = c_3 - 1 - e^\pi \quad .$$

If we choose $c_3 = 1$, $c_4 = 0$ we arrive at the solution

$$y_A(k) = e^x + (\pi-1)\frac{e^\pi}{2} - e^\pi x + \sin x,$$

whereas if we let $c_3 = 0$, $c_4 = 1$, we find another linearly independent solution

$$y_B(x) = e^x + (\pi-1)\frac{e^\pi}{2} + \frac{\pi}{2} - (1 + e^\pi)x + \cos x.$$

Thus, y_A and y_B form a linear independent family of solutions of the nonhomogeneous system.

EXERCISE 1

1. Given the nonhomogeneous system shown below, write the related homogeneous system

   (a) $x^2y'' + xy' + y = x^2$
   $y(0) + y'(0) = 2$
   $y(0) - 3y(5) = -1$
   $y'(0) = 4$

   (b) $y^{(iv)} + 6y'' + 9y = \sin x + \cos x$
   $y(a) - 2y'(a) + 3y(b) = 3$
   $y'''(b) = 1$
   $y''(a) - y'''(a) - 2y(b) - 3y'(b) = 4$

2. What is largest number of boundary conditions permitted if the order of the given differential equation is 5?

3. Find the index of compatibility and the general solution of

   (a) $y'' + 4y = 0;\quad y(0) + y'\left[\frac{\pi}{2}\right] = 0$

   (b) $y'' + 4y = 0;\quad y(0) + y'\left[\frac{\pi}{2}\right] = 0,\ y\left[\frac{\pi}{2}\right] = 0$

   (c) $y'' - 9y = 0;\quad y(0) + y(1) = 0,$
   $y'(0) - y'(1) = 0$

(d) $y''' - y'' + y' - y = 0$;

$y(0) = 0,\ y(0) + y''(\pi) = 0,\ y''(\pi) = 0$

(e) $y^{(iv)} - y''' = 0$; $y(0) + y(1) = 0$

(f) $y^{(iv)} = 0$; $y(0) - y'(1) = 0,\ y''(0) = 0$

$y'(0) + y''(1) = 0,\ y'''(0) - y'''(1) = 0$

(g) $x^2y'' + xy' + y = 0,\ y(1) + y'(1) = 0$.

4. Find thye index of compatibility and the general solution of

(a) $y' = 1$; $y(0) = 0,\ y(2) = 3$

(b) $y'' + 16y = x$; $y(0) + y'(\pi) = 0,\ y'(0) = 2$

(c) $y'' + 9y = \sin x$; $y(0) + y'(0) = 1$

(d) $y'' - 2y' - 3y = e^x$; $y(0) = 1,\ y'(0) = 0,$ $y(1) + y'(1) = 2.$

2. ADJOINT DIFFERENTIAL EQUATIONS AND BOUNDARY CONDITIONS

Let L be the nth order differential operator of the form

$$L = a_0(x)\frac{d^n}{dx^n} + a(x)\frac{d^{n-1}}{dx^n} + \ldots. + a_{n-1}(x)\frac{d}{dx} + a_n.$$

where $a_0, \ldots, a_n$ are continuous and $a_0 \neq 0$ on $[a,b]$. In particular if L operates on a function u we can write

$$Lu = a_0(x)u^{(n)}(x) + a_1(x)u^{(n-1)}(x) + \ldots. + a_{n-1}(x)u'(x) + a_nu.$$

The differential operator L^* whose form is given by

$$L^* = (-1)^n\frac{d^n}{dx^n}(a_0(x)) + (-1)^{n-1}\frac{d^{n-1}}{dx^{n-1}}(a_1(x)) + \ldots - \frac{d}{dx}(a_{n-1}(x)) + a_n \tag{2.1}$$

is called the *adjoint* of the differential operator L. If L^* operates on the function v we write

$$\begin{aligned} L^{*}v = {} & (-1)^{n}(a_0(x)v(x))^{(n)} \\ & + (-1)^{n-1}(a_1(x)v(x))^{(n-1)} \\ & + \dots -(a_{n-1}(x)v(x)) + a_n(x)v(x). \end{aligned}$$

Example 1: If $Ly = r(x)y'' + q(x)y' + p(x)y$ then $L^{*}z = (r(x)z)'' - (q(x)z)' + p(x)z$.

If $L^{*} = L$ then the operators are said to be *formally self-adjoint*.

Example 2: Let $Ly = y'' + 4y$, then $L^{*}z = (-1)^{2}z'' + 4z = z'' + 4z$ thus $L^{*} = L$ are self-adjoint.

Earlier in paragraph we discussed the adjoint form of an ordinary linear differential equation. The idea can be carried further to apply not only to the differential equation but also the boundary conditions.

Consider the system

$L(u) = 0$, where L is nth order operator and

$U_i(u) = 0, \quad 0 \le i \le 2n$,

then $\displaystyle\int_a^b (vL(u) - uL^{*}(v))dx = P(u,v)\Big|_a^b$

where

$$P(u,v) = \begin{cases} u[a_{n-1}v - (a_{n-2}v)' +' \dots +(-1)^{n-1}(a_0v)^{(n-1)}] \\ + u'[a_{n-2}v - (a_{n-3}v) + \dots +(-1)^{n-2}(a_0v)^{(n-2)}] \\ + \qquad \vdots \\ + u^{n-2}[a_1v - (a_0v)'] \\ + u^{n-1}[a_0v] \end{cases} .$$

$P(u,v)$ is known as the *bilinear concomitant*
To see, in general, how we arrive at such a result, let us look at the second order case.

If

$L(u) = a_0(x)u''(x) + a_1(x)u'(x) + a_2(x)u(x)$

then by multiplying $L(u)$ by a function $v(x)\epsilon C^2$ and integrating by parts over the interval $[a,b]$, we find that

$$\int_a^b vL(u)dx = u'(a_0v) + u[a_1v - (a_0v)]\Big|_a^b + \int_a^b u(a_0v)'' - (a_1v)' + a_2v]dx.$$

Combining the integrals on the lefthand side of the equation we see that

$$\int_a^b [vL(u) - uL^*(v)]dx = u'(a_0v) + u[a_1v - (a_0v)']\Big|_a^b = P(u,v)\Big|_a^b .$$

This equation is known as *Green's formula*. From Green's formula it is easy to derive the *Lagrange identity* which is given by

$$vL(u) - uL^*(v) = \frac{d}{dx} P(u,v) .$$

We now turn to the boundary conditions $U_i(x) = 0$, $0 \leq i = m \leq 2n$, where we assume that all m equations are linearly independent. If $m < 2n$ we enlarge the set $U_i(x) = 0$ to $2n$ linearly independent equations by attaching $2n - m$ essentially arbitrary equations. In more advanced texts it is shown that there exists a set $V_1, V_2, \ldots, V_{2n}$ such that

$$U_1V_{2n} + U_2V_{2n-1} + \ldots + U_{2n-1}V_2 + U_{2n}V_1 = P(u,v)\Big|_a^b \tag{2.2}$$

From (2.2) we can determine a set of boundary conditions to go along with the adjoint equation. These ideas are summarized in the next paragraph.

If the system

$L(u) = 0$ of order n, and

$U_i(u) = 0 \qquad 1 \leq i = m \leq 2n$,

then the *adjoint system* is given by

$L^*(v) = 0$ of order n, and

$V_i(v) = 0 \qquad 1 \leq i = 2n - m \leq 2n.$

Example 3:

Given the system

$$L(u) = u'' + 4u = 0$$
$$U_1 = u(0) + u'(L) = 0$$
$$U_2 = u(L) = 0 \qquad m = 3 \leq 2n = 4$$
$$U_3 = u'(0) = 0,$$

find the adjoint system.

First write out

$$P(u,v)\Big|_0^L = u'(L)v(L) + u(L)[4v(L) - v'(L)]$$
$$- u'(0)v(0) - u(0)[4v(0) - v'(0)].$$

Next pick any expression for U_4 such that the set $\{U_i(u),\ 0 \leq i \leq 4\}$ is linear independent. We shall choose $U_4 = u(0)$.

Substitute the expressions $U_1,\dots,U_4$ in the equation (2.2). It follows that

$$[u(0) + u'(L)]V_4 + u(L)V_3 + u'(0)V_2 + u(0)V_1$$
$$= P(u,v)\Big|_0^L = u'(L)v(L) + u(L)[4v(L) - v'(L)]$$
$$- u'(0)v(0) - u(0)[4v(0) - v'(0)].$$

This equation holds if the individual coefficients of $u(0)$, $u'(0)$, $u(L)$, $u'(L)$ are equal from which we find

$$\begin{aligned} V_1 + V_4 &= -4v(0) + v'(0) \\ V_2 &= -v(0) \\ V_3 &= 4v(L) - v'(L) \\ V_4 &= v(L) \quad . \end{aligned} \tag{2.3}$$

Solving for V_1 we find

$$V_1 = -4v(0) + v'(0) - v(L) \quad .$$

The adjoint system takes the form

$$L^*(v) = v'' + v = 0$$

$$V_1 = -4v(0) + v'(0) - v(L) = 0.$$

EXERCISE 2

1. Write out the bilinear concomitant $P(u,v)$ for the differential equation $L(u)$ and write out Green's formula.

   (a) $L(u) = u' = 0 \qquad a = 1,\ b = 2$

   (b) $L(u) = u'' + 4u = 0 \qquad a = 0,\ b = 1$

   (c) $L(u) = x^2u''(x) + 5xu'(x) + 4u(1) = 0$
   $a = 2,\ b = 4$

   (d) $L(u) = 3u''' - 7u'' + u' - 4u = 0$
   $a = 0,\ b = 1$

   (e) $L(u) = x^3u'''(x) - 3xu'(x) + 4u(x) = 0$
   $a = 1,\ b = 5$

   (f) $L(u) = u^{(iv)} - 2u'' + u = 0 \qquad a = 0,\ b = 2$.

2. Show that Lagrange's identity is true for

   (a) $L(u) = x^2u''(x) + xu'(x) + u(x)$

   (b) $L(u) = u''' - 3u' + u$.

3. Given the system

   $L(u) = u'' - 2u' + u = 0$

   $U_1 = u(0) + u(L) = 0$

   $U_2 = u(0) + u'(0) = 0$

   $U_3 = u'(L) = 0$,

   find the adjoint system.

4. Given the system

   $L(u) = u'' + xu = 0$

   $U_1 = u(0) = 0$

$U_2 = u(0) + u'(L) = 0$,

find the adjoint system.

5. Given the system

$L(u) = u'' + u = 0$

$U_1 = u(0) + 2u'(0) + u(L) = 0$,

find the adjoint system.

6. Given the system

$L(u) = u'' + u = 0$

$U_1 = u(0) + u(L) = 0$

$U_2 = u(0) = 0$

$U_3 = u'(L) = 0$

$U_4 = u'(0) + u(L) = 0$,

find the adjoint system.

7. Given the system

$L(u) = u''' + 9u = 0$

$U_1 = u(0) + u''(L) = 0$

$U_2 = u'(0) = 0$

$U_3 = u''(0) = 0$,

find the adjoint system.

3. SELF-ADJOINT SYSTEMS

Probably the most studied systems are *self-adjoint systems* where $L = L^*$ and $V_i(u)$ are linear combinations of $U_i(u)$ and vice versa, that is $U_i(v)$ are linear combinations of $V_i(v)$.

Consider the system

$L(u) = 0$

$U_i(u) = 0 \quad 0 \leq i = m \leq 2n$

where L is a linear differential operator. In order

that the system be self-adjoint the following conditions must hold.

1. L must be a formally self-adjoint operator, i.e. $L = L^*$.
2. The number of boundary conditions $U_i(u) = 0$ must equal n, i.e. $m = n$.
3. The number of boundary conditions $V_i(v) = 0$ must also equal n.
4. $V_1(v), \ldots, V_n(v)$ must be linear combinations of $U_1(v), \ldots, U_n(v)$ and vice versa.

The adjoint system for this case can be written

$$L(v) = L^*(v) = 0$$

$$V_i(v) = 0 \qquad i = n .$$

Example 1: Show that the system

$$u'' + u = 0$$

$$u(a) = 0$$

$$u(b) = 0$$

is self-adjoint.

1. $Lu = \dfrac{d^2u}{dx^2} + u$ and $L^*v = (-1)^2 \dfrac{d^2v}{dx^2} + v$

$$= \frac{d^2v}{dx^2} + v .$$

Therefore $L = L^*$, i.e. formally self-adjoint.

2. Number of boundary conditions on u is 2.
3. Now

$$\int_a^b \{v(u'' + u) - u(v'' + v)\}dx$$

$$= P(u,v)\Big|_a^b = u'v - uv'\Big|_a^b \tag{3.1}$$

$$= u'(b)v(b) - u(b)v'(b) - u'(a)v(a) + u(a)v'(a).$$

This final expression must equal

$$U_1V_4 + U_2V_3 + U_3V_2 + U_4V_1$$

$$= u(a)V_4 + u(b)V_3 + u'(a)V_2 + u'(b)V_1, \qquad (3.2)$$

where we have chosen U_3, U_4 arbitrarily except that the set $U_1, \ldots, U_4$ must be linearly independent.

Comparing the righthand side of (3.1) and (3.2) we find

$$V_1 = v(b)$$

$$V_2 = -v(a)$$

$$V_3 = -v'(b)$$

$$V_4 = v'(a) \quad .$$

Notice we only need the first two conditions V_1, V_2, i.e.

$$V_1 = v(b) = 0$$

$$V_2 = v(a) = 0.$$

4. Obviously, V_1, V_2 are linear combinations of $U_1(v)$, $U_2(v)$ and U_1, U_2 are linear combinations of $V_1(u)$, $V_2(u)$. The system is self-adjoint and can be written as

$$v'' + v = 0$$

$$v(a) = 0$$

$$v(b) = 0 \quad .$$

EXERCISE 3

1. Given the homogeneous linear second order equation

$$a_0(x)y'' + a_1(x)y' + a_2(x)y = 0 \quad . \qquad (3.3)$$

   (a) Show that in order for the differential equation above to be self-adjoint

   $$a_0'(x) = a_1(x).$$

(b) If the differential equation is not self-adjoint, find a function $v(x)$ which can be used to multiply both sides of equation (3.3) so that the lefthand side becomes self-adjoint.

(c) Show that a homogeneous linear second order self-adjoint equation can be written in the form

$$[r(x)y'(x)]' + p(x)y = 0 .$$

2. Show the following systems are self-adjoint.

(a) $L(u) = u'' + u = 0$

$U_1 = u(a) + u(b) = 0$

$U_2 = u'(a) = 0$

(b) $L(u) = u'' + 4u = 0$

$U_1 = u(a) - u'(b) = 0$

$U_2 = u'(a) + u(b) = 0$

(c) $L(u) = (xu')' + 4u \qquad 0 < x$

$U_1 = u(1) + 2u'(2) = 0$

$U_2 = u(1) = 0$

(d) $L(u) = (xu')' + 4u \qquad 0 < x$

$U_1 = u'(1) - u(2) = 0$

$U_2 = u(1) + 2u'(2) = 0$

(e) $L(u) = e^{2x}u'' + 2e^{2x}u' + e^{2x}u = 0$

$U_1 = 2u(b) + u'(b) = 0$

$U_2 = u(b) + u'(b) = 0$

3. Show that the system

$L(u) = u'' + u = 0$

$U_1 = u(0) + u(\pi) = 0$

$U_2 = u(0) - 2u'(\pi) = 0$

is not a self-adjoint system.

4. A BROADER APPROACH TO SELF-ADJOINT SYSTEMS.

In the theoretical study of self-adjoint differential systems it has been found useful to write the ordinary differential equation in *canonical* form rather than *classical* form.

Example 1: In classical form Bessel's equation looks like

$$x^2y'' + xy' + (x^2 - \nu^2)y = 0$$

whereas in canonical form (self-adjoint) it is written

$$(xy')' + \left[x - \frac{\nu^2}{x}\right]y = 0.$$

Returning to the operator (2.1) for writing the adjoint of the operator L, we see that it is defined for all $n > 0$. However, when we investigate the property of an operator being self-adjoint, we run into trouble for odd values of n. In these cases we see that the term containing the highest order of derivative of L and L^* are equal in magnitude but opposite in sign, that is

$$a_0(x)y^{(n)} = - a_0(x)y^{(n)}, \quad n \text{ odd} \quad .$$

Therefore, if n is odd it is impossible for L to be self-adjoint. Only even powers can be self-adjoint.

Sometimes we speak of odd order differential equations as being antiself-adjoint.

Definition 1: If $L = a_0y^{(2n+1)} + a_1y^{(2n)}+...+a_ny$ $(n = 1,2,...)$, and if $L^* = -L$, the operator L is said to be *antiself-adjoint*.

Because of the awkward nature of odd order differential equations, we shall confine our remarks in what follows to even order differential equations.

In recent theoretical work in differential equations it has become customary to write an even order self-adjoint operator in the form,

$$L(y) = \{\ldots[(\langle p_0 y' \rangle' + p_1 y^{(n-1)})' + p_2 y^{(n-2)}]' + \ldots + p_{n-1} y'\}' + p_n y$$

where $n = 1,2,\ldots$. The advantage of writing the differential operator in this way is that the coefficients $p_0, p_1, \ldots, p_n$ can be taken functions of C^0 and $p_0 > 0$ in the interval $[a,b]$.

Example 2: The canonical forms of a 2nd, 4th, and 6th order homogeneous linear self-adjoint equations are

$$(qy')' + py = 0 \quad q,p \in C^0,\ q > 0 \quad \text{on} \quad [a,b]$$

$$[ry'')' + qy']' + py = 0 \quad r,q,p \in C^0,\ r > 0 \quad \text{on} \quad [a,b]$$

$$\{[(sy''')' + ry'']' + qy'\}' + py = 0 \quad s,r,q,p \in C^0,\ s > 0 \quad \text{on} \quad [a,b].$$

Furthermore, when it comes to writing initial or boundary conditions it is more convenient to use quasi-derivatives rather than ordinary derivatives.

Definition 2: The $2n$ *quasi-derivatives* for use with a $2n$ order differential equation are

$$D_i y = y^{(i)} \qquad i = 0,1,\ldots,n-1$$

$$D_n y = p_0 y^{(n)}$$

$$D_{n+1} y = (p_0 y^{(n)})' + p_1 y^{(n-1)}$$

$$\vdots$$

$$D_{2n-1} y = \left\{\ldots\left[\left[\langle p_0 y^{(n)} \rangle' + p_1 y^{(n-1)}\right]' + p_2 y^{(n-2)}\right]' + \ldots\right\}' + p_{n-1} y' \quad .$$

Example 3: The 4th order self-adjoint linear differential equation initial value problem can be written

$$[(ry'')' + qy']' + py = 0$$

$$D_0 y(x_0) = y(x_0) = \alpha_0$$

$$D_1y(x_0) = y'(x_0) = \alpha_1$$
$$D_2y(x_0) = r(x_0)y''(x_0) = \alpha_2$$
$$D_3y(x_0) = [r(x_0)y''(x_0)]' + q(x_0)y'(x_0) = \alpha_3$$

where $\alpha_0,\ldots,\alpha_3$ are constants.

EXERCISE 4

1. Given the differential equation as shown below, write a set of initial conditions at $x = x_0$ in quasi-derivative form as in Example 3.

   (a) $y'' + 4y = 0$

   (b) $(e^x y')' + e^{-x}y = 0$

   (c) $x^2y'' + 2xy' + 3y = 0$ Hint: Rewrite equation in canonical form.

   (d) $[(4y'')' + 2y']' + 8y = 0$

   (e) $(x^2y'')'' + 4y = 0$

   (f) $\{[(x^3y''')' + x^2y'']' + xy'\}' + y = 0$.

5. STURM-LIOUVILLE THEORY.

Sturm-Liouville theory can be developed for linear differential equations of order higher than the second but historically the theory was applied principally to the 2nd order.

When analyzing many physical problems, we are often led to the differential system,

$$\begin{aligned} &[r(x)y'(x)]' + [p(x) + \lambda s(x)]y(x) = 0 \\ &a_1y(a) + a_2y'(a) = 0 \\ &b_1y(b) + b_2y'(b) = 0 \ . \end{aligned} \tag{5.1}$$

Such a problem is called a Sturm-Liouville problem. We assume that the following conditions hold in this problem.

(a) r, p, s are continuously differentiable on $[a,b]$

(b) $r(x) > 0$ and $s(x) > 0$ on $[a,b]$

(c) λ is a number possibly complex.

(d) a_1, a_2, b_1, b_2 are real

(e) both a_1, a_2 are not zero

(f) both b_1, b_2 are not zero.

The first interesting result concerning the Sturm-Liouville problem is the fact that under the above conditions all eigenvalues are real. This property is contained in the following theorem.

Theorem 1: If $\varphi(x)$ is a solution of the Sturm-Liouville problem corresponding to the eigenvalue λ, then λ is real.

Proof: Suppose λ is a complex number (i.e. $\lambda = \alpha + i\beta$). Under these circumstances φ could be a complex function of a real variable. From the theory of such functions we know that $\overline{\varphi'(x)} = \overline{\varphi}'(x)$. If we now take the conjugate of the differential equation and the boundary conditions, we have

$$\overline{[r(x)\varphi'(x)]' + [p(x) + \lambda s(x)]\varphi(x)} =$$

$$[r(x)\overline{\varphi}'(x)]' + [p(x) + \overline{\lambda}s(x)]\overline{\varphi} = 0, \quad \text{and} \tag{5.2}$$

$$a_1\overline{\varphi}(a) + a_2\overline{\varphi}'(a) = 0$$

$$b_1\overline{\varphi}(b) + b_2\overline{\varphi}'(b) = 0 \quad .$$

Thus, if φ is a solution of the Sturm-Liouville problem corresponding to the eigenvalue λ, then $\overline{\varphi}$ is a solution to the problem corresponding to the eigenvalue $\overline{\lambda}$.

Now if we multiply equations (5.1) and (5.2) by $\overline{\varphi}(x)$ and $\varphi(x)$ respectively and subtract one from the other, we find

$$(\lambda-\bar{\lambda})s(x)\varphi(x)\bar{\varphi}(x) = \varphi(x)[r(x)\bar{\varphi}'(x)]' - \bar{\varphi}(x)[r(x)\varphi'(x)]'.$$

The next step is to integrate both sides from a to b using integration by parts on the righthand side which yields

$$(\lambda-\bar{\lambda})\int_a^b s(x)\varphi(x)\bar{\varphi}(x)dx = r(x)[\varphi(x)\bar{\varphi}'(x) - \bar{\varphi}(x)\varphi'(x)]_a^b = r(b)[\varphi(b)\bar{\varphi}'(b) - \bar{\varphi}(b)\varphi'(b)] - r(a)[\varphi(a)\bar{\varphi}'(a) - \bar{\varphi}(a)\varphi'(a)].$$

Upon substituting the boundary conditions given in (5.1) and (5.2), we find that the righthand side is zero, and therefore

$$(\lambda-\bar{\lambda})\int_a^b s(x)\varphi(x)\bar{\varphi}(x)dx = 0 \quad .$$

Since eigenfunctions are non-trivial, the integral cannot be zero and we can write

$$\lambda - \bar{\lambda} = 0$$

or

$$\lambda = \bar{\lambda},$$

and, therefore, the eigenvalues are real.

In a similar way we can prove the orthogonality of solutions of the Sturm-Liouville problem. This idea is summarized in the next theorem.

Theorem 2: If φ_m and φ_n are solutions of the Sturm-Liouville problem whose respective eigenvalues λ_m, λ_n are not equal, then φ_m and φ_n satisfy the orthogonality condition

$$\int_a^b s(x)\varphi_m(x)\varphi_n dx = 0 \quad m \neq n.$$

Both of these theorems can be extended so that they hold for less stringent conditions. These ideas are summarized in Table 1.

	Differential Equation	Boundary Condition	Comment
(1)	$r(a) = 0$	$b_1 y(b)+b_2 y'(b)=0$	singularity at $x = a$
(2)	$r(b) = 0$	$a_1 y(a)+a_2 y'(a)=0$	singularity at $x = b$
(3)	$r(a)=r(b)=0$	None needed	singularity at $x = a$ and $x = b$.
(4)	$r(a)=r(b)$	$y(a)=y(b)$, $y'(a)=y'(b)$	periodic on (a,b)

TABLE 1

When we investigate problems using a set of orthogonal functions it is to our advantage to normalize each function.

Definition 1: The function f is said to be *normal* over the interval $[a,b]$ with respect to weight function s if

$$||f||^2 = \int_a^b s(x)f^2(x)dx = 1.$$

Theorem 3: Any set of non-zero orthogonal functions $\{\varphi_n\}$ can be easily normalized by letting

$$\psi_n = \frac{\varphi_n}{||\varphi_n||} .$$

Proof: The integral

$$\int_a^b s(x)\psi_n^2(x)dx = \int_a^b s(x) \frac{\varphi_n^2(x)}{||\varphi_n||^2} dx$$

$$= \frac{1}{||\varphi_n||^2} \int_a^b s(x)\varphi_n^2(x)dx = \frac{||\varphi_n||^2}{||\varphi_n||^2} = 1.$$

Furthermore, it is easy to show that ψ_n satisfies the same Sturm-Liouville problem as φ_n and the set $\{\psi_n\}$ is orthogonal.

Another important property concerning eigenvalues and eigenfunctions is contained in the following theorem:

Theorem 4: If along with the conditions normally assigned to the Sturm-Liouville problem on $[a,b]$, the coefficients of the Sturm-Liouville equation and its boundary conditions are independent of λ, then there exists an infinite set of eigenvalues $\lambda_0, \lambda_1, \ldots, \lambda_n, \ldots$ such that the $\lim_{n\to+\infty} \lambda_n = +\infty$ and the corresponding eigenfunctions $y_1, y_1, \ldots, y_n, \ldots$ have exactly n zeros on $a < x \leq b$.

Example 1: Let us solve the Sturm-Liouville problem.

$$y'' + \lambda y = 0$$
$$y(0) = y(\pi) = 0 \quad .$$

The general solution of the differential equation is

$$y(x) = A \cos \sqrt{\lambda}\, x + B \sin \sqrt{\lambda}\, x \quad .$$

Applying the first boundary condition we have

$$y(0) = A = 0.$$

The second boundary condition yields

$$y(\pi) = B \sin \sqrt{\lambda}\, \pi = 0$$

from which obtain the eigenvalues

$$\lambda_n = n^2 . \quad n = 1,2,\ldots \quad .$$

Notice $\lim_{n\to+\infty} \lambda_n = \lim_{n\to+\infty} n^2 = +\infty$.

Furthermore, if we choose an eigenfunction, say $y_3 = \sin 3x$, corresponding to $\lambda_3 = 9$, we know from trigonometry that y_3 has exactly 3 zeros on $0 < x \leq \pi$. See Figure 1.

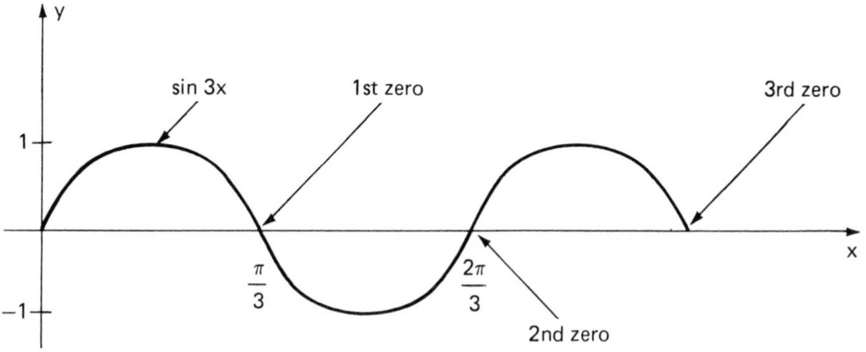

Figure 1

EXERCISE 5

1. Given the Sturm-Liouville problems listed below, show that $\lim_{n\to+\infty} \lambda_n = +\infty$ and that the particular eigenfunction y_n (i.e., eigenfunction corresponding to eigenvalue λ_n) has exactly n zeros on the interval indicated.

	Sturm-Liouville Problem	Eigenfunction	Interval
(a)	$y''+\lambda y=0, .y(0)=y'(\pi)=0$	y_2	$(0,\pi]$
(b)	$y''+\lambda y=0,\ y'(0)=y'(\frac{\pi}{2})=0$	y_4	$(0,\frac{\pi}{2}]$
(c)	$y''+\lambda y=0,\ y(0)=0,\ y(\pi)-y'(\pi)=0$	y_4	$(0,\pi]$
(d)	$y''+(4+\lambda)y=0,\ y(0)=y(1)=0$	y_3	$(0,1]$
(e)	$x^2y''+xy'+\lambda y=0,\ y(1)=y(e)=0$	y_4	$(1,e]$
(f)	$x^2y''+xy'+2\lambda y=0,\ y'(1)=y'(e^{\pi})=0$	y_2	$(1,2^{\pi}]$

2. In solving boundary value problems we are often confronted by the system
$y'' + \lambda y = 0;\ \ y(0) = y(L) = 0.$
How do we know all the eigenvalues are real?

3. (a) Write Bessel's equation $x^2y'' + xy' + (x^2 - \lambda)y = 0$ in self-adjoint form.
   (b) If we let $a = 0$ and $b = L$ and $y(L) + y'(L) = 0$, how do we know all the eigenvalues are real.
4. (a) Write Legendre's equation $(1-x^2)y'' - 2xy' + \lambda y = 0$ in self-adjoint form.
   (b) If $a = -1$, $b = 1$, will all the eigenvalues be real?

6. INTRODUCTION TO ORTHOGONALITY AND COMPLETENESS

In Section 5 we saw that the number of eigenfunctions defined over $(a,b]$ of a Sturm-Liouville problem is infinite. Let the eigenfunctions be represented by the orthonormal set of functions $\{\psi_n\}$. In the study of solutions of boundary value problems we are often faced with the following question. Given a function f is it possible to express this function f as an infinite linear combination of functions ψ_n, i.e. does

$$f(x) = \sum_{i=1}^{\infty} c_i\psi_i$$

over $[a,b]$? In looking for the answer to this question, we must first look into the matter of convergence.

6.1 Convergence in the mean

In the study of analysis there are many types of convergence. Probably the most well-known type is pointwise convergence (or convergence) followed by uniform convergence which is a "stronger" type of convergence.

However, for the study of Fourier series we find it convenient to use a third type of convergence known as *convergence in the mean* which roughly falls between convergence and uniform convergence.

Definition 1: A function $f(x)$ is said to be *piecewise continuous* over $[a,b]$ if there are at most a finite number if points on $[a,b]$ at which f is discontinuous and at each such point the righthand and lefthand limits of f exist. If a or b is a point at which f is discontinuous only the righthand limit must exist at a and only the lefthand limit at b. If f is piecewise continuous we write $f \in C_p$.

Example 1: (a) $$f(x) = \begin{cases} 0 & x = a,b \\ 1 & a < x < b \end{cases}$$

is piecewise continuous on $[a,b]$.

(b) $$f(x) = \begin{cases} 0 & x = 0 \\ \frac{1}{x} & 0 < x \leq 1 \end{cases}$$

is not piecewise continuous on $[0,1]$ since $\lim_{x \to 0^+} f(x)$ does not exist.

It can be shown that if f is piecewise continuous on $[a,b]$, it is integrable on $[a,b]$.

Example 2: If $$f(x) = \begin{cases} x & 0 \quad x < 1 \\ \frac{1}{2} & x \quad 1 \\ 3 & 1 < x \leq 4 , \end{cases}$$

then $f(x)$ is piecewise continuous on $[0,4]$ and

$$\int_0^4 f(x)dx = \int_0^1 x\ dx + \int_1^1 \frac{1}{2}\ dx + \int_1^4 3dx$$
$$= \frac{1}{2} + 0 + 12 - 3 = 9\,\frac{1}{2}.$$

See Figure 2

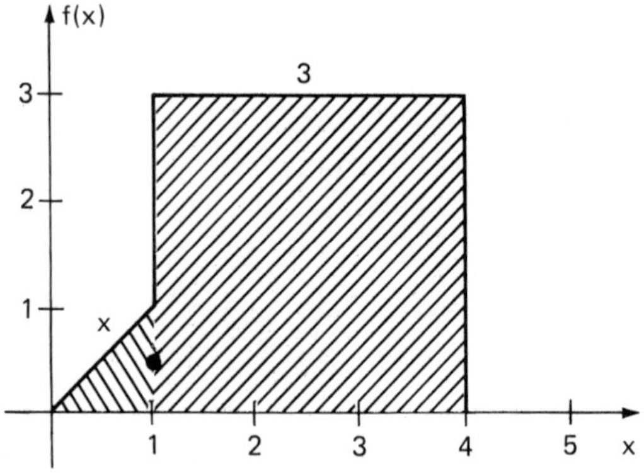

Figure 2

Definition 2: If $f \in C_p$ and $s(x) > 0$ and continuous on $[a,b]$,

$$||f|| = \sqrt{\int_a^b s(x)[f(x)]^2 dx}$$ is the *norm* of f over $[a,b]$

Definition 3: If f and $g \in C_p$ and $s(x) > 0$ and continuous on $[a,b]$

$$||f-g|| = \sqrt{\int_a^b s(x)[f(x)-g(x)]^2 dx}$$ is *distance* between two functions.

Suppose $\sum_{i=1}^{\infty} f_i(x)$ is a series of piecewise continuous functions $f_i(x)$ on $[a,b]$. Let $S_n(x) = \sum_{i=1}^{n} f_i(x)$ be the sequence of partial sums $\{S_n\}$.

Definition 4: Given $\epsilon > 0$, if there exists an integer $N > 0$ such that $||f(x) - S_n(x)|| < \epsilon$ whenever $N \leq n$, then the series $\sum_{i=1}^{\infty} f_i(x)$ is said to *converge in the mean* to $f(x)$ on $[a,b]$.

Example 3: We shall show the series $\sum_{i=0}^{\infty} x^i$ is convergent in the mean to $f(x) = \frac{1}{1-x}$ on $[0,r]$ where $r < 1$, and $s(x) = 1$. We write $S_{n-2} = \frac{1-x^{n-1}}{1-x}$ since the series is geometric. We are using S_{n-2} instead of S_n for convenience. Given $\epsilon > 0$, we see that

$$\left\|\frac{1}{1-x} - \frac{1-x^{n-1}}{1-x}\right\| = \sqrt{\int_0^r \left[\frac{x^{n-1}}{1-x}\right]^2 dx} \leq$$

$$\sqrt{\int_0^r \left[\frac{r^{n-1}}{1-r}\right]^2 dx} = \frac{r^n}{1-r} < \epsilon .$$

The inequality $\frac{x^{n-1}}{1-x} \leq \frac{r^{n-1}}{1-r}$ follows from the fact $0 \leq x \leq r$. Now $\left\|\frac{1}{1-x} - \frac{1-x^{n-1}}{1-x}\right\| < \epsilon$ if $\frac{r^n}{1-r} < \epsilon$ which implies

$$r^n < (1 - r)\epsilon \quad .$$

Taking the natural log of both sides we have

$$n \ln r < \ln(1 - r)\epsilon$$

and if we assume $\epsilon < 1$

$$\frac{\ln(1-r)\epsilon}{\ln r} < n .$$

By choosing N equal the next integer greater than $\frac{\ln(1-r)\epsilon}{\ln r}$, $\left\|\frac{1}{1-x} - \frac{1-x^{n-1}}{1-x}\right\|$ will be less than ϵ.

Therefore, $\sum_{i=1}^{\infty} x^i$ converges in the mean to $\frac{1}{1-x}$ on $[0,r]$.

Theorem 1: If $\sum_{i=1}^{\infty} u_i$ converges in the mean on $[0,r]$,

$$u_j \sum_{i=1}^{\infty} u_i = \sum_{i=1}^{\infty} u_i u_j \quad \text{on} \quad [a,b] \quad .$$

Proof: We write

$$\left\|u_j \sum_{i=1}^{\infty} u_i - \sum_{i=1}^{n} u_i u_j\right\| = \left\{\int_a^b s(x)\left[u_j \sum_{i=1}^{\infty} u_i - \sum_{i=1}^{n} u_i u_j\right]^2 dx\right\}^{1/2} .$$

$$= \left\{\int_a^b s(x) u_j^2 \left[\sum_{i=1}^{\infty} u_i - \sum_{i=1}^{n} u_i\right]^2 dx\right\}^{1/2} .$$

Since u_j is bounded on $[a,b]$, i.e $u_j^2 \le M$.

then

$$\left\{\int_a^b s(x)\ u_j^2 \left[\sum_{i=1}^{\infty} u_i - \sum_{i=1}^{\infty} u_i\right]^2 dx \right\}^{1/2} \le M \left\|\sum_{i=1}^{\infty} u_i - \sum_{i=1}^{n} u_i\right\|.$$

By hypothesis $\left\|\sum_{i=1}^{\infty} u_i - \sum_{i=1}^{n} u_i\right\| < \frac{\epsilon}{M}$ for $N \le n$, and $\left\|u_j \sum_{i=1}^{\infty} u_i - \sum_{i=1}^{n} u_i u_j\right\| < M \frac{\epsilon}{M} = \epsilon$ for $N \le n$. The theorem is proved.

Theorem 2: If $\sum_{i=1}^{\infty} u_i$ converges in the mean on $[a,b]$ and if the terms u_i are piecewise continuous on $[a,b]$, then

$$\int_a^b \left[\sum_{i=1}^{\infty} u_i(x)\right] dx = \sum_{i=1}^{\infty} \left[\int_a^b u_i(x) dx\right] .$$

Proof: We can write for t in $[a,b]$

$$\left|\int_a^t \left[\sum_{i=1}^{\infty} u_i(x)\right] dx - \sum_{i=1}^{n} \int_a^t u_i(x) dx\right|^2 =$$

$$= \left|\int_a^t \left[\sum_{i=1}^{\infty} u_i(x) - \sum_{i=1}^{n} u_i(x)\right] dx\right|^2 \tag{6.1}$$

$$\le \left[\int_a^t \left|\sum_{i=1}^{\infty} u_i(x) - \sum_{i=1}^{n} u_i(x)\right| dx\right]^2 .$$

Since $s(x) > 0$, we can rewrite the integral in (6.1) in the form

$$\int_a^t \left[\frac{\sqrt{s(x)}}{\sqrt{s(x)}}\right] \left| \sum_{i=1}^{\infty} u_i(x) - \sum_{i=1}^{n} u_i(x) \right| dx \quad . \tag{6.2}$$

Therefore for all t in $[a,b]$

$$\left| \int_a^t \left[\sum_{i=1}^{\infty} u_i(x) \right] dx - \sum_{i=1}^{n} \int_a^t u_i(x)dx \right|^2 \leq$$

$$\left[\int_a^t \left[\frac{1}{\sqrt{s(x)}} \right] \left[\sqrt{s(x)} \left| \sum_{i=1}^{\infty} u_i(x) - \sum_{i=1}^{n} u_i(x) \right| \right] dx \right]^2 \tag{6.3}$$

$$\leq \int_a^t \frac{dx}{s(x)} \cdot \int_a^t s(x) \left| \sum_{i=1}^{\infty} u_i(x) - \sum_{i=1}^{n} u_i(x) \right|^2 dx$$

where the last expression follows from using the Schwarz inequality.

Let $\int_a^b \frac{dx}{s(x)} = M$ and since $\sum_{i=1}^{\infty} u_i(x)$ converges in the mean, we can choose N so large that for all $N \leq n$

$$\int_a^b s(x) \left| \sum_{i=1}^{\infty} u_i(x) - \sum_{i=1}^{n} u_i(x) \right|^2 dx$$

$$= \left\| \sum_{i=1}^{\infty} u_i(x) - \sum_{i=1}^{n} u_i(x) \right\| < \frac{\epsilon^2}{M} \quad . \tag{6.4}$$

Substituting the result from (6.4) into (6.3) we see that for $\epsilon > 0$, there exists an $N > 0$ such that

$$\left| \int_a^b \sum_{i=1}^{\infty} u_i(x)dx - \sum_{i=1}^{n} \int_a^b u_i(x)dx \right| < \epsilon$$

for $N \leq n$.

Thus,

$$\int_a^b \sum_{i=1}^{\infty} u_i(x)dx = \sum_{i=1}^{\infty} \int_a^b u_i(x)dx \quad .$$

This result can be stated in the following way.

If $\sum_{i=1}^{\infty} u_i(x)$ converges in the mean on $[a,b]$

and the terms u_i are piecewise continuous on $[a,b]$, then the integral of the series over $[a,b]$ equals the series of the integrals

6.2 Generalized Fourier Series

Let $\{\psi_i\}$ be an infinite sequence of orthonormal functions with respect to the weight function $s(x)$ defined over the interval $[a,b]$. Suppose we wish to represent a piecewise continuous function $f(x)$ on $[a,b]$ by an infinite series $\sum_{i=1}^{\infty} c_i\psi_i$, that is

$$f(x) = \sum_{i=1}^{\infty} c_i\psi_i \quad .$$

Assuming this series converges in the mean to f, we multiply both sides of equation by $s(x)\psi_j(x)$, i.e.

$$s(x)f(x)\psi_j(x) = s(x)\psi_j(x) \sum_{i=1}^{\infty} c_i\psi_i(x)$$

$= \sum_{i=1}^{\infty} c_i s(x)\psi_i(x)\psi_j(x)$, and then integrate both sides from a to b. We find

$$\int_a^b s(x)f(x)\psi_j(x)dx = \int_a^b \sum_{i=1}^{\infty} c_i s(x)\psi_i(x)\psi_j(x)dx$$

$$= \sum_{i=1}^{\infty} c_i \int_a^b s(x)\psi_i(x)\psi_j(x)dx. \tag{6.5}$$

The operation of multiplication by ψ_j and integrating term by term follow from the fact that $\Sigma\, c_i\psi_i$ converges in the mean to $f(x)$ as we have shown in paragraph 6.1.

Now since $\{\psi_i\}$ is an orthonormal set, we have from (6.5) that

$$c_j = \int_a^b s(x)f(x)\psi_j(x)dx \tag{6.6}$$

or since j is a dummy index we can write (6.6) as

$$c_i = \int_a^b s(x)f(x)\psi_i(x)dx \quad . \tag{6.7}$$

Whenever c_i is calculated by the equation (6.7) it is called a *Fourier coefficient* and the corresponding series $\sum_{i=1}^{\infty} c_i\psi_i$ is called a *Fourier series*.

The key supposition is deriving the formula for finding c_i in the previous argument depended on knowing that the series $\sum_{i=1}^{\infty} c_i\psi_i$ converges in the mean to $f(x)$. Unfortunately, we are usually given a function f and asked to find its Fourier series without knowing whether or not the series converges in the mean or if it converges, does it converge to f. Under these circumstances we use the $\sim$ sign and write

$$f(x) \sim \sum_{i=1}^{\infty} c_i\psi_i \quad .$$

This notation is often called the formal representation of the Fourier series for f. If we can deduce in some way that the series converges in the mean to f, then we can replace the $\sim$ sign by the $=$ sign.

6.3 Best Approximation

Suppose we wish to represent $f(x)$ by a series of orthonormal functions ψ_i which can be written $f(x) = \sum_{i=1}^{\infty} a_i\psi_i(x)$. Which choice of constants a_i yields the best approximation in the mean to f?

Let

$$E_n = ||f - T_n||$$

where T_n is the nth partial sum of the series $\sum_{i=1}^{\infty} a_i\psi_i$, i.e. $T_n = \sum_{i=1}^{n} a_i\psi_i$.

The best approximation in the mean occurs when the a_i's are chosen so that E_n is minimized. To find this set of a_i's we see that

$$E_n^2 = ||f - T_n||^2 = \int_a^b s(x)\left[f(x) - T_n(x)\right]^2 dx$$

$$= \int_a^b s(x)f^2(x)dx - 2\int_a^b s(x)f(x)T_n(x)dx$$

$$+ \int_a^b s(x)T_n^2(x)dx \quad .$$

Replacing T_n by $\sum_{i=1}^{n} a_i\psi_i$ we see

$$E_n^2 = ||f||^2 - 2\int_a^b s(x)f(x)\sum_{i=1}^{n} a_i\psi_i dx$$

$$+ \int_a^b s(x)\left[\sum_{i=1}^{n} a_i\psi_i\right]^2 dx = ||f||^2$$

$$- \int_a^b s(x)\left[2\sum_{i=1}^{n} a_i f(x)\psi_i - \sum_{i=1}^{n} a_i^2\psi_i^2\right]dx$$

where we used the fact that $\left[\sum_{i=1}^{n} a_i\psi_i\right]^2 = \sum_{i=1}^{n} a_i^2\psi_i^2$ because the ψ_i are orthogonal functions over [a,b]. Since ψ_i are also normalized

$$E_n^2 = ||f||^2 - 2\sum_{i=1}^{n} a_i c_i + \sum_{i=1}^{n} a_i^2$$

where $c_i = \int_a^b s(x)f(x)\psi_i(x)dx$ are the Fourier coefficients of f. Then, by completing the square

$$E_n^2 = ||f||^2 + \sum_{i=1}^{n}(a_i - c_i)^2 - \sum_{i=1}^{n} c_i^2 \quad . \qquad (6.8)$$

Since f and c_i are fixed numbers, E_n^2 will be minimized when $a_i = c_i$. We summarize these ideas in the following theorem.

Theorem 3: The best approximation in the mean of the series of orthonormal functions $\sum_{i=1}^{\infty} a_i\psi_i$ to f over $[a,b]$ occurs when

$$a_i = c_i = \int_a^b s(x)f(x)\psi_i(x)dx \quad ,$$

that is when the a_i's are Fourier coefficients and the series is a Fourier series.

Returning to equation (6.8) we see that if $a_i = c_i$

$$E_n^2 = ||f||^2 - \sum_{i=1}^{n} c_i^2 \ .$$

But since $E_n^2 \geq 0$, we can write

$$\sum_{i=1}^{n} c_i^2 \leq ||f||^2 \quad .$$

Letting $n \to +\infty$ we arrive at *Bessel's inequality*

$$\sum_{i=1}^{\infty} c_i^2 \leq ||f||^2 \quad .$$

Theorem 4: If f is piecewise continuous on $[a,b]$, then $\sum_{i=1}^{\infty} c_i^2$ converges absolutely.

Proof: Let $\sigma_n = \sum_{i=1}^{n} c_i^2$. Then the sequence of partial sums $\{\sigma_n\}$ is increasing and is bounded above by $||f||^2$. Thus, the series $\sum_{i=1}^{\infty} c_i^2$ is convergent but a positive term series is absolutely convergent.

Example 4: It can be shown that the Fourier series of $f(x) = x - x^2$ on $[-1,1]$ is given by

$$x - x^2 = \frac{8}{\pi^3} \sum_{i=1,3}^{\infty} \frac{\sin i\pi t}{i^3} \quad \text{where} \quad c_i = \frac{8}{(\pi i)^3} .$$

Using Bessel's inequality and recalling $s(x) = 1$ for a sine series

$$\frac{8}{\pi^3} \sum_{i=1}^{\infty} \frac{1}{i^3} \le \int_{-1}^{1} (x - x^2)^2 dx = \frac{16}{15} .$$

The series $\sum_{i=1}^{\infty} \frac{1}{i^3} \le \frac{2}{15} \pi^3$.

An interesting observation about Bessel's inequality is that it holds whether $\sum_{i=1}^{\infty} c_i\psi_i$ converges to f on $[a,b]$ or not. This result occurs because in our earlier argument it was just necessary to minimize E_n^2. If $E_n \neq 0$, then $\sum_{i=1}^{\infty} c_i\psi_i$ does not converge in the mean to f.

From Theorem 4 we can deduce the following corollary.

Corollary 1: If c_i are Fourier coefficients of a piecewise continuous function f, then $\lim_{i\to\infty} c_i \to 0$.

Proof: Since $\sum_{i=1}^{\infty} c_i^2$ is convergent, the $\lim_{i\to 0} c_i^2 = 0.$ Therefore, $\lim_{i\to\infty} c_i = 0.$

Example 5: The Fourier sries of

$$f(x) = \begin{cases} h & 0 < x < \pi \\ -h & -\pi < x < 0 \end{cases}$$

is given by

$$\frac{4h}{\pi} \sum_{i=1,3,5}^{\infty} \frac{\sin ix}{i} .$$

Therefore, $c_i = \frac{4h}{\pi i}$ and the $\lim_{i\to\infty} c_i = 0.$

6.4 Completeness.

Let $\{\psi_i\}$ be an infinite orthonormal set of eigenfunctions which are piecewise continuous on $[a,b]$. Then the following question arises. Given any piecewise continuous function f on $[a,b]$, can f be represented by a Fourier series $\sum_{i=1}^{\infty} c_i\psi_i$? We are assuming the Fourier series converges in the mean to f so we can write $f(x) = \sum_{i=1}^{\infty} c_i\psi_i(x)$.

Definition 5: If for any piecewise continuous function f defined on $[a,b]$, $f = \sum_{i=1}^{\infty} c_i\psi_i$, then the orthonormal set $\{\psi_i\}$ is said to be *complete*.

Theorem 5: If $\{\psi_i\}$ is a complete orthonormal set of eigenfunctions on $[a,b]$ and f is piecewise continuous on $[a,b]$, $\sum_{i=1}^{\infty} c_i\psi_i$ converges in the mean to f if and only if

$$||f||^2 = \sum_{i=1}^{n} c_i^2.$$

Proof: Assume $\sum_{i=1}^{\infty} c_i\psi_i$ converges in the mean to f.

Let $S_n = \sum_{i=1}^{n} c_i\psi_i$ and we can write $\lim_{n\to\infty} ||f-S_n|| = 0$.

Thus,

$$\lim_{n\to\infty} \int_a^b s(x)\left[f(x) - \sum_{i=1}^{n} c_i\psi_i\right]^2 dx = 0.$$

Squaring the term inside the brackets we can write

$$\lim_{n\to\infty}\left\{\int_a^b s(x)f(x)^2dx - 2\int_a^b s(x)f(x)\sum_{i=1}^{n} c_i\psi_i dx + \int_a^b s(x)\left[\sum_{i=1}^{n} c_i\psi_i\right]^2 dx\right.$$

$$= ||f||^2 - \lim_{n\to\infty} 2\sum_{i=1}^{n} c_i^2 + \lim_{n\to\infty}\int_a^b s(x)\sum_{i=1}^{n} c_i^2\psi_i^2\, dx$$

$$= ||f||^2 - \lim_{n\to\infty} 2\sum_{i=1}^{n} c_i^2 + \lim \sum_{i=1}^{n} c_i^2 = 0,$$

where we used the fact that

$$\int_a^b s(x)\sum_{i=1}^{n} c_i^2\psi_i^2 = \sum_{i=1}^{n} c_i^2\int_a^b s(x)\psi_i^2dx = \sum_{i=1}^{n} c_i^2 .$$

it follows that

$$||f||^2 = \sum_{i=1}^{\infty} c_i^2 .$$

Assume $||f||^2 = \sum_{i=1}^{\infty} c_i^2$. Since by hypothesis the orthonormal set $\{\psi_i\}$ is complete, we can use the expansion found in the first half of this theorem to write

$$\lim_{n\to\infty} ||f - S_n||^2 = ||f||^2 - \lim_{n\to\infty}\sum_{i=1}^{n} c_i^2$$

$$= ||f||^2 - \sum_{i=1}^{\infty} c_i^2 = 0 ,$$

which shows us the series $\sum_{i=1}^{\infty} c_i\psi_i$ converges in the mean to f on $[a,b]$ and the theorem is proved.

Definition 6: The equality $||f||^2 = \sum_{i=1}^{\infty} c_i^2$ in Theorem 5 is known as *Parseval's equality*.

When a set of orthonormal eigenfunctions $\{\psi_i\}$ is complete, no more nontrivial piecewise continuous functions can be added to the set. A trivial piecewise continuous function is zero in $[a,b]$ except at a finite number of points. Suppose we attempt to add the function f to the set $\{\psi_i\}$. Then since the members of the set must be orthogonal

$$c_i = \int_c^b s(x)f(x)\psi_i(x)dx = 0 \quad .$$

But by Parseval's equality

$$||f||^2 = \sum_{i=1}^{\infty} c_i^2 = 0$$

from which it follows that

$$f(x) = 0$$

at all but a finite number of points in $[a,b]$.

Example 6: The set of functions $\{\sin nx\}$ is not complete over $[-\pi,\pi]$. If we attempt to write the Fourier sine series of $f(x) = 1$ over $[-\pi,\pi]$ we know that a_0, a_n are zero since $f(x)$ is an even function. Now

$$b_n = \frac{1}{\pi}\int_{-\pi}^{\pi} \sin nx\, dx = -\frac{1}{\pi n}\cos nx\Big|_{-\pi}^{\pi}$$

$$= -\frac{1}{\pi n}[\cos n\pi - \cos n(-\pi)] = 0.$$

The Fourier series $\sum_{n=1}^{\infty} b_n \sin nx = 0$ which is certainly not equal to $f(x)$.

6.5. Comments on convergence of Fourier series

We recall that in much of our discussion in the previous paragraphs, we did not know whether a Fourier series converged or not and if it did converge, did it converge to the right value.

The answer to these questions is known for certain classes of functions but their derivation is beyond the scope of this text. For your information we quote one such result.

Definition 7: A function f is said to be *piecewise differentiable* on $[a,b]$ if f is piecewise continuous on $[a,b]$ and if there are at most a finite number of points on $[a,b]$ at which f' does not exist and at each such point the righthand and lefthand limits of f' exist. If a or b is a point at which f' does not exist, only the righthand limit of f' must exist at a and only the lefthand limit at b.

Theorem 6: (A Fourier Convergence Theorem)

Suppose f is a piecewise differentiable function on $[a,b]$, then at any point x in $[a,b]$ the Fourier series $\sum_{i=1}^{\infty} c_i\psi_i$ where

$$c_i = \int_a^b s(x)f(x)\psi_i(x)dx,$$ converges (pointwise) to

$$\frac{f(x^+)+f(x^-)}{2}.$$

Notice that at a point x_1 where f has a jump discontinuity, the Fourier series converges to a point halfway between $f(x_1^+)$ and $f(x_1^-)$.

Example 7: Suppose $f(x)$ on $[a,b]$ is given by graph in Figure 3(b). Then the Fourier series of $f(x)$ converges to the function shown in 3(b).

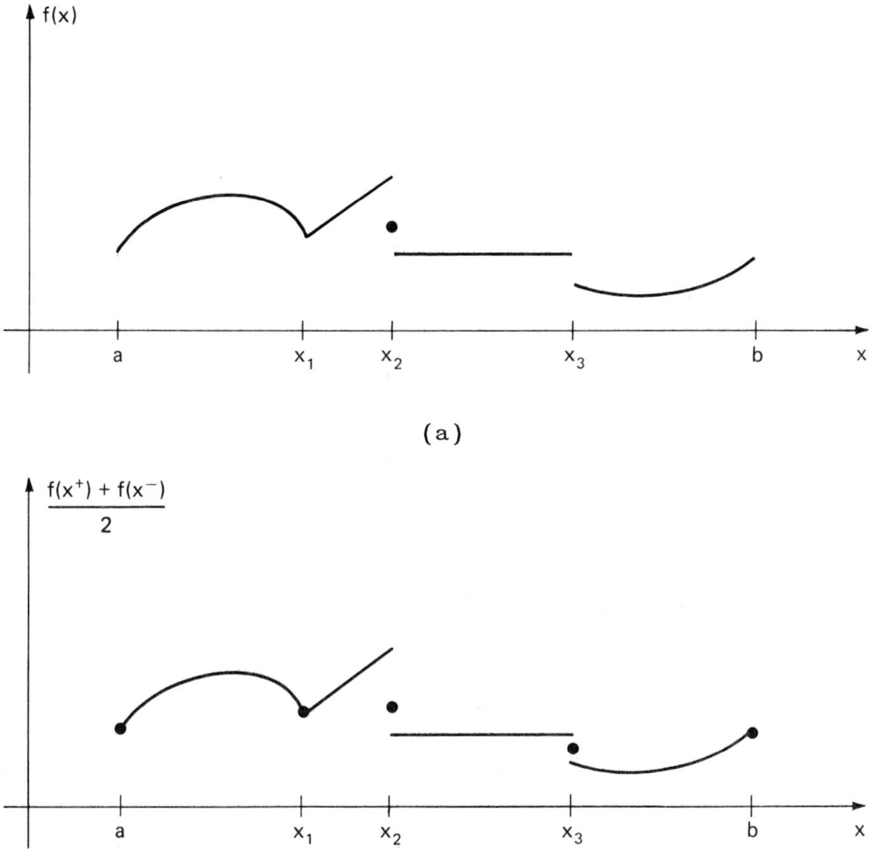

(b)

Figure 3

Notice the Fourier series converges to $f(x)$ at all points of $[a,b]$ except $x = a, x_3, b$.

EXERCISE 6

1. Using Bessel's inequality, find an upper bound for the series $\sum_{i=1}^{\infty} c_i^2$ where the c_i's are the

Fourier coefficients of the given Fourier series defined over the given interval.

(a) $\sum_{i=1}^{\infty} (-1)^{i+1} \frac{2\sin ix}{i} = x \qquad -\pi < x < \pi.$

(b) $\sum_{i=0}^{\infty} \frac{\sin(2i+1)x}{2i+1} = \frac{\pi}{4} \qquad 0 < x < \pi$

(c) $\frac{\pi^3}{3} + 4 \sum_{i=1}^{\infty} (-1)^i \frac{\cos ix}{i^2} = x^2 \qquad -\pi \leq x \leq \pi$

(d) $\frac{\pi}{2} - \frac{4}{\pi} \sum_{i=0}^{\infty} \frac{\cos(2i+1)x}{(2i+1)^2} = |x| \qquad -\pi \leq x \leq \pi$

(e) $1 - \frac{4}{\pi^2} \sum_{i=0}^{\infty} \frac{\cos(2i+1)\pi x}{(2i+1)^2} - \frac{2}{\pi} \sum_{i=1}^{\infty} \frac{\sin 2i\pi x}{2i}$

$= f(x)$

where $f(x) = \begin{cases} 1 & -1 < x < 0 \\ 2x & 0 < x < 1 \end{cases}$.

2. Show that Corollary 1 holds in the Fourier series in Problem 1.

3. Let $f(x) = \frac{a_0}{2} + \sum_{i=1}^{\infty} (a_i \cos ix + b_i \sin ix)$

over $(-\pi,\pi)$. Write out Bessel's inequality for $f(x)$.

4. Using Parseval's equation and the Fourier series

$$1 = \frac{4}{\pi} \sum_{i=0}^{\infty} \frac{\sin(2i+1)x}{2i+1} \qquad 0 < x < \pi$$

show that

$$\sum_{i=0}^{\infty} \frac{1}{(2i+1)^2} = \frac{\pi^2}{8} .$$

Hint: Remember set of eigenfunmctions must be orthonormal.

5. Apply problem 4 to the Fourier series

$$x = 2 \sum_{i=1}^{\infty} (-1)^{i+1} \frac{\sin ix}{i} \qquad -\pi < x < \pi$$

to show that

$$\sum_{i=1}^{\infty} \frac{1}{i^2} = \frac{\pi^2}{6} \quad .$$

6. Apply problem 4 to the Fourier series

$$x^2 = \frac{1}{3} + \frac{4}{\pi^2} \sum_{i=1}^{\infty} (-1)^i \frac{\cos i\pi x}{i^2} \qquad -1 < x < 1$$

to show that

$$\sum_{i=1}^{\infty} \frac{1}{i^4} = \frac{\pi^4}{90} \quad .$$

7. We know the set of functions $\{\cos nx\}$, $n = 0,1,\ldots$ is complete over the interval $[0,\pi]$. Show that it is not complete over the interval $[-\pi,\pi]$.

   Hint: Try to express $f(x) = x$ over $(-\pi,\pi)$ in terms of $\cos nx$.

8. Without integrating, why would we say the set of Legendre polynomials $\{P_{2n}\}$, $n = 0,1,\ldots$ defined over $(-1,1)$ is not complete?

9. Why does the Fourier series of $\sin x$ converge (pointwise) at all points to $\sin x$ in $(-\infty,\infty)$.

10. Define a square wave so that the Fourier series converges at every point in $(-\infty,\infty)$.

11. Why does $f(x) = \tan x$ on $\left[-\frac{\pi}{2}, \frac{\pi}{2}\right]$ not meet the hypothesis of Theorem 5?

12. Sketch the graphs of the Fourier series for the functions given below.

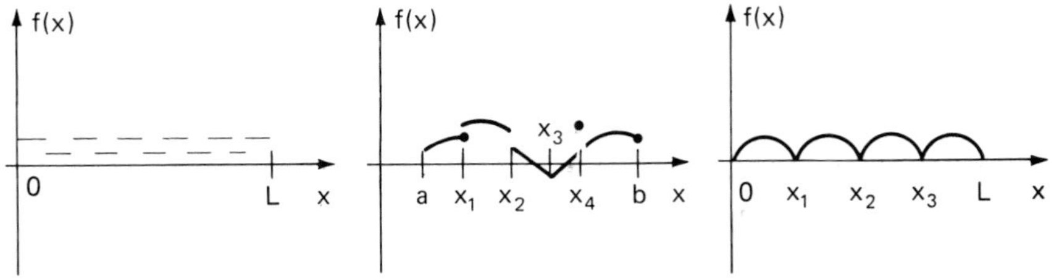

Figure 4

CHAPTER 2 SPECIAL FUNCTIONS

1. HYPERGEOMETRIC SERIES

1.1 INTRODUCTION

In the year 1812 Johann Frederick Carl Gauss published a comprehensive work studying the *hypergeometric series* which has the form

$$1 + \frac{\alpha\beta}{\gamma} x + \frac{\alpha(1+\alpha)\beta(1+\beta)}{2!\gamma(1+\gamma)} x^2$$

$$+ \frac{\alpha(1+\alpha)(2+\alpha)\beta(1+\beta)(2+\beta)}{3!\gamma(1+\gamma)(2+\gamma)} x^3 + \dots \quad (1.1)$$

where α, β, γ, x are real numbers. The name hypergeometric comes from the fact that this series is a generalization of the geometric series. In fact, if we set $\alpha = 1$ and $\beta = \gamma$ in (1.1) above we have

$$1 + x + x^2 + x^3 + \dots$$

which is the well-known geometric series.

In his work Gauss determined the restrictions which must be placed on α, β, γ, x so that the series converges. As it turned out with proper choice of α, β, γ, x this series can be converted into many of the important series in analysis. The logarithmic function, trigonometric functions, binomial series, and Legendre polynomials are just a few of very well-known and much used functions which can be expressed in terms of the hypergeometric series. To quote Eric Bell in his article *The Prince of Mathematics** we see the importance of the study of this series. "By disposing of this series (hypergeometric) in its general form Gauss slew a multitude at one smash. From this work developed many applications to the differential equations of physics in the nineteenth century."

1.2 HYPERGEOMETRIC DIFFERENTIAL EQUATION - HYPERGEOMETRIC SERIES

The second order differential equation

$$x(1-x)y'' + [\gamma-(\alpha+\beta+1)x]y' - \alpha\beta y = 0 \qquad (1.2)$$

is known as Gauss' or the hypergeometric equation.

It can easily be seen that $x = 0$ and $x = 1$ are singular points, and since

$$\lim_{x\to\infty} x\, \frac{\gamma-(\alpha+\beta+1)x}{x(1-x)} = \gamma$$

and

$$\lim_{x\to\infty} x^2\, \frac{-\alpha\beta}{x(1-x)} = 0 \quad ,$$

$x = 0$ is a *regular* singular point.

Similarly, since

$$\lim_{x\to 1} (1-x)\, \frac{\gamma-(\alpha+\beta+1)x}{x(1-x)} = \gamma-(\alpha+\beta+1)$$

and

$$\lim_{x\to 1} (1-x)^2\, \frac{-\alpha\beta}{x(1-x)} = 0 \quad ,$$

$x = 1$ is also a regular singular point.

We desire to find the general solution to Gauss' differential equation in powers of x. Since $x = 0$

*E.T. Bell, *Men of Mathematics*, Simon Schuster (1937)

is a regular singular point, we assume the solution of the differential equation is given by the Frobenius series

$$y = x^r \sum_{n=0}^{\infty} c_n x^n = \sum_{n=0}^{\infty} c_n x^{n+r} \quad .$$

Differentiating term by term we can write

$$y' = \sum_{n=0}^{\infty} (n+r) c_n x^{n+r-1}$$

and

$$y'' = \sum_{n=0}^{\infty} (n+r)(n+r-1) c_n x^{n+r-2} \quad .$$

Substituting the series representations of y'', y', y into (1.2) we find

$$(x-x^2) \sum_{n=0}^{\infty} (n+r)(n+r-1) c_n x^{n+r-2} + [\gamma-(\alpha+\beta+1)x]\cdot$$
$$\sum_{n=0}^{\infty} (n+r) c_n x^{n+r-1} - \alpha\beta \sum_{n=0}^{\infty} c_n x^{n+r} = 0 \quad .$$

Combining monomials of like powers, the equation above can be written in the more compact form

$$\sum_{n=0}^{\infty} (n+r)(n+r-1+\gamma) c_n x^{n+r-1}$$
$$- \sum_{n=0}^{\infty} \left[(n+r)^2 + (\alpha+\beta)(n+r) + \alpha\beta\right] c_n x^{n+r} = 0. \quad (1.3)$$

If we reindex the first series on the lefthand side by setting $n = n + 1$ equation (1.3) becomes

$$\sum_{n=-1}^{\infty} (n+r+1)(n+r+\gamma) c_{n+1} x^{n+r}$$
$$- \sum_{n=0}^{\infty} \left[(n+r)^2 + (\alpha+\beta)(n+r) + \alpha\beta\right] c_n x^{n+r} = 0. \quad (1.4)$$

By removing the $n = -1$ term from the first series on the lefthand side we can write (1.4) in the form

$$r(r-1+\gamma) c_0 x^{r-1} + \sum_{n=0}^{\infty} \{(n+r+1)(n+r+\gamma) c_{n+1}$$
$$- (n+r+\alpha)(n+r+\beta) c_n\} x^{n+r} = 0. \quad (1.5)$$

If we pick $c_o \neq 0$, and since the coefficient $r(r-1+\gamma)c_o$ must equal 0 and we see that the *indicial equation* is

$$r(r-1+\gamma) = 0$$

from which we get two roots

$$r_1 = 0 \quad \text{and} \quad r_2 = 1 - \gamma \quad .$$

Since every coefficient must equal zero we see from (1.5) that

$$(n+r+1)(n+r+\gamma)c_{n+1} - (n+r+\alpha)(n+r+\beta)c_n = 0$$

or

$$c_{n+1} = \frac{(n+r+\alpha)(n+r+\beta)}{(n+r+1)(n+r+\gamma)} c_n, \quad n = 0,1,2,\ldots \qquad (1.6)$$

which is the recurrence relation for Gauss' differential equation.

For $r_1 = 0$ and $r_2 = 1 - \gamma < 0$, ($1-\gamma$ integer or not), the recurrence relation becomes

$$c_{n+1} = \frac{(n+\alpha)(n+\beta)}{(n+1)(n+\gamma)} c_n \quad .$$

Setting $c_0 = 1$ and repeatedly using this formula we arrive at the solution

$$y_1 = 1 + \frac{\alpha\beta}{\gamma} x + \frac{\alpha(1+\alpha)\beta(1+\beta)}{2!\gamma(1+\gamma)} x^2 + \ldots$$
$$+ \frac{\alpha(1+\alpha)\ldots(n-1+\alpha)\beta(1+\beta)\ldots(n-1+\beta)}{n!\gamma(1+\gamma)\ldots(n-1+\gamma)} x^n + \ldots \quad . \qquad (1.7)$$

We see immediately that this solution is the hypergeometric series. It is also referred to as the hypergeometric function.

Before we find another linearly independent solution we wish to write the hypergeometric series in a more convenient form. To do so we introduce the following notation known as the *Pochhammer symbol*.

Definition 1: $(\alpha)_0 = 1$ if $\alpha \neq 0$.

$$(\alpha)_k = \alpha(1+\alpha)\ldots(k-1+\alpha) \quad , \quad k = 1,2,\ldots \quad .$$

Example 1: Find $(2)_k$, $k = 0,1,2,3$.

Using Definition 1, we find

$$(2)_0 = 1$$

$$(2)_1 = 2$$

$$(2)_2 = 2(1+2) = 2\cdot 3 = 6$$

$$(2)_3 = 2(1+2)(2+2) = 2\cdot 3\cdot 4 = 24$$

Example 2: Show $(1)_n = n!$

From Definition 1, we can write

$$(1)_0 = 1$$

and $(1)_n = 1(1+1)(2+1)\ldots[(n-1) + 1]$, $n = 1,2,\ldots$

$$= 1\cdot 2\cdot 3 \ldots . n$$

$$= n! .$$

In fact, it is easy to show $(k)_n = \dfrac{(n+k-1)!}{(k-1)!}$ where k and n are positive integers.

Example 3: How is the expression $(\alpha)_n$, $\alpha \neq 0$, related to the Gamma function? (See Section 4.)

We know that a key identity in dealing with Gamma functions is given by

$$\Gamma(1+\alpha) = \alpha\Gamma(\alpha) \ , \ \alpha \neq 0,-1,-2, \ \ldots \ .$$

By repeated use of this identity we can derive the following relation where $\alpha \neq 0,-1,-2, \ \ldots$ and n is a positive integer, i.e.

$$\Gamma(\alpha+n) = (\alpha+n-1)\Gamma(\alpha+n-1) = (\alpha+n-1)(\alpha+n-2)\ldots\alpha\Gamma(\alpha).$$

But the factor multiplying $\Gamma(\alpha)$ on the righthand side equals $(\alpha)_n$ and, therefore, we can write the identity

$$(\alpha)_n = \frac{\Gamma(\alpha+n)}{\Gamma(\alpha)} \ .$$

Returning to our solution (1.7) of Gauss' equation we see that the hypergeometric series in (1.7) can be written as

$$_2F_1(\alpha,\beta;\gamma;x) = \sum_{n=0}^{\infty} \frac{(\alpha)_n(\beta)_n}{n!(\gamma)_n} x^n$$

where the subscript 2 indicates the two expressions $(\alpha)_n(\beta)_n$ are in the numerator while the subscript 1 indicates the single expression $(\gamma)_n$ in the denominator. In section 7 we shall have more to say about this notation.

Let us return to our search for the general solution of Gauss' differential equation. The second solution is found by using the root $r_2 = 1 - \gamma$ where $1 - \gamma$ is not an integer. The recurrence relation becomes

$$c_{n+1} = \frac{(n+1-\gamma+\alpha)(n+1-\gamma+\beta)}{(n+1-\gamma+1)(n+1-\gamma+\gamma)} c_n$$

$$= \frac{(n+1-\gamma+\alpha)(n+1-\gamma+\beta)}{(n+2-\gamma)(n+1)} c_n, \quad n = 0,1,2,\ldots .$$

Specifically, we see that

$$c_1 = \frac{(1-\gamma+\alpha)(1-\gamma+\beta)}{(2-\gamma)} c_0$$

$$c_2 = \frac{(2-\gamma+\alpha)(2-\gamma+\beta)}{(3-\gamma)2} c_1$$

and eventually

$$c_n = \frac{(n-\gamma+\alpha)(n-1-\gamma+\alpha)\cdots(1-\gamma+\alpha)(n-\gamma+\beta)\cdots(1-\gamma+\beta)}{(n+1-\gamma)(n-\gamma)\cdots(2-\gamma)n!} c_0 .$$

Once again, setting $c_0 = 1$, a linearly independent solution is given by

$$y_2(x) = x^{1-\gamma}\left[1 + \frac{(1-\gamma+\alpha)(1-\gamma+\beta)}{2-\gamma} x + \cdots \right.$$

$$\left. + \frac{(n-\gamma+\alpha)\cdots(1-\gamma+\alpha)(n-\gamma+\beta)\cdots(1-\gamma+\beta)}{(n+1-\gamma)(n-\gamma)\cdots(2-\gamma)n!} x^n + \cdots\right].$$

This solution can be expressed in terms of the hypergeometric series as follows:

$$y_2(x) = x^{1-\gamma}\,{}_2F_1(1-\gamma+\alpha, 1-\gamma+\beta; 2-\gamma; x) .$$

The general solution for Gauss' equation for $1 - \gamma$ not equal to an integer is

$$y(x) = A\,{}_2F_1(\alpha,\beta;\gamma;x) + B\,{}_2F_1(1-\gamma+\alpha,1-\gamma+\beta;2-\gamma;\alpha).$$

It can be shown that the solution for $y_2(x)$ is also true for γ an integer and $r_2 = 1 - \gamma \geq 0$.

3. CONVERGENCE OF HYPERGEOMETRIC SERIES

First we observe that if α or β is equal to zero or a negative integer, the hypergeometric series becomes the hypergeometric polynomial and there is no problem of convergence.

Example 4: Write out the hypergeometric series for $\alpha = 1$, $\beta = -3$, $\gamma = 2$.

$$ {}_2F_1(1,-3;2;x) = 1 + \frac{1(-3)}{1!2}\,x + \frac{1(1+1)(-3)(-3+1)}{2!2(2+1)}\,x^2 + \ldots + \frac{1(2)(3)4(-3)(-2)(-1)(0)}{4!2(3)(4)(5)}\,x^4 $$

$$ = 1 - \frac{3}{2}x + x^2 - \frac{1}{4}x^3 $$

which is a third degree polynomial.

If α,β are not equal to zero or a negative integer, we can find the radius of convergence by applying the ratio test to the hypergeometric series. From the recurrence relation (1.6) we see that the

$$\lim_{n\to+\infty} \frac{c_{n+1}}{c_n}|x| = \lim_{n\to+\infty} \frac{(n+r+\alpha)(n+r+\beta)}{(n+r+1)(n+r+\gamma)}|x| = |x|.$$

Therefore, the radius of convergence is given by $R = 1$.

Example 5: Write out in series form the hypergeometric function ${}_2F_1(1,1;2;x)$.

$$ {}_2F_1(1,1;2;x) = \sum_{n=0}^{\infty} \frac{(1)_n(1)_n}{n!(2)_n}\,x^n = \sum_{n=0}^{\infty} \frac{n!n!}{n!(n+1)!}\,x^n $$

$$ = \sum_{n=0}^{\infty} \frac{x^n}{n+1} \,. \tag{1.8} $$

We know $\frac{1}{1-x} = \sum_{n=0}^{\infty} x^n$ for $|x| < 1$,

and therefore

$$\int \frac{dx}{1-x} = \sum_{n=0}^{\infty} \frac{x^{n+1}}{n+1} = -\ln(1-x).$$

Comparing this series with (1.8) we see that

$$\ln(1-x) = x \sum_{n=0}^{\infty} \frac{x^n}{n+1} = -\, x\,{}_2F_1(1,1;2;x) \quad \text{for} \quad |x|<1.$$

In particular we have represented the log function in terms of the hypergeometric series.

4. SOLUTIONS TO GAUSS' EQUATION FOR LARGE X.

We saw in section 2 that we were able to find the general solution to Gauss' equation in the interval $|x| < 1$. The question arises as to whether we can find a solution for Gauss' equation which converges in the intervals $1 < |x|$. This can be done since we shall demonstrate that $x = \infty$ is also a regular singular point. In order to investigate x at infinity we consider the transformation $x = \frac{1}{t}$ or $t = \frac{1}{x}$ which carries the point $\pm\infty$ into the point zero. If $y(x)$ is a solution of Gauss' equation we shall use the following notation:

$$y(x) = y\left[\frac{1}{t}\right] = \bar{y}(t).$$

Using the fact that $\frac{dt}{dx} = -\frac{1}{x^2} = -t^2$, the derivatives and

$$\frac{dy}{dx} = \frac{d\bar{y}}{dt}\frac{dt}{dx} = -\frac{d\bar{y}}{dt}t^2$$

and

$$\frac{d^2y}{dx^2} = \frac{d}{dx}\left[\frac{dy}{dx}\right] = \frac{dt}{dx}\frac{d}{dt}\left[-t^2\frac{d\bar{y}}{dt}\right]$$

$$= -t^2\left[-t^2\frac{d^2\bar{y}}{dt^2} - 2t\frac{d\bar{y}}{dt}\right] = t^2\left[t^2\frac{d^2\bar{y}}{dt^2} + 2t\frac{d\bar{y}}{dt}\right].$$

Substituting these results into Gauss' equation (1.2) we find

$$\frac{1}{t}\left[1-\frac{1}{t}\right]t^2\left[t^2\,\frac{d^2\bar{y}}{dt^2} + 2t\,\frac{d\bar{y}}{dt}\right]$$

$$+\left[\gamma - (\alpha+\beta+1)\frac{1}{t}\right]\left[-t^2\,\frac{d\bar{y}}{dt}\right] - \alpha\beta\bar{y} = 0 \ .$$

Simplifying this equation we arrive at the transformed Gauss' equation

$$t^2(1-t)\frac{d^2\bar{y}}{dt^2} - \left[(\alpha+\beta-1)t + (2-\gamma)t^2\right]\frac{d\bar{y}}{dt} + \alpha\beta\bar{y} = 0. \tag{1.9}$$

This equation has singular points at $t = 0$ and $t = 1$. Furthermore, by examining the limits

$$\lim_{t\to 0} t \cdot \frac{-[(\alpha+\beta-1)t+(2-\gamma)t^2]}{t^2(1-t)} = 1 - \alpha - \beta \quad \text{replace}$$

above expression.

and

$$\lim_{t\to 0} t^2 \cdot \frac{\alpha\beta}{t^2(1-t)} = \alpha\beta \ ,$$

we see easily that $t = 0$ is a regular singular point of the transformed equation. Since $t = 0$ corresponds to $x = \infty$ in the original Gauss' equation, we say $x = \infty$ is a regular singular point of the original Gauss' equation (1.2). Thus Gauss' equation has 3 regular singular points $x = 0, 1, \infty$.

To find solutions of Gauss' equation for large x we use Frobenius' method around $t = 0$ to solve equation 1.9.

We assume a solution of the form

$$\bar{y}(t) = t^r \sum_{n=0}^{\infty} c_n t^n \ .$$

Proceeding in a way similar to solving equation (1.2) we arrive at the series equation

$$c_0(r-\alpha)(r-\beta)t^r + \sum_{n=1}^{\infty}\Big\{[(n+r)(n+r-1)-(\alpha+\beta-1)(n+r)+\alpha\beta]c_n - [(n+r+1)(n+r-2)+(2-\gamma)(n+r+1)]c_{n+1}\Big\}t^{n+r} = 0.$$

If we set $c_0 \neq 0$, then we arrive at the indicial equation

$$(r-\alpha)(r-\beta) = 0$$

whose solution is

$$r_1 = \alpha, \quad r_2 = \beta$$

which are the indices of the equation.

If we set the coefficients under the sigma sign equal to zero we arrive at the recurrence relation

$$c_n = \frac{(n+r-1)(n+r-\gamma)}{(n+r-\alpha)(n+r-\beta)}\, c_{n-1}, \quad n = 1,2,\ldots \quad .$$

To find the solution to the differential equation corresponding to $r_1 = \alpha$, the recurrence relation becomes

$$c_n = \frac{(n+\alpha-1)(n+\alpha-\gamma)}{n(n+\alpha-\beta)}\, c_{n-1} \quad .$$

By repeated use of the recurrence relation we arrive at the value of the nth coefficient c_n.

$$c_n = \frac{(\alpha)_n(1+\alpha-\gamma)_n}{n!(1+\alpha-\beta)_n}\, c_0 \ , \quad c_0 \neq 0.$$

Therefore, the solution corresponding to $r_1 = \alpha$ is

$$\bar{y}_1(t) = t^{\alpha} \sum_{n=0}^{\infty} \frac{(\alpha)_n(1+\alpha-\gamma)_n}{n!(1+\alpha-\beta)_n}\, t^n$$

from which we get the corresponding solution for Gauss' equation (1.2) which is

$$y_1(x) = x^{-\alpha} \sum_{n=0}^{\infty} \frac{(\alpha)_n(1+\alpha-\gamma)_n}{n!(1+\alpha-\beta)_n} \left[\frac{1}{x}\right]^n \quad .$$

notice this answer can be written in the hypergeometric notation

$$y_1(x) = x^{-\alpha}\, {}_2F_1\left[\alpha, 1+\alpha-\gamma; 1+\alpha-\beta; \frac{1}{x}\right].$$

Since we already know the hypergeometric series

${}_2F_1\left[\alpha, 1+\alpha-\gamma; \alpha-\beta; \frac{1}{x}\right]$ converges for $\frac{1}{|x|} < 1$, it is clear that $y_1(x)$ converges for $1 < |x|$. Because of the symmetry of α and β in the equation, the other linear independent solution is found to equal

$$y_2(x) = x^{-\beta} {}_2F_1\left[\beta, 1+\beta-\gamma; 1+\beta-\alpha; \frac{1}{x}\right]$$

provided $\alpha + \beta \neq$ an integer. The general solution to Gauss' equation for $1 < |x|$ takes the form

$$y(x) = A {}_2F_1\left[\alpha, 1+\alpha-\gamma; 1+\alpha-\beta; \frac{1}{x}\right]$$

$$+ B {}_2F_1\left[\beta, 1+\beta-\gamma; 1+\beta-\alpha; \frac{1}{x}\right].$$

5. CONFLUENT HYPERGEOMETRIC DIFFERENTIAL EQUATION

We have just examined the solutions to Gauss' equation. In particular we found that the equation admits three regular singular points $x = 0, 1, \infty$. If we replace the independent variable x by $\frac{u}{\beta}$, Gauss' equation looks like

$$u\left[1-\frac{u}{\beta}\right] \frac{d^2\bar{y}}{du^2} + \left[\gamma - \left[\frac{\alpha}{\beta} + 1 + \frac{1}{\beta}\right]u\right]\frac{d\bar{y}}{du} - \alpha\bar{y} = 0 \quad (1.10)$$

where

$$y(x) = y\left[\frac{u}{\beta}\right] = \bar{y}(u).$$

This equation can be shown to have regular singular points at $x = 0, \beta, \infty$. Now if we let $\beta \to +\infty$, then, equation (1.10) is formally transformed into the equation

$$u\bar{y}'' + (\gamma-u)\bar{y}' - \alpha\bar{y} = 0 .$$

This differential equation is called the *confluent hypergeometric equation* because it results when the regular singular point $x = \beta$ "flows together with" the regular singular point at infinity. For purposes of the discussion which follows we shall rewrite the hypergeometric differential equation in terms of x and y that is

$$xy'' + (\gamma-x)y' - \alpha y = 0. \tag{1.11}$$

This equation is also known as Kummer's equation and has a regular singular point at $x = 0$ but an *irregular* singular point at $x = \infty$.

We shall find the solution to equation (1.11) around 0 by using the method of Frobenius. If we let

$$y(x) = x^r \sum_{n=0}^{\infty} c_n x^n$$

we find, after substituting into equation (1.11) the series equation

$$[r(r-1)+\gamma r]c_0 x^{r-1} +$$

$$\sum_{n=0}^{\infty} \Big[(n+r+1)(n+r+\gamma)c_{n+1} - (n+r+\alpha)c_n\Big]x^{n+r} = 0 .$$

Assuming $c_0 \neq 0$ we arrive at the indicial equation

$$r(r-1) + \gamma r = 0$$

from which we find the indicies $r_1 = 0$ and $r_2 = 1 - \gamma$. Setting the coefficient of the general series term under the sigma sign equal to zero we are led to the recurrence relation

$$c_{n+1} = \frac{n+r+\alpha}{(n+r+1)(n+r+\gamma)} c_n \quad , \quad n = 0,1,2,\ldots .$$

To find one solution to the differential equation we set $r = 0$ in the recurrence relation and observe that

$$c_{n+1} = \frac{n+\alpha}{(n+1)(n+\gamma)} c_n$$

which by repeated use yields the result

$$c_n = \frac{(\alpha)_n}{n!(\gamma)_n} c_0 .$$

The corresponding solution takes the form upon letting $c_0 = 1$

$$y_1(x) = \sum_{n=0}^{\infty} \frac{(\alpha)_n}{n!(\gamma)_n} x^n \quad . \tag{1.12}$$

Just as we found it useful to represent the hypergeometric series by the symbol ${}_2F_1$, we define the series in (1.12) as

$$ {}_1F_1(\alpha;\gamma;x) = \sum_{n=0}^{\infty} \frac{(\alpha)_n}{n!(\gamma)_n} x^n \quad . $$

This particular series is called *Kummer's series* or the *confluent hypergeometric series*.

If $1 - \gamma$ is not an integer we can easily find a second linearly independent solution to the confluent hypergeometric equation. It is given by

$$ y_2(x) = x^{1-\gamma} \sum_{n=0}^{\infty} \frac{(1-\gamma+\alpha)_n}{n!(2-\gamma)_n} x^n \quad . $$

But $y_2(x)$ can be written in terms of ${}_1F_1$. It is

$$ y_2(x) = x^{1-\gamma}{}_1F_1(1-\gamma+\alpha;2-\gamma;x) $$

and the general solution to the confluent hypergeometric equation is given as

$$ y(x) = A{}_1F_1(\alpha;\gamma;x) + Bx^{1-\gamma}{}_1F_1(1-\gamma+\alpha;2-\gamma;x) $$

if $1 - \gamma \neq$ an integer.

It is a straightforward application of the ratio test to show that the confluent hypergeometric series converges for all x, (i.e. $R = \infty$).

6. RELATION WITH OTHER SPECIAL FUNCTIONS.

In section 1 we saw that the solution to Gauss' differential equation

$$ x(1-x)y'' + [\gamma - (\alpha + \beta + 1)x]y' - \alpha\beta y = 0 $$

is given by

$$ y(x) = A{}_2F_1(\alpha,\beta;\gamma;x) $$
$$ + Bx^{1-\gamma}{}_2F_1(1-\gamma+\alpha,1-\gamma+\beta;2-\gamma;x) . $$

Unfortunately, many of the special classic differential equations have the coefficient of the y'' term in the form $(1-x^2)$ rather than $x(1-x)$. We shall now show how to transform Gauss' equation into this form.

If we let $x = \frac{1-u}{2}$, then

$$x(1-x) = \frac{1-u}{2}\left[1 - \frac{1-u}{2}\right] = \frac{1-u^2}{4} .$$

This transformation converts the coefficient of y'' into the desirable form. Let us see what Gauss' equation looks like if we use the transformation $x = \frac{1-u}{2}$ or $u = 1 - 2x$ and

$$y(x) = y\left[\frac{1-u}{2}\right] = \bar{y}(u) .$$

The derivatives

$$y'(x) = \frac{dy}{dx} = \frac{dy}{du}\frac{du}{dx} = -2\frac{d\bar{y}}{du}$$

and

$$y''(x) = \frac{d^2y}{dx^2} = \frac{d\left[\frac{d\bar{y}}{du}\right]}{du}\frac{du}{dx}(-2) = 4\frac{d^2\bar{y}}{du^2} .$$

Substituting these derivatives into the differential equation (1.13) we arrive at the transformed differential equation

$$\frac{1-u^2}{4} \cdot 4\frac{d^2\bar{y}}{du^2} + \left[\gamma - (\alpha+\beta+1)\frac{1-u}{2}\right](-2)\frac{d\bar{y}}{du} - \alpha\beta\bar{y} = 0$$

or

$$(1-u^2)\frac{d^2\bar{y}}{du^2} + [\alpha+\beta+1-2\gamma-(\alpha+\beta+1)u]\frac{d\bar{y}}{du}$$
$$- \alpha\beta\bar{y} = 0 . \qquad (1.14)$$

This equation is an alternate form of Gauss' differential equation and its solution is

$$\bar{y}(u) = A\,{}_2F_1\left[\alpha,\beta;\gamma;\frac{1-u}{2}\right]$$
$$+ B\left[\frac{1-u}{2}\right]^{1-\gamma}{}_2F_1\left[1-\gamma+\alpha,1-\gamma+\beta;2-\gamma;\frac{1-u}{2}\right].$$

Example 6: Show that the Legendre polynomials $P_n(u)$ satisfies

$$P_n(u) = {}_2F_1\left[-n, n+1; 1; \frac{1-u}{2}\right].$$

Legendre's equation is given by

$$(1-u^2)\bar{y}''(u) - 2u\bar{y}(u) + n(n+1)\bar{y} = 0. \qquad (1.15)$$

This equation is a special case of the differential equation (1.14) if $\alpha = -n$, $\beta = n + 1$, $\gamma = 1$, $x = \frac{1-u}{2}$.

The particular solution of (1.15) satisfying the initial conditions $y(1) = 1$ and bounded on $(-1,1)$ is $y(u) = P_n(u)$. Under these conditions the solution of Gauss' equation is

$$\bar{y}(u) = y\left[\frac{1-u}{2}\right] = A\,{}_2F_1\left[-n, n+1; 1; \frac{1-u}{2}\right].$$

At $u = 1$

$$\bar{y}(1) = 1 = y(0) = A\,{}_2F_1(-n, n+1; 1; 0) = A\ .$$

Therefore, $A = 1$ and

$$P_n(u) = {}_2F_1\left[-n, n+1; 1; \frac{1-u}{2}\right].$$

Once we have transformed Gauss' equation into the form (1.14) it is relatively easy to convert many of the well-known classic differential equations into Gauss' hypergeometric equation (1.13). See Chart (1).

Example 7: Show that the Chebyshev polynomial $T_3(u) = 4u^3 - 3u = {}_2F_1\left[-3, 3; \frac{1}{2}; \frac{1-u}{2}\right]$.

We know $${}_2F_1\left[-3, 3; \frac{1}{2}; \frac{1-u}{2}\right] = \sum_{n=0}^{\infty} \frac{(-3)_n (3)_n}{n!\left[\frac{1}{2}\right]_n} \left[\frac{1-u}{2}\right]^n$$

$$= 1 - \frac{3\cdot 3}{\frac{1}{2}}\left[\frac{1-u}{2}\right] + \frac{(-3)(-2)(3)(4)}{2!\ \frac{1}{2}\cdot\frac{3}{2}}\left[\frac{1-u}{2}\right]^2$$

$$+ \frac{(-3)(-2)(-1)(3)(4)(5)}{3!\frac{1}{2}\cdot\frac{3}{2}\cdot\frac{5}{2}}\left[\frac{1-u}{2}\right]^3$$

$$= 1 - 9(1-u) + 12(1-u)^2 - 4(1-u)^3$$

$$= 4u^3 - 3u\ .$$

Example 8: Show that the Jacobi polynomial

$$P_2^{(1,1)}(u) = \frac{15u^2}{4} - \frac{3}{4} = 3\,{}_2F_1\left[-2, 5; 2; \frac{1-u}{2}\right].$$

	Standard Form	Equivalent Gauss' Form
Legendre	$(1-u^2)\overline{y}''-2u\overline{y}'+n(n+1)\overline{y} = 0$	$x(1-x)y''+[1-2x]y'+n(n+1)y = 0$
Jacobi	$(1-u^2)\overline{y}''+[b-a-(a+b+2)u]\overline{y}'+n(n+a+b+1)\overline{y} = 0$	$x(1-x)y''+[a+1-(a+b+2)x]y'+n(n+a+b+1)y = 0$
Chebyshev #1	$(1-u^2)\overline{y}''-u\overline{y}'+n^2\overline{y} = 0$	$x(1-x)y''+[\frac{1}{2} - x]y'+n^2y = 0$
Chebyshev #2	$(1-u^2)\overline{y}''-3u\overline{y}+n(n+2)\overline{y} = 0$	$x(1-x)y''+[\frac{3}{2} - 3x]y'+n(n+2)y = 0$
Gegenbauer	$(1-u^2)y''-(2\lambda+1)u\overline{y}'+n(n+2\lambda)\overline{y} = 0$	$x(1-x)y''+[\lambda+\frac{1}{2} -(2\lambda+1)x]y'+n(n+2\lambda)y = 0$

Note: In all cases change of independent variable is given by $x = \frac{1-u}{2}$

CHART 1

We know $3\left[{}_2F_1\left[-2,5;2;\frac{1-u}{2}\right]\right]$

$$= \sum_{n=0}^{\infty} \frac{(-2)_n(5)_n}{n!(2)_n} \left[\frac{1-u}{2}\right]^n$$

$$= 3\left[1 + \frac{(-2)5}{2}\left[\frac{1-u}{2}\right] + \frac{(-2)(-1)(5)(6)}{2!2\cdot3}\left[\frac{1-u}{2}\right]^2\right.$$

$$= \frac{15u^2}{4} - \frac{3}{4} \quad .$$

7. SOME PROPERTIES OF THE HYPERGEOMETRIC FUNCTION.

In the same way that trigonometric functions, hyperbolic functions, etc. obey certain relationships, we can derive several useful properties concerning hypergeometric functions.

Recall from subsection 2 that the hypergeometric function ${}_2F_1$ is defined as follows:

$${}_2F_1(\alpha,\beta;\gamma;x) = \sum_{n=0}^{\infty} \frac{(\alpha)_n(\beta)_n}{n!(\gamma)_n} x^n \qquad |x| < 1. \qquad (1.16)$$

Theorem 1: ${}_2F_1(\alpha,\beta;\gamma;0) = 1.$

Proof: When we set $x = 0$ in (1.16), the series has only one nonzero term (the first) and its value is 1.

Theorem 2: The hypergeometric function is commutative with respect to α and β, i.e.

$${}_2F_1(\alpha,\beta;\gamma;x) = {}_2F_1(\beta,\alpha;\gamma;x) \; .$$

Proof: Since $(\alpha)_n(\beta)_n = (\beta)_n(\alpha)_n$,

$${}_2F_1(\alpha,\beta;\gamma;x) = \sum_{n=0}^{\infty} \frac{(\alpha)_n(\beta)_n}{n!(\gamma)_n} x^n = \sum_{n=0}^{\infty} \frac{(\beta)_n(\alpha)_n}{n!(\gamma)_n} x^n$$

$$= {}_2F_1(\beta,\alpha;\gamma;x).$$

There is a relationship between the hypergeometric function and the confluent hypergeometric function. If in Gauss' equation (1.13) we replace the independent variable x by $\frac{u}{\beta}$. It is easy to show that $u = \beta$ is a regular singular point and ${}_2F_1\left[\alpha,\beta;\gamma;\frac{u}{\beta}\right]$ is a solution to the differential equation.

Now suppose we allow β to slide up toward $+\infty$. Then

$$\lim_{\beta\to+\infty} {}_2F_1\left[\alpha,\beta;\gamma;\frac{u}{\beta}\right] = \lim_{\beta\to+\infty} \sum_{n=0}^{\infty} \frac{(\alpha)_n(\beta)_n}{n!(\gamma)_n}\left[\frac{u}{\beta}\right]^n$$

$$= \lim_{\beta\to+\infty} \sum_{n=0}^{\infty} \frac{(\alpha)_n}{n!(\gamma)_n}\frac{(\beta)_n}{\beta^n}u^n$$

$$= \sum_{n=0}^{\infty} \frac{(\alpha)_n}{n!(\gamma)_n}\left[\lim_{\beta\to+\infty}\frac{(\beta)_n}{\beta^n}\right]u^n . \tag{1.17}$$

Now

$$\lim_{\beta\to+\infty}\frac{(\beta)_n}{\beta^n} = \lim_{\beta\to+\infty}\frac{\beta(\beta+1)\ldots(\beta+k-1)}{\underbrace{\beta\cdot\beta\ \ldots\ \beta}_{k \text{ factors}}}$$

$$= \lim_{\beta\to+\infty} 1\left[1+\frac{1}{\beta}\right]\ \ldots\ \left[1-\frac{k-1}{\beta}\right] = 1.$$

Therefore, we see from (1.17)

$$\lim_{\beta\to+\infty} {}_2F_1\left[\alpha,\beta;\gamma;\frac{u}{\beta}\right] = \sum_{n=0}^{\infty}\frac{(\alpha)_n}{n!(\gamma)_n}u^n = {}_1F_1(\alpha;\gamma;u),$$

which is the confluent hypergeometric function.

It is from this technique that the name confluent hypergeometric function or confluent differential equation is derived. If we recall we modified Gauss' equation so that instead of having regular singular points at $0,1,\infty$, our transformed equation had its regular singular points at $0,\beta,\infty$. Furthermore, we allowed the singular point β to move toward $+\infty$ so that the two singular points flowed together or became *confluent*.

The hypergeometric function satisfies several differentiation and integration relations. Of prime importance is the following theorem

Theorem 3: For $|x| < 1$,

$$\frac{d}{dx}\left\{{}_2F_1(\alpha,\beta;\gamma;x)\right\} = \frac{\alpha\beta}{\gamma}\ {}_2F_1(\alpha+1,\beta+1;\gamma+1;x).$$

Proof: Using the definition of ${}_2F_1$ we can write

$$\frac{d}{dx}\left\{{}_2F_1(\alpha,\beta;\gamma;x)\right\} = \frac{d}{dx}\sum_{n=0}^{\infty}\frac{(\alpha)_n(\beta)_n}{n!(\gamma)_n}x^n$$

$$= \sum_{n=0}^{\infty} n\,\frac{(\alpha)_n(\beta)_n}{n!(\gamma)_n}x^{n-1} = \sum_{n=0}^{\infty}\frac{\alpha(\alpha+1)_{n-1}\beta(\beta+1)_{n-1}}{(n-1)!\gamma(\gamma+1)_{n-1}}x^{n-1}$$

$$= \frac{\alpha\beta}{\gamma}\ {}_2F_1(\alpha+1,\beta+1;\gamma+1;x)\quad .$$

By repeated use of this theorem we arrive at the general result.

Theorem 4: For $1 < |x|$,

$$\frac{d^n}{dx^n}\left\{{}_2F_1(\alpha,\beta;\gamma;x)\right\} = \frac{(\alpha)_n(\beta)_n}{(\gamma)_n}\ {}_2F_1(\alpha+n,\beta+n;\gamma+n;x)\quad .$$

Using the result of Theorem 3, we easily arrive at the following integral relationship.

Corollary 1: If $|x| < 1$ and $\alpha,\beta,\gamma \neq 1$, then

$$\int {}_2F_1(\alpha,\beta;\gamma;x)dx$$

$$= \frac{\gamma-1}{(\alpha-1)(\beta-1)}\ {}_2F_1(\alpha-1,\beta-1;\gamma-1;x) + c.$$

Gauss defined six functions called *contiguous functions* ("touching functions") associated with ${}_2F_1(\alpha,\beta;\gamma;x)$ in the following way. If we replace the indice α by $\alpha + 1$ or $\alpha - 1$ and leave all other indices unchanged, i.e. ${}_2F_1(\alpha+1,\beta;\gamma;x)$ or

${}_2F_1(\alpha-1,\beta;\gamma;x)$, we define two of the contiguous functions. Doing the same reindexing with β and γ, gives us the other 4 contiguous functions.

Having defined these functions it is possible to show that the hypergeometric function ${}_2F_1(\alpha,\beta;\gamma;x)$ can be expressed in terms of any two contiguous functions whose coefficients are functions of x.

Example 9: Show that

$$(\alpha-\gamma+1)\,{}_2F_1(\alpha,\beta;\gamma;x)$$
$$= \alpha\,{}_2F_1(\alpha+1,\beta;\gamma;x) - (\gamma-1)\,{}_2F_1(\alpha,\beta;\gamma-1;x) \quad . \qquad (1.18)$$

By writing the two terms on the righthand side of (1.18) we have

$$\alpha\,{}_2F_1(\alpha+1,\beta;\gamma;x) - (\gamma-1)\,{}_2F_1(\alpha,\beta;\gamma-1;x)$$

$$= \sum_{n=0}^{\infty} \frac{\alpha(\alpha+1)_n(\beta)_n}{n!(\gamma)_n} x^n - \sum_{n=0}^{\infty} \frac{(\gamma-1)(\alpha)_n(\beta)_n}{n!(\gamma-1)_n} x^n \quad .$$

But from our knowledge of Pochhammer's symbols we can write

$$\alpha(\alpha+1)_n = (\alpha+n)(\alpha)_n \quad \text{and} \quad \frac{\gamma-1}{(\gamma-1)_n} = \frac{\gamma+n-1}{(\gamma)_n} \quad .$$

Substituting these identities in the appropriate series above we have

$$\alpha\,{}_2F_1(\alpha+1,\beta;\gamma;x) - (\gamma-1)\,{}_2F_1(\alpha,\beta;\gamma-1;x)$$

$$= \sum_{n=0}^{\infty} \frac{(\alpha+n)(\alpha)_n(\beta)_n}{n!(\gamma)_n} x^n - \sum_{n=0}^{\infty} \frac{(\gamma+n-1)(\alpha)_n(\beta)_n}{n!(\gamma)_n} x^n$$

$$= (\alpha-\gamma-1) \sum_{n=0}^{\infty} \frac{(\alpha)_n(\beta)_n}{n!(\gamma)_n} x^n = (\alpha-\gamma+1)\,{}_2F_1(\alpha,\beta;\gamma;x).$$

Unfortunately, from our knowledge of combinations, six contiguous functions taken two at a time yields 15 recurrence relations (i.e. ${}_6C_2 = 15$). Some of these recurrence relations are left as exercises but for a complete listing see Rainville, *Special Functions* MacMillan Co. 1960, p. 71, Ex. 21.

The hypergeometric series is usually defined in terms of a MacLaurin series, but it can also be written in terms of an integral which becomes another useful tool when seeking properties of the hypergeometric series. We shall show how this integral representation can be found in the next theorem.

Theorem 5: The hypergeometric series ${}_2F_1(\alpha,\beta;\gamma;x)$ can be represented in integral form as follows:

$${}_2F_1(\alpha,\beta;\gamma;x) = \frac{\Gamma(\gamma)}{\Gamma(\beta)\Gamma(\gamma-\beta)} \int_0^1 \frac{t^{\beta-1}(1-t)^{\gamma-\beta-1}}{(1-xt)^{\alpha}}\, dt$$

provided $\gamma > \beta > 0$.

Proof: Earlier we showed that Pochhammer's symbol can be written as

$$(\alpha)_n = \frac{\Gamma(\alpha+n)}{\Gamma(\alpha)} .$$

The factor $\frac{(\beta)_n}{(\gamma)_n}$ can be written in terms of the Gamma function as follows:

$$\begin{aligned}\frac{(\beta)_n}{(\gamma)_n} &= \frac{\Gamma(\beta+n)\Gamma(\gamma)}{\Gamma(\beta)\Gamma(\gamma+n)} \\ &= \frac{\Gamma(\beta+n)\Gamma(\gamma-\beta)}{\Gamma(\gamma+n)} \cdot \frac{\Gamma(\gamma)}{\Gamma(\beta)\Gamma(\gamma-\beta)} \qquad (1.19)\end{aligned}$$

where we have multiplied the expression on the righthand side by $\frac{\Gamma(\gamma-\beta)}{\Gamma(\gamma-\beta)}$.

Now the factor

$$\frac{\Gamma(\beta+n)\Gamma(\gamma-\beta)}{\Gamma(\gamma+n)} = B(\beta+n,\gamma-\beta) = \int_0^1 t^{\beta+n-1}(1-t)^{\gamma-\beta-1}dt \qquad (1.20)$$

provided $\gamma > \beta > 0$ where $B(x,y)$ is the Beta function.

Substituting into the series representation of the hypergeometric series the results from (1.19) and (1.20) we have

$$ {}_2F_1(\alpha,\beta;\gamma;x) = \sum_{n=0}^{\infty} \frac{(\alpha)_n(\beta)_n}{n!(\gamma)_n} x^n $$

$$ = \sum_{n=0}^{\infty} \frac{(\alpha)_n}{n!}\left[\frac{\Gamma(\gamma)}{\Gamma(\beta)\Gamma(\gamma-\beta)}\right]\int_0^1 t^{\beta+n-1}(1-t)^{\gamma-\beta-1}dt. $$

Interchanging the order of the summation and integration we can write

$$ {}_2F_1(\alpha,\beta;\gamma;x) $$

$$ = \frac{\Gamma(\gamma)}{\Gamma(\beta)\Gamma(\gamma-\beta)}\left[\int_0^1 \sum_{n=0}^{\infty}\frac{(\alpha)_n}{n!}\, t^{\beta+n-1}(1-t)^{\gamma-\beta-1}dt\right]x^n $$

$$ = \frac{\Gamma(\gamma)}{\Gamma(\beta)\Gamma(\gamma-\beta)}\int_0^1 t^{\beta-1}\,(1-t)^{\gamma-\beta-1}\sum_{n=0}^{\infty}\frac{(\alpha)_n}{n!}\,(xt)^n dt. \tag{1.21} $$

The series in (1.21) is the binomial expansion of the function

$$ \frac{1}{(1-xt)^{\alpha}}, \quad -1 < xt < 1 \quad . $$

Replacing the series in (1.21) by this function we arrive at an integral representation for the hypergeometric series which is

$$ {}_2F_1(\alpha,\beta;\gamma;x) = \frac{\Gamma(\gamma)}{\Gamma(\beta)\Gamma(\gamma-\beta)}\int_0^1 \frac{t^{\beta-1}(1-t)^{\gamma-\beta-1}}{(1-xt)^{\alpha}}\,dt \quad . \tag{1.22} $$

From this result we get the particularly simple evaluation of the hypergeometric series at $x = 1$. We find that

$$ {}_2F_1(\alpha,\beta;\gamma;1) = \frac{\Gamma(\gamma)}{\Gamma(\beta)\Gamma(\gamma-\beta)}\int_0^1 t^{\beta-1}(1-t)^{\gamma-\alpha-\beta-1}dt. $$

But as we have seen earlier in (1.20) the integral in (1.22) can be written in terms of Gamma functions. Thus

$$ {}_2F_1(\alpha,\beta;\gamma;1) = \frac{\Gamma(\gamma)}{\Gamma(\beta)\Gamma(\gamma-\beta)} \cdot \frac{\Gamma(\beta)\Gamma(\gamma-\alpha-\beta)}{\Gamma(\gamma-\alpha)} $$

$$ = \frac{\Gamma(\gamma)\Gamma(\gamma-\alpha-\beta)}{\Gamma(\gamma-\alpha)\Gamma(\gamma-\beta)} . $$

8. PROPERTIES OF THE CONFLUENT HYPERGEOMETRIC FUNCTION.

Many of the properties derived in the last section for Gauss' hypergeometric series also apply to the confluent hypergeometric function or Kummer's function

$$ {}_1F_1(\alpha;\beta;x) = \sum_{n=0}^{\infty} \frac{(\alpha)_n}{n!(\gamma)_n} x^n , \quad |x| < \infty . $$

The result

$$ {}_1F_1(\alpha;\beta;0) = 1 $$

follows from the definition of ${}_1F_1(\alpha;\beta;x)$.

Furthermore, many well-known functions are included among the confluent hypergeometric functions, for example,

$$ e^x = \sum_{n=0}^{\infty} \frac{x^n}{n!} = {}_1F_1(\beta;\beta;x) $$

$$ e^{-x} = \sum_{n=0}^{\infty} \frac{(-x)^n}{n!} = {}_1F_1(\beta;\beta;-x) $$

Theorem 6: For all x

$$ \frac{d}{dx}\{{}_1F_1(\alpha;\beta;x)\} = \frac{\alpha}{\beta}\, {}_1F_1(\alpha+1;\beta+1;x) $$

and in general

$$ \frac{d^n}{dx^n}\{{}_1F_1(\alpha;\beta;x)\} = \frac{(\alpha)_n}{(\beta)_n}\, {}_1F_1(\alpha+n;\beta+n;x). $$

Proof: The proof is similar to Theorem 3 and Theorem 4.

We recall from Theorem 5 that the hypergeometric series can be written as an integral. This type of representation is also available to the confluent hypergeometric series.

Theorem 7: If $\gamma > \alpha > 0$, then

$$ {}_1F_1(\alpha;\gamma;x) = \frac{\Gamma(\gamma)}{\Gamma(\gamma)\Gamma(\gamma-\alpha)} \int_0^1 t^{\alpha-1}(1-t)^{\gamma-\alpha-1}e^{xt}dt \quad . $$

The confluent hypergeometric series has four contiguous functions namely, ${}_1F_1(\alpha+1;\beta;x)$, ${}_1F_1(\alpha-1;\beta;x)$, ${}_1F_1(\alpha;\beta+1;x)$, ${}_1F_1(\alpha,\beta-1;x)$. From these functions we are able to generate 6 recurrence relations $\left[{}_4C_2 = 6\right]$ between any two distinct contiguous functions and the confluent hypergeometric series.

For example one of these recurrence relations is given as

$$ (2\alpha-\beta+x)\,{}_1F_1(\alpha;\beta;x) = \alpha\,{}_1F_1(\alpha+1;\beta;x) - (\beta-\alpha)\,{}_1F_1(\alpha-1;\beta;x). \qquad (1.23) $$

Example 10: Derive the recurrence relation for the Laguerre polynomial L_n

$$ (n+1)L_{n+1}(x) = (2n-1-x)L_n(x) - nL_{n-1}(x). $$

We write the Laguerre polynomial as a confluent hypergeometric series as follows:

$$ L_n(x) = {}_1F_1(-n;1;x) \qquad (1.24) $$

If we apply the recurrence relation for contiguous functions we derived in (1.23), we can write

$$ (-2n-1+x)\,{}_1F_1(-n;1;x) = -n\,{}_1F_1(-(n-1);1;x) + (1+n)\,{}_1F_1(-(n+1);1;x). \qquad (1.25) $$

Using the identity (1.24), the recurrence relation (1.25) in terms of Laguerre polynomials becomes

$$ (n+1)L_{n+1}(x) = (2n-1-x)L_n(x) - nL_{n-1}(x). $$

Although the recurrence relation for the contiguous confluent hypergeometric series was easily converted into a recurrence relation for the Laguerre polynomials, this approach is rarely used because as we shall see there are more direct ways to find recurrence relations for orthogonal polynomials and Bessel functions.

9. GENERATING FUNCTIONS.

Back in section (7) we showed that although it is a relatively straightforward matter to derive recurrence relations for hypergeometric functions, it is not easy to adapt them to specific orthogonal polynomials.

Generating functions not only permit us to derive recurrence relations directly but also yield other properties related to orthogonal polynomials.

Definition 2: A function $G(x,t) = \sum_{n=0}^{\infty} c_n g_n(x) t^n$ is called a generating function for the set of functions $\{g_n(x)\}$.

As an example of a generating function we recall that

$$\frac{1}{1-xt} = \sum_{n=0}^{\infty} x^n t^n \quad , \quad |xt| < 1.$$

Therefore, $(1-xt)^{-1}$ is a *generating function* for the set of functions $\{x^n\}$.

Unfortunately, there appears to be no straight forward systematic approach to finding generating functions and although a great number of generating functions have been found, they are attained by individual methods.

A number of generating functions are made up from the expression $2xt-t^2$. Let $G(u)$ be a function that has a power series expansion and let

$$G(2xt-t^2) = \sum_{n=0}^{\infty} g_n(x)t^n \quad .$$

If we let $F(x,t) = G(2xt-t^2)$ and compute the two derivatives

$$\frac{\partial F}{\partial x} = 2t\frac{dG}{du}$$

and

$$\frac{\partial F}{\partial t} = (2x-2t)\frac{dG}{du}$$

we arrive, after eliminating $\frac{dG}{du}$, at the differential equation

$$(x-t)\frac{\partial F}{\partial x} - t\frac{\partial F}{\partial t} = 0 \quad . \tag{1.26}$$

Now since

$$F(x,t) = \sum_{n=0}^{\infty} g_n(x)t^n \quad ,$$

$$\frac{\partial F}{\partial x} = \sum_{n=0}^{\infty} g_n'(x)t^n$$

and

$$\frac{\partial F}{\partial t} = \sum_{n=0}^{\infty} ng_n(x)t^{n-1} = \sum_{n=1}^{\infty} ng_n(x)t^{n-1} \quad ,$$

we have after substituting these derivatives into (1.26)

$$\sum_{n=0}^{\infty} xg_n'(x)t^n - \sum_{n=0}^{\infty} g_n'(x)t^{n+1} - \sum_{n=1}^{\infty} ng_n(x)t^n = 0 \quad .$$

Reindexing the middle series we can write

$$\sum_{n=0}^{\infty} xg_n'(x)t^n - \sum_{n=1}^{\infty} g_{n-1}'(x)t^n - \sum_{n=1}^{\infty} ng_n(x)t^n = 0$$

or

$$xg_0'(x) + \sum_{n=1}^{\infty} \Big[xg_n'(x)-g_{n-1}'(x)-ng_n(x)\Big]t^n = 0 \quad .$$

Since the series must hold for arbitrary t, this can only happen if each coefficient of t^n is zero.

Therefore, we have

$$xg_0'(x) = 0$$

or

$$g_0'(x) = 0 \qquad \text{if} \quad x \neq 0$$

and

$$xg_n'(x) - g_{n-1}'(x) - ng_n(x) = 0, \quad n = 1,2,\ldots \tag{1.27}$$

The equation (1.27) is a recurrence relation between the functions

$$g_n'(x),\ g_{n-1}'(x),\ g_n(x) .$$

Example 11: The generating function for the Hermite polynomials $H_n(x)$ is given by

$$\exp(2xt-t^2) = \sum_{n=0}^{\infty} \frac{H_n(x)}{n!} t^n .$$

Using the relation (1.27) we find that

$$x\frac{H_n'(x)}{n!} - \frac{H_{n-1}'(x)}{(n-1)!} - n\frac{H_n(x)}{n!} = 0$$

or

$$xH_n'(x) - nH_n(x) = nH_{n-1}'(x)$$

which is a recurrence relation for the Hermite polynomials.

EXERCISE 1

1. Write out ${}_2F_1(\alpha-\gamma+1,\beta-\gamma+1;2-\gamma;x)$ as an infinite series. Show that $x^{1-\gamma}{}_2F_1(\alpha-\gamma+1,\beta-\gamma+1;2-\gamma;x)$ is a solution of Gauss' differential equation, if $1-\gamma$ is not an integer or γ is an integer such that $1-\gamma \geq 0$.

2. (a) Show that the hypergeometric series becomes the geometric series when $\alpha=1,\beta=\gamma$ in ${}_2F_1(\alpha,\beta;\gamma;x)$.

(b) Can α, β, γ be chosen in other ways to yield the geometric series?

3. (a) Show that the Legendre polynomial $P_2(x)$ can be expressed in the form ${}_2F_1\left[-2,3;1;\frac{1-x}{2}\right]$.

(b) Show that $P_n(x) = {}_2F_1\left[-n,n+1;1;\frac{1-x}{2}\right]$.

4. Show that

(a) $\mathrm{Sin}^{-1}x = x\,{}_2F_1\left[\frac{1}{2},\frac{1}{2};\frac{3}{2};x^2\right]$

(b) $\mathrm{Tan}^{-1}x = x\,{}_2F_1\left[\frac{1}{2},1;\frac{3}{2};-x^2\right]$.

5. Prove

(a) $(\alpha)_n = (\alpha)_{n-1}(\alpha+n-1)$, $n = 1,2,\ldots$, $\alpha \neq 0$

(b) $(\alpha)_n = \alpha(\alpha+1)_{n-1}$, $n = 1,2,\ldots$, $\alpha \neq 0$.

6. Prove

$$(\alpha)_{2n} = 2^{2n}\left[\frac{a}{2}\right]_n \left[\frac{\alpha+1}{2}\right]_n .$$

7. Show that the Bessel function $J_p(x)$ can be expressed in terms of the confluent hypergeometric function as

$$J_p(z) = \frac{1}{\Gamma(p+1)}\left[\frac{z}{2}\right]^p e^{-iz}\;{}_1F_1\left[\frac{1}{2}+p,1+2p;2iz\right] .$$

Hint: $J_p(z)$ satisfies

$$z^2J_p'' + zJ_p' + \left[z^2-p^2\right]J_p = 0$$

and is given explicitly as

$$J_p(z) = \sum_{k=0}^{\infty} \frac{(-1)^k}{k!\,\Gamma(k+p+1)}\left[\frac{z}{2}\right]^{p+2k} .$$

8. The equation (1.28) is obtained in the process of solving for the Quantum mechanical motion of a particle in a spherical oscillator well, i.e.

$$\left[\frac{d^2}{dx^2} - x^2 - \frac{L(L+1)}{x^2} + 2\epsilon\right]R(x) = 0. \qquad (1.28)$$

where $L = 0,1,2,...$ and ϵ = constant.

(a) Show that this equation can be reduced to the confluent hypergeometic equation

$$\left[z\frac{d^2}{dz^2} + (2s + \frac{1}{2} - z)\frac{d}{dz} + n\right]W(z) = 0$$

using the transformation

$$R(x) = e^{-z/2}z^s W(z)$$

where $z = x^2$

$$\epsilon = 2\left[n+s+\frac{1}{4}\right]$$

$$L(L+1) = 4s\left[s-\frac{1}{2}\right]$$

(b) For what values of ϵ will $\lim_{z\to+\infty} W(z) = 0$?

9. Show that

$${}_1F_1(a;c;x) = e^x {}_1F_1(c-a,,c;-x) \quad .$$

10. Show that when $a = -n$ where n is a positive integer, ${}_1F_1(a;c;x)$ is a polynomial. These polynomials are called (up to a normalization constant) the generalized Laguerre polynomials.

11. The most general second order equation with three regular singular points at $0,1,\infty$ is Riemann's equation

$$y'' + \left[\frac{A}{x} + \frac{B}{1-x}\right]y' + \left[\frac{C}{x^2} + \frac{D}{(1-x)^2} + \frac{E}{x(1-x)}\right]y = 0 \ .$$

Show that any Riemann equation can be transformed into the hypergeometric equation (1.2) by a transformation of the form

$$y(x) = x^\lambda(1-x)^\mu Y(x)$$

by a proper choice of λ,μ.

2. BESSEL FUNCTIONS

2.1 INTRODUCTION

Although Bessel functions are an example of a hypergeometric series, they are most easily understood if we start with Bessel's differential equation

$$x^2y''(x) + xy'(x) + (x^2 - \nu^2)y(x) = 0$$

and derive the properties of Bessel functions through solving this equation.

Probably no other special function has received so much attention in the last hundred years. Whole books have been devoted to a detailed search of its properties. We propose to look at some of the more important results over the next few pages.

2.2 BESSEL'S DIFFERENTIAL EQUATION AND ITS SOLUTION

The second order ordinary differential equation

$$x^2y''(x) + xy'(x) + (x^2 - \nu^2)y(x) = 0 \qquad (2.1)$$

is known as Bessel's equation of order ν. Although ν itself can be taken as any number, real or complex, it is normally taken as zero or a positive number if ν is real.

In order to solve this equation we note that $x = 0$ is a regular singular point and, therefore, we can solve the differential equation by using Frobenius' method. In that regard we assume a series solution of the form

$$y(x) = x^r \sum_{n=0}^{\infty} c_n x^n = \sum_{n=0}^{\infty} c_n x^{n+r} \quad .$$

Upon differentiating $y(x)$ twice term by term and substituting these results into the differential equation (2.1) we arrive at the series equation

$$c_0(r^2 - \nu^2)x^r + c_1\left[(r+1)^2 - \nu^2\right]x^{r+1}$$
$$+ \sum_{n=2}^{\infty} \left\{c_n\left[(n+r)^2 - \nu^2\right] + c_{n-2}\right\}x^{n+r} = 0 \quad . \tag{2.2}$$

Normally we proceed by setting $c_0 \neq 0$ from which we get the indicial equation

$$r^2 - \nu^2 = 0 \quad .$$

Upon solving this equation we find the two indices

$$r = \pm \nu \quad . \tag{2.3}$$

Once these indices are selected it is obvious that in order to satisfy the condition

$$c_1\left[(r+1)^2 - \nu^2\right] = c_1[1 \pm 2\nu] = 0 \quad ,$$

we must choose $c_1 = 0$ unless $\nu = \pm \frac{1}{2}$. In the situation $\nu = \pm \frac{1}{2}$, since c_1 can be any number, we choose to set it to zero.

Returning to equation (2.2) we observe that the coefficient of x^{n+r} must equal zero, that is

$$c_n\left[(n+r)^2 - \nu^2\right] + c_{n-2} = 0 \quad .$$

Solving for c_n we arrive at the recurrence relation

$$c_n = - \frac{c_{n-2}}{(n+r+\nu)(n+r-\nu)} \qquad n = 2,3,\ldots \quad . \tag{2.4}$$

From the general theory of solving differential equations by Frobenius' method we know that if the indicies $r = \pm \nu$ are such that their difference is <u>not</u> zero or an integer then we can find two linearly

independent solutions at once. The first solution is
found by taking $r = \upsilon$, the second by using $r = -\upsilon$.
On the other hand, if the difference between the roots
is zero or an integer, then both roots yield
essentially the same solution and we are forced to find
our second linearly independent solution elsewhere.

   We shall first examine the relations to Bessel's
differential equation for the first case listed above,
(i.e., the difference of the indicies is _not_ zero or an
integer).

   Returning to the recurrence relation (2.4) and
setting $r = \upsilon$ we obtain

$$c_n = - \frac{c_{n-2}}{n(n+2\upsilon)} \qquad n = 2,3,\ldots \quad . \qquad\qquad (2.5)$$

By repeated use of this recurrence relation we see that

$$c_2 = - \frac{c_0}{2(2+2\upsilon)}$$

$$c_4 = - \frac{c_2}{4(4+2\upsilon)} = \frac{c_0}{2.4.(2+2\upsilon)(4+2\upsilon)}$$

and eventually

$$c_{2n} = (-1)^n \frac{c_0 \Gamma(\upsilon+1)}{2^{2n} n! \Gamma(\upsilon+n+1)}$$

where we have used the identity from Gamma functions
$\Gamma(\upsilon+n+1) = (\upsilon+n)(\upsilon+n-1) \ldots (\upsilon+1)\Gamma(\upsilon+1)$. (See Section
4).

   A solution of Bessel's differential equations is
given by

$$y_1(x) = c_0 \Gamma(\upsilon+1) x^\upsilon \sum_{n=0}^{\infty} (-1)^n \frac{x^{2n}}{2^{2n} n! \Gamma(\upsilon+n+1)} \quad . \quad (2.6)$$

Multiplying numerator and denominator by 2^υ and
bringing the factor x^υ under the summation sign
equation (2.6) can be written as

$$y_1(x) = c_0 \Gamma(\upsilon+1) \sum_{n=0}^{\infty} \frac{(-1)^n}{n! \Gamma(\upsilon+n+1)} \left[\frac{x}{2}\right]^{\upsilon+2n} \quad .$$

Since the summation alone is also a solution to

Bessel's differential equation, we define the Bessel function of the first kind of order υ, denoted by $J_\upsilon(x)$, as

$$J_\upsilon(x) = \sum_{n=0}^{\infty} \frac{(-1)^n}{n!\Gamma(\upsilon+n+1)} \left[\frac{x}{2}\right]^{\upsilon+2n} .$$

By using the index $r = -\upsilon$ we can find a second linearly independent solution

$$J_{-\upsilon}(x) = \sum_{n=0}^{\infty} \frac{(-1)^n}{n!\Gamma(-\upsilon+n+1)} \left[\frac{x}{2}\right]^{-\upsilon+2n} .$$

It so happens in the case of Bessel's differential equation that although the general theory of solutions around regular singular points indicates we may have trouble if the difference between the indices is an integer, we only have exceptional solutions when the difference of indices is an even integer.

Therefore, the general solution of Bessel's equation of order $\upsilon(\upsilon \neq 0,1,2,\ldots)$ is given by

$$y(x) = AJ_\upsilon(x) + BJ_{-\upsilon}(x) .$$

Although our derivation of the general solution of Bessel's differential equation is purely formal, it can be shown by using the ratio test that both solutions $J_\upsilon(x)$ and $J_{-\upsilon}(x)$ are absolutely convergent for all x.

Applying the ratio test to $J_\upsilon(x)$ we have

$$\lim_{n\to+\infty} \left| \frac{\dfrac{\left[\frac{x}{2}\right]^{2(n+1)+\upsilon}}{(n+1)!\Gamma(2+\upsilon+n)}}{\dfrac{\left[\frac{x}{2}\right]^{2n+\upsilon}}{n!\Gamma(1+\upsilon+n)}} \right| = \lim_{n\to+\infty} \frac{\Gamma(1+\upsilon+n)}{(n+1)\Gamma(2+\upsilon+n)} \left[\frac{x}{2}\right]^2$$

$$= \lim_{n\to+\infty} \frac{1}{(n+1)(1+\upsilon+n)} \left[\frac{x}{2}\right]^2 = 0\cdot \left[\frac{x}{2}\right]^2 < 1 .$$

This last inequality is true for all values of x.

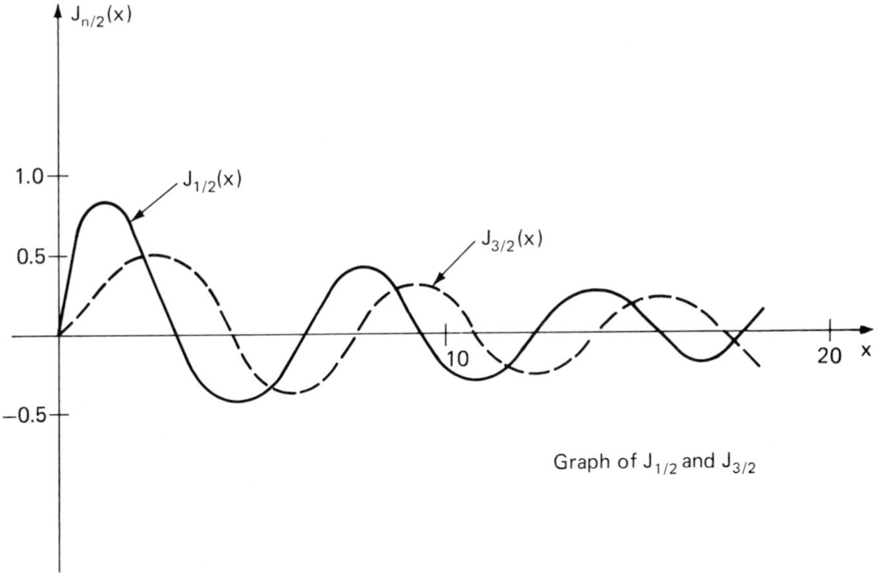

Figure 1

2.3 SOLUTIONS OF BESSEL'S DIFFERENTIAL EQUATION OF INTEGER ORDER

When the parameter $\upsilon = n$ is an integer the method of Frobenius yields only one series solution directly. In fact the method developed in the previous section works perfectly well for $\upsilon = n$ where $n = 0,1,...$ is the larger of the two indices. This solution which is valid for all x takes the form

$$J_n(x) = \sum_{k=0}^{\infty} \frac{(-1)^k}{k!\Gamma(n+k+1)}\left[\frac{x}{2}\right]^{k+2n}$$

$$= \sum_{k=0}^{\infty} \frac{(-1)^k}{k!(n+k)!}\left[\frac{x}{2}\right]^{k+2n}, \quad n = 0,1,2,... \quad .$$

The second summation follows easily from the first since

$$\Gamma(n+k+1) = (n+k)!$$

when n is an integer.

Example 1: In expanded form

$$J_0(x) = 1 - \left[\frac{x}{2}\right]^2 + \frac{1}{4}\left[\frac{x}{2}\right]^4 - \frac{1}{36}\left[\frac{x}{2}\right]^6 + \cdots$$

$$J_1(x) = \frac{x}{2} - \frac{1}{2}\left[\frac{x}{2}\right]^3 + \frac{1}{12}\left[\frac{x}{2}\right]^5 - \frac{1}{144}\left[\frac{x}{2}\right]^7 + \cdots .$$

Since n is an integer we cannot find the other linearly independent solution by the previous method. There are a number of ways in which the second solution may be found but we shall choose an approach generally covered in elementary texts on ordinary differential equations, the method of reduction of order.

We recall that if we have one solution $y_1(x)$ to the differential equation

$$y'' + P(x)y' + Q(x)y = 0 ,$$

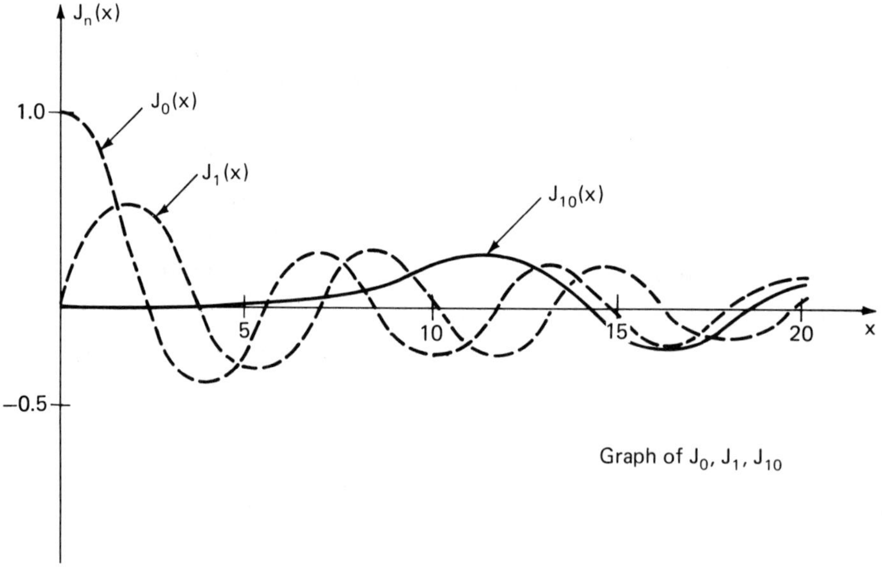

Figure 2

a linearly independent solution $y_2(x)$ is given in integral form by

$$y_2(x) = y_1(x)\int\frac{e^{-\int P(x)dx}}{y_1^2(x)}\,dx \quad .$$

Applying this technique to Bessel's differential equation, a solution linearly independent to $J_n(x)$ has the form

$$\begin{aligned} y_2 &= J_n(x)\int\frac{e^{-dx/x}}{J_n^2(x)}\,dx \\ &= J_n(x)\int\frac{dx}{xJ_n^2(x)} \quad , \end{aligned} \tag{2.7}$$

provided $x \neq 0$.

Since (2.7) is an independent solution

$$y(x) = AJ_n(x) + BJ_n(x)\int\frac{dx}{xJ_n^2(x)}$$

is the general solution to Bessel's differential equation. The normally accepted form of the linearly independent solution for $x > 0$ is denoted by $Y_n(x)$ and is called the Bessel function of the second kind of order n. They are expressed as follows:

$$\begin{aligned} Y_0(x) &= \frac{2}{\pi}\left[\ln\frac{x}{2} + \gamma\right]J_0(x) \\ &- \frac{2}{\pi}\sum_{r=1}^{\infty}\frac{(-1)^r}{(r!)^2}\left[\frac{x}{2}\right]^{2r}\left[1 + \frac{1}{2} + \frac{1}{3} + \ldots + \frac{1}{r}\right] \end{aligned} \tag{2.8}$$

and

$$Y_n(x) = \frac{2}{\pi}\left[\ln\frac{x}{2} + \gamma\right] J_n(x)$$

$$- \frac{1}{\pi}\sum_{r=0}^{n-1}\frac{(n-r-1)!}{r!}\left[\frac{x}{2}\right]^{2r-n} \tag{2.9}$$

$$- \frac{1}{\pi}\sum_{r=0}^{\infty}\frac{(-1)^r}{r!(n+r)!}\left[\frac{x}{2}\right]^{2r+n}$$

$$\left\{\left[1+\frac{1}{2}+\frac{1}{3}+\ \ldots+\frac{1}{r}\right] + \left[1+\frac{1}{2}+\frac{1}{3}+\ldots+\frac{1}{n+r}\right]\right\}$$

where the term $\left[1 + \frac{1}{2} + \frac{1}{3} + \ldots + \frac{1}{r}\right]$ is zero for $r = 0$. In both definitions $\gamma = 0.5772$ is Euler's constant which is defined as

$$\gamma = \lim_{n\to+\infty}\left\{\left[\sum_{k=1}^{n}\frac{1}{k}\right] - \ln n\right\} .$$

It is important to note from equations (2.8) and (2.9) that

$$\lim_{x\to 0^+} Y_n = -\infty , \quad n = 0,1,\ldots$$

which tells us that the solution $Y_n(x)$ is unbounded on $(0,c)$.

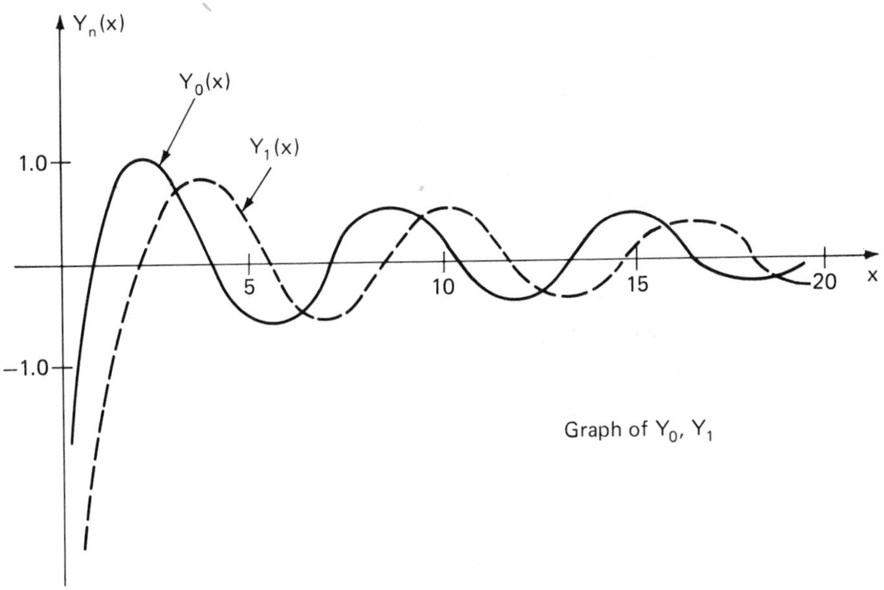

Figure 3

2.4. PROPERTIES OF BESSEL FUNCTIONS

Now that we have found the solutions to Bessel's differential equation, we shall investigate more fully their properties. Before proceeding let us summarize our notation.

If $\nu = n$ a positive integer or zero, the solution $J_n(x)$ is called Bessel function of the first kind of order n while a linear independent solution $Y_n(x)$ is called Bessel function of the second kind of order n.

If $\nu \neq n$, then a linearly independent set of solutions is denoted by $J_\nu(x)$ and $J_{-\nu}(x)$ both of which are Bessel functions of the first kind of order ν.

Joseph Krauskopf Memorial Library

Bessel functions of the first kind are continuous and bounded for all x although in most applications this domain is some set of the positive x axis. When drawn on a graph they remind us of a damped sinusoidal. In fact

$$J_{\frac{1}{2}}(x) = \sqrt{\frac{2}{\pi x}} \sin x$$

is actually a damped sinusoidal. (See Figure 1).

On the other hand we notice that Bessel functions of the second kind are discontinuous at $x = 0$ since

$$\lim_{x \to 0} Y_n = -\infty$$

because of the $\ln x$ term. But other than $x = 0$, Y_n is a well-behaved function and it too looks like a damped sinusoidal. (See Figure 3).

In the same way as trigonometric functions, one needs tables to find the value of a Bessel function for a given x. The Handbook of Mathematical Functions* contains a large number of such tables up to order 9. Tables up to J_{100} have been constructed.

The Bessel functions of the first kind also possess derivatives of all orders for any x. Unfortunately the expression for $\frac{d}{dx} J_\nu(x)$ is not easily expressed so it is customary to establish the following differentiation formula

Theorem 1: $$\frac{d}{dx}\left[x^\nu J_\nu(x)\right] = x^\nu J_{\nu-1}(x) \quad . \qquad (2.10)$$

Proof: We see from the definition of $J_\nu(x)$ that

$$x^\nu J_\nu(x) = \sum_{k=0}^{\infty} \frac{(-1)^k}{k!\Gamma(\nu+k+1)} \frac{x^{2\nu+2k}}{2^{\nu+2k}} \quad .$$

Then

$$\frac{d}{dx}\left[x^{\nu}J_{\nu}(x)\right] = \sum_{k=0}^{\infty} \frac{(-1)^k 2(\nu+k)}{k!\Gamma(\nu+k+1)} \frac{x^{2\nu+2k-1}}{2^{\nu+2k}}$$

$$= x^{\nu} \sum_{k=0}^{\infty} \frac{(-1)^k(\nu+k)}{k!\Gamma(\nu+k+1)} \left[\frac{x}{2}\right]^{\nu-1+2k} \qquad (2.11)$$

*Abramowitz, M. and I. A. Stegon (Eds.) "Handbook of Mathematical Fucntions" U.S. Department of Commerce, 1964.

Now from our knowledge of the Gamma function we see that

$$\frac{(\nu+k)}{\Gamma(\nu+k+1)} = \frac{1}{\Gamma(\nu+k)} \quad .$$

Therefore, substituting this result in (2.11) we obtain

$$\frac{d}{dx}\left[x^{\nu}J_{\nu}(x)\right] = x^{\nu} \sum_{k=0}^{\infty} \frac{(-1)^k}{k!\Gamma(\nu+k)} \left[\frac{x}{2}\right]^{\nu-1+2k}$$

$$= x^{\nu}J_{\nu-1}(x).$$

In a similar way we can show

$$\frac{d}{dx}\, x^{-\nu}J_{\nu}(x) = -x^{-\nu}J_{\nu+1}(x). \qquad (2.12)$$

Example 2: Using formula (2.10) we see that

$$\frac{d}{dx}\, x^2 J_2(x) = x^2 J_1(x) \quad .$$

We recall from our earlier discussion concerning Frobenius' solutions that $J_n(x)$ and $J_{-n}(x)$ are linearly dependent solutions. We proceed now to find that exact relation.

Theorem 2: If $n = 0,1,2,\ldots$ then

$$J_{-n}(x) = (-1)J_n(x) \quad . \qquad (2.13)$$

Proof: Starting with the definition of $J_{-n}(x)$ we can write

$$J_{-n}(x) = \sum_{k=0}^{\infty} \frac{(-1)^n}{k!\Gamma(-n+k+1)} \left[\frac{x}{2}\right]^{-n+2k} \quad .$$

Notice that we can start summing at $k = n$ rather than

zero because $\frac{1}{\Gamma(-n+k+1)} = 0$ for $k = 0,1,\ldots,n-1$.

Therefore,

$$J_{-n}(x) = \sum_{k=n}^{\infty} \frac{(-1)^n}{k!\Gamma(-n+k+1)} \left[\frac{x}{2}\right]^{-n+2k}$$

which upon reindexing by setting $k = \ell + n$ becomes

$$J_{-n}(x) = \sum_{\ell=0}^{\infty} \frac{(-1)^{\ell+n}}{(\ell+n)!\Gamma(\ell+1)} \left[\frac{x}{2}\right]^{n+2\ell} . \qquad (2.14)$$

But $(\ell+n)!\Gamma(\ell+1) = \ell!(\ell+n)(\ell+n-1)\ldots(\ell+1)\Gamma(\ell+1)$ $= \ell!\Gamma(n+\ell+1)$. Substituting this result into (2.14) yields

$$J_{-n}(x) = (-1)^n \sum_{\ell=0}^{\infty} \frac{(-1)^{\ell}}{\ell!\Gamma(n+\ell+1)} \left[\frac{x}{2}\right]^{n+2\ell}$$

$$= (-1)^n J_n(x).$$

Using Theorem 1 and Theorem 2 we can derive the useful formula

$$\frac{d}{dx} J_0(x) = \frac{d}{dx} x^0 J_0(x) = x^0 J_{-1}(x) = -J_1(x). \qquad (2.15)$$

Once we have derived the basic differentiation formulas (2.10) and (2.12), it is a relatively straightforward matter to derive some recurrence relations concerning derivatives.

Applying the product rule to (2.10) and (2.12) we have

$$x^{\nu} \frac{d}{dx} J_{\nu}(x) + \nu x^{\nu-1} J_{\nu}(x) = x^{\nu} J_{\nu-1}(x) \qquad (2.16)$$

$$x^{-\nu} \frac{d}{dx} J_{\nu}(x) - \nu x^{-\nu-1} J_{\nu}(x) = -x^{-\nu} J_{\nu+1}(x) . \qquad (2.17)$$

Dividing equation (2.16) by x^{ν} and (2.17) by $x^{-\nu}$ we get two slightly different formulas for evaluating the derivative of a Bessel function in terms of Bessel functions. Thus, we have

$$\frac{d}{dx} J_{\nu}(x) = J_{\nu-1}(x) - \frac{\nu}{x} J_{\nu}(x) \qquad (2.18)$$

or

$$\frac{d}{dx} J_{\nu}(x) = \frac{\nu}{x} J_{\nu}(x) - J_{\nu+1}(x) . \qquad (2.19)$$

Example 3: What is $\frac{d}{dx} J_3(4)$

$$\frac{d}{dx} J_3(4) = J_2(4) - \frac{3}{4} J_3(4)$$
$$= .3641 - \frac{3}{4} (.4302) = .0415 \quad .$$

Example 4: What is $\frac{d^2}{dx^2} J_\nu(x)$?

$$\frac{d^2}{dx^2} J_\nu(x) = \frac{d}{dx}\left[\frac{d}{dx} J_\nu(x)\right] = \frac{d}{dx} J_{\nu-1}(x) - \frac{d}{dx}\left[\frac{\nu}{x} J_\nu(x)\right]$$
$$= J_{\nu-2}(x) - \frac{\nu-1}{x} J_{\nu-1}(x)$$
$$- \nu \frac{x(J_{\nu-1}(x) - \frac{\nu}{x} J_\nu) - J_\nu(x)}{x^2}$$
$$= J_{\nu-2}(x) + \frac{(1-2\nu)}{x} J_{\nu-1}(x) + \frac{\nu(\nu+1)}{x^2} J_\nu(x) \quad .$$

We derive a very important recurrence relation by subtracting equation (2.19) from equation (2.18) which yields

$$xJ_{\nu+1}(x) = 2\nu J_\nu(x) - xJ_{\nu-1}(x) \quad .$$

With this relation we are able to evaluate a Bessel function in terms of Bessel functions of lower order.

Example 5: What is the value of $J_2(5.5)$?

We write

$$5.5J_2(5.5) = 2J_1(5.5) - 5.5J_0(5.5)$$

and, therefore,

$$J_2(5.5) = \frac{2}{5.5} (-0.3414) + 0.0068 = -0.1173 \quad .$$

As we recall from our study of integrals, once we have found a differentiation formula we can easily turn it into an integral formula. Using equations (2.10) and (2.12) we can easily write two important integral formulas,

$$\int x^\nu J_{\nu-1}(x)dx = x^\nu J_\nu(x) + C \tag{2.20}$$

and

$$\int x^{-\nu} J_{\nu+1}(x)dx = -x^{-\nu} J_\nu(x) + C \quad . \tag{2.21}$$

Example 6: As a special case of formula (2.20) we see

$$\int xJ_0(x)dx = xJ_1(x) + C.$$

Sometimes it is difficult to decide which integral formula to use. If you are trying to write an integral expression in terms of lower order Bessel functions then we would probably use equation (2.21). Often times some subtle mathematical manipulations must be made in order to use formulas (2.20) and (2.21).

Example 7: Integrate $\int x^5 J_2(x)dx$.

We factor the integral as follows:

$$\int x^5 J_2(x)dx = \int x^2 \cdot x^3 J_2(x)dx \quad .$$

Now using integration by parts we set

$$u = x^2 \quad \text{and} \quad dv = x^3 J_2(x)dx$$

$$du = 2xdx \qquad v = x^3 J_3(x) \quad .$$

Therefore,

$$\int x^5 J_2(x)dx = x^5 J_3(x) - 2\int x \cdot x^3 J_3(x)dx$$

$$= x^5 J_3(x) - 2x^4 J_4(x) + C.$$

Example 8: Show $\int_0^a xJ_2(x)dx = 2 - aJ_1(a) - 2J_0(a)$.

From equation (2.21) with $\upsilon = 1$ we find using integration by parts that

$$\int_0^a xJ_2(x)dx = \int_0^a x^2 x^{-1} J_2(x)dx = -xJ_1(x)\Big|_0^a$$

$$+ 2\int_0^a J_1(x)dx = -aJ_1(a) - 2J_0(x)\Big|_0^a$$

$$= 2 - aJ_1(a) - 2J_0(a) \quad .$$

Although an emphasis has been on Bessel functions of the first kind, Bessel functions of the second kind satisfy similar properties. For example, wherever $Y_n(x)$ is differentiable

$$\frac{d}{dx} x^n Y_n(x) = x^n Y_{n-1}(x)$$

and

$$\frac{d}{dx} Y_n(x) = Y_{n-1}(x) - \frac{n}{x} Y_n(x) \quad .$$

Many more of these formulas and recurrence relations can be found in the Handbook of Mathematical Functions.

2.5 ZEROS OF BESSEL FUNCTIONS

It is important in what follows that we show Bessel functions have an infinite number of values of x which make them zero. Unfortunately, since a Bessel function is defined by a Frobenius series and it is usually very difficult to find those values of x which make a Frobenius series zero, we are forced to look elsewhere in mathematics for our answer. The answer to our problems is found in a segment of the study of differential equations called Sturmian theory.

Definition 1: Suppose $x = x_0$ is a number such that $f(x_0) = 0$. The $x = x_0$ is said to be *zero* of $f(x)$.

Example 9: The zeros of $\sin bx$ are the values $x = \frac{n\pi}{b}$, $n = 0, \pm 1 \ldots$. The function $f(x) = 1-e^x$ has only one zero, that is $x = 0$.

In the discussion which follows it is necessary to write our given second order linear differential equation in self-adjoint form which is written

$$(r(x)y'(x))' + p(x)y = 0 \quad .$$

Every second order linear differential equation can be written this way and so it is known that Bessel's equation looks like

$$(xy')' + \left[x - \frac{\nu^2}{x}\right]y = 0$$

in self-adjoint form.

The following theorem called an oscillation theorem is taken from Sturmian theory.

Theorem 3: Given the differential equation $(r(x)y')' + p(x)y = 0$ where $r(x) > 0$ and $r(x)$ and $p(x)$ are continuous on the interval $0 < x < +\infty$. If the two integrals

$$\int_1^{+\infty} \frac{dx}{r(x)} = +\infty$$

and

$$\int_1^{+\infty} p(x)dx = +\infty$$

then every solution $y(x)$ has an infinite number of zeros on the interval $(1,+\infty)$.

Example 10: We already know that $\sin bx$ has an infinite number of zeros from our knowledge of trigonometry but let us justify the fact by applying Theorem 3.

Sin bx is a solution to the differential equation $y'' + b^2y = 0$ which is in self-adjoint form. From Theorem 3 we see

$$\int_1^{+\infty} 1dx = +\infty.$$

and

$$\int_1^{+\infty} b^2dx = +\infty.$$

Therefore, $\sin bx$ has an infinite number of zeros on $(1,+\infty)$. See Figure 4.

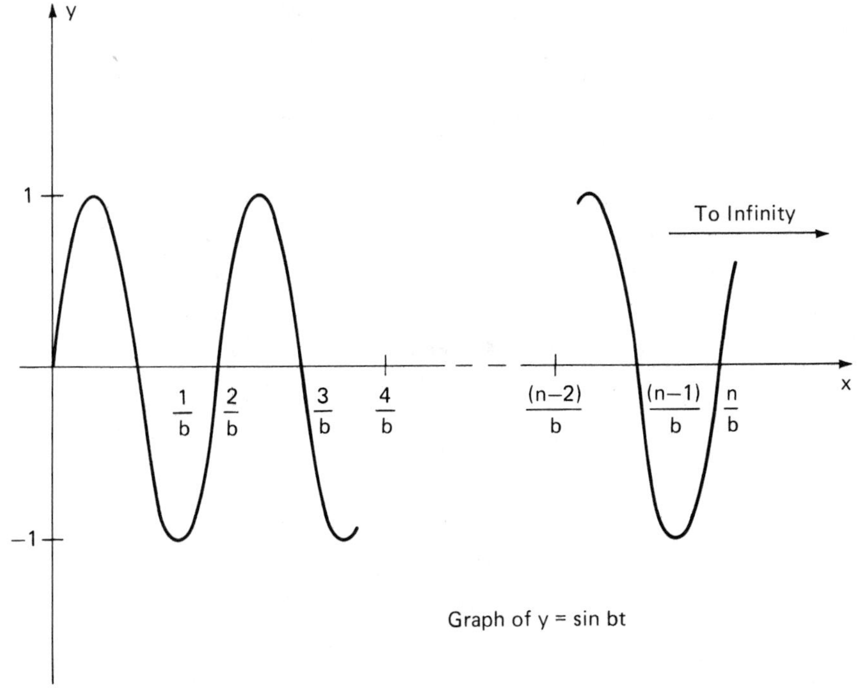

Figure 4

We apply Theorem 3 to Bessel's differential equation $(xy')' + \left[x - \frac{\nu^2}{x}\right]y = 0$. Therefore,

$$r(x) = x \quad \text{and} \quad p(x) = x - \frac{\nu^2}{x} \quad .$$

The integral

$$\int_1^{+\infty} \frac{dx}{x} = \lim_{c\to+\infty} \ln c = +\infty$$

while the integral

$$\int_1^{+\infty} \left[x - \frac{\nu^2}{x}\right]dx \geq \int_1^{+\infty} (x - \nu^2)dx$$

$$= \lim_{c\to+\infty} x\left[\frac{x}{2} - \nu^2\right]\bigg|_1^c = +\infty.$$

Therefore, the solutions of any Bessel's equation have an infinite number of zeros on $(1,+\infty)$. Some of these zeros are shown in Figure 2.

Unlike the sine and cosine function the zeros of a Bessel function must be found by approximate method. Tables of the zeros of certain Bessel functions are listed in a number of handbooks. Shown below in Table 1 is a brief list showing some zeros of certain Bessel functions.

Location of Zeros for $x > 0$

Bessel Function	1st	2nd	3rd	4th
J_0	2.41	5.52	8.65	11.79
J_1	3.83	7.02	10.17	13.32
$J_{\frac{1}{2}}$	3.14	6.28	9.42	12.57
J_8	12.20	16.04	19.60	22.90
Y_0	0.89	3.96	7.09	10.22
Y_1	2.20	5.43	8.60	11.75

TABLE 1

It is important that we identify these zeros of a Bessel function $J_\upsilon(x)$ (or $Y_n(x)$) for if we let x_j equal the jth positive root of $J_\upsilon(x) = 0$, then we can show that the infinite set of functions $\left\{J_\upsilon\left[\frac{x_j x}{c}\right]\right\}$, $j = 1,2,3,\ldots$ and c = constant is particularly useful when desiring to express a given function in terms of an infinite series of Bessel functions.

2.6 STURM-LIOUVILLE PROBLEMS LEADING TO ORTHOGONAL SETS OF BESSEL FUNCTIONS

Very often when we are solving boundary value problems whose domain is a circular disk or cylinder, we are lead to the following Sturm-Liouville problem

$$\frac{d}{dx}\left[x\,\frac{dy(x)}{dx}\right] + \left[\lambda x - \frac{\nu^2}{x}\right]y(x) = 0 \tag{2.22}$$

where ν is a real number normally taken positive or zero along with the boundary condition

$$b_1 y(c) + b_2 y'(c) = 0$$

where b_1 and b_2 are not both zero simultaneously.

Unfortunately the differential equation (2.22) is "not quite" Bessel's differential equation because of the parameter λ which may be some other value than one. In what follows we shall find it convenient to replace λ by α^2 where $\alpha \geq 0$. We can convert equation (2.22) into Bessel's equation making the change of independent variable $x = \frac{u}{\alpha}$.

Using the notation

$$y(x) = y\left[\frac{u}{\alpha}\right] = \bar{y}(u)$$

we see that the derivatives

$$\frac{dy}{dx} = \frac{d\bar{y}}{du}\,\frac{du}{dx} = \frac{d\bar{y}}{du}\,\alpha$$

and

$$\frac{d}{dx}\left[x\,\frac{dy}{dx}\right] = \alpha\,\frac{d}{du}\left[u\,\frac{d\bar{y}}{du}\right] . \tag{2.23}$$

Substituting derivatives (2.23) into (2.22), equation (2.22) becomes

$$\alpha\,\frac{d}{du}\left[u\,\frac{d\bar{y}}{du}\right] + \left[\alpha^2\,\frac{u}{\alpha} - \alpha\,\frac{\nu^2}{u}\right]\bar{y} = 0$$

or

$$\frac{d}{du}\left[u\,\frac{d\bar{y}}{du}\right] + \left[u - \frac{\nu^2}{u}\right]\bar{y} = 0 \tag{2.24}$$

which is Bessel's equation of order ν. If $J_\nu(u)$ is a solution to equation (2.24), then $J_\nu(\alpha x)$ is a solution to (2.22). $J_{-\nu}(\alpha x)$ is a linearly indepedent solution if $\nu \neq$ an integer while $Y_n(\alpha x)$ is a linearly independent solution if $\nu = n$, an integer.

In the discussion which follows we shall be using the derivative $\frac{d}{dx} J_\nu(\alpha x)$. Care must be taken to use the chain rule when evaluating this derivative.

Consider the function $J_\nu(u)$ where $u = \alpha x$. Then, from our knowledge of the chain rule,

$$\frac{d}{dx} J(\alpha x) = \frac{d}{du} J_\nu(u) \frac{du}{dx} = \alpha \frac{d}{du} J_\nu(u).$$

Example 11: Evaluate $\frac{d}{dx} J_0(2.41x)$ at $x = 2$.

$$\frac{d}{dx} J_0(2.41x)\Big|_{x=2} = 2.41 \frac{d}{du} J_0(u)\Big|_{u=4.82}$$

$$= -2.41 J_1(u)\Big|_{u=4.82} = (-2.41)(-0.234)$$

$$= 0.564.$$

Using this notation

$$\frac{d}{dx} J_\nu(\alpha c) = \frac{d}{dx} J_\nu(\alpha x)\Big|_{x=c} = \alpha \frac{d}{du} J_\nu(\alpha c) \quad .$$

Let us search for bounded solutions of the Sturm-Liouville boundary value problem over the interval $[0, c]$ given by

$$\frac{d}{dx}\left[x \frac{dy}{dx}\right] + \left[\alpha^2 x - \frac{\nu^2}{x}\right] y = 0 \tag{2.25}$$

$$b_1 y(c) + b_2 y'(c) = 0 \quad . \tag{2.26}$$

From our previous discussions we know that the general solution of the differential equation is given by either

$$y(x) = AJ_\nu(\alpha x) + BJ_{-\nu}(\alpha x), \ \nu \neq \text{an integer}$$

or

$$y(x) = AJ_n(\alpha x) + BY_n(\alpha x), \quad n = \text{nonnegative integer}.$$

But both $J_{-\upsilon}(\alpha x)$ and $Y_n(\alpha x)$ are unbounded on the interval $[0,c]$. Since we are seeking only bounded solutions to our problem we must take $B = 0$, and the solution has the form

$$y(x) = AJ_\upsilon(\alpha x) \quad \text{for} \quad 0 \leq \upsilon .$$

In order to meet the boundary condition (2.26)

$$b_1 J_\upsilon(\alpha c) + b_2 \frac{d}{dx} J_\upsilon(\alpha c) = 0 \quad . \tag{2.27}$$

For convenience it will be helpful to look at two possible interpretations of boundary condition (2.27). They are

$$\begin{aligned} &(a) \quad J_\upsilon(\alpha c) = 0 \quad , \quad b_2 = 0 \\ &(b) \quad hJ_\upsilon(\alpha c) + c \frac{d}{dx} J_\upsilon(\alpha c) = 0 \end{aligned} \tag{2.28}$$

where $h \geq 0$, $c > 0$, $\frac{d}{dx} J_\upsilon(\alpha c) = \alpha \frac{d}{du} J_\upsilon(\alpha c)$. If the boundary condition is of the form (2.28a), i.e. $J_\upsilon(\alpha c) = 0$, then because $J_\upsilon(x)$ has an infinite number of isolated zeros

$$\alpha c = x_i \qquad i = 1,2,\ldots$$

where $J_\upsilon(x_i) = 0$. Therefore,

$$\alpha_i = \frac{x_i}{c}$$

or

$$\lambda_i = \alpha_i^2 = \frac{x_i^2}{c^2}$$

are the eigenvalues associated with the problem and $y_i(x) = J_\upsilon(\alpha_i x)$ are the eigenfunctions.

Our Sturm-Liouville problem has an infinite number of solutions $\{J_\upsilon(\alpha_i x)\}$. In the next paragraph we shall show that members of this set satisfy an orthogonal property.

For the boundary conditions (2.28b), we observe that

$$hJ_\nu(\alpha c) + \alpha c \frac{d}{du} J_\nu(\alpha c) = 0 \quad .$$

From Sturmian theory it can be shown that there are an infinite number of isolated roots x_i of the equation

$$hJ_\nu(x) + \alpha c \frac{d}{dx} J_\nu(x) = 0 \quad .$$

Once again we see that $\alpha_i = \frac{x_i}{c}$ and $\lambda_i = \frac{x_i^2}{c^2}$ are the eigenvalues and $J_\nu(\alpha_i x)$ are the eigenfunctions associated with boundary value problem (2.28b).

Example 12: Find the first three eigenvalues and eigenfunctions of the boundary value problem

$$\frac{d}{dx}\left[x \frac{dy}{dx}\right] + \left[\alpha^2 x - \frac{4}{x}\right] y = 0$$

$$\frac{d}{dx} J_2(\alpha c) = 0 \quad .$$

First we notice

$$\frac{d}{dx} J_2(\alpha c) = \alpha \frac{d}{du} J_2(u)\bigg|_{u=\alpha c} = 0$$

which implies

$$\frac{d}{du} J_2(u) = 0 \quad . \tag{2.29}$$

The first three positive roots of equation (2.29) can be found in a handbook and are $u_1 = 3.05$, $u_2 = 6.71$, $u_3 = 9.97$. The first three eigenvalues are

$$\lambda_1 = \alpha_1^2 = \left[\frac{3.05}{c}\right]^2 = \frac{9.30}{c^2}$$

$$\lambda_2 = \alpha_2^2 = \left[\frac{6.71}{c}\right]^2 = \frac{45.02}{c^2}$$

$$\lambda_3 = \alpha_3^2 = \left[\frac{9.97}{c}\right]^2 = \frac{99.40}{c^2}$$

while $J_2\left[\frac{3.05x}{c}\right]$, $J_2\left[\frac{6.71x}{c}\right]$, $J_2\left[\frac{9.97x}{c}\right]$ are the corresponding eigenfunctions.

2.7 ORTHOGONAL SETS

The two infinite sets of Bessel functions $\{J_\nu(\alpha_i x)\}$ corresponding to boundary conditions (2.28a) and (2.28b) are extremely useful in solving boundary value problems since they possess the property of being orthogonal sets. Because of this fact we shall show in the next paragraph that many functions defined over an interval $[0,c]$ may be expressed as an infinite series of Bessel functions.

Assume $J_\nu(\alpha_i x)$ and $J_\nu(\alpha_j x)$ are solutions of the differential equation

$$\frac{d}{dx}\left[x\frac{dy}{dx}\right] + \left[\lambda x - \frac{\nu^2}{x}\right] y = 0 \quad \text{where} \quad \lambda = \alpha^2.$$

We express this fact in the two equations

$$\frac{d}{dx}\left[x\frac{d}{dx} J_\nu(\alpha_i x)\right] + \left[\lambda_i x - \frac{\nu^2}{x}\right] J_\nu(\alpha_i x) = 0 \qquad (2.30)$$

$$\frac{d}{dx}\left[x\frac{d}{dx} J_\nu(\alpha_j x)\right] + \left[\lambda_j x - \frac{\nu^2}{x}\right] J_\nu(\alpha_j x) = 0 \qquad (2.31)$$

where $\lambda_i = \alpha_i^2$ and $\lambda_j = \alpha_j^2$.

We multiply (2.30) by $J_\nu(\alpha_j x)$ and (2.31) by $J_\nu(\alpha_i x)$ and subtract one equation from the other. Upon integrating both sides of this difference from 0 to c we can write

$$(\lambda_i - \lambda_j)\int_0^c xJ_\nu(\alpha_i x)J_\nu(\alpha_j x)dx$$
$$= x\left[J_\nu(\alpha_j x\frac{d}{dx} J_\nu(\alpha_i x)\right.$$
$$\left.- J_\nu(\alpha_i x)\frac{d}{dx} J_\nu(\alpha_j x)\right]\Bigg|_0^c = c\left[J_\nu(\alpha_j c)\frac{d}{dx} J_\nu(\alpha_i c)\right.$$
$$\left.- J_\nu(\alpha_i c)\frac{d}{dx} J_\nu(\alpha_j c)\right] \qquad . \qquad (2.32)$$

By examining the righthand side of (2.32) we notice that the boundary conditions

$$\text{(a)} \quad J_\nu(\alpha c) = 0$$

$$\text{(b)} \quad hJ_\nu(\alpha c) + c\frac{d}{dx} J_\nu(\alpha c) = 0 \qquad (2.33)$$

make the expression on the righthand side zero. Now if $\lambda_i \neq \lambda_j$ and our set of Bessel functions satisfies one of the conditions in (2.28), we see that

$$\int_0^c xJ_\nu(\alpha_i x)J_\nu(\alpha_j x)dx = 0 \quad ,$$

that is $J_\nu(\alpha_i x)$ is orthogonal to $J_\nu(\alpha_j x)$ with respect to the weight function x over the interval $[0,c]$.

Another integral we shall need in order to evaluate a Fourier-Bessel series is the integral

$$\int_0^c xJ_\nu^2\,(\alpha_i x)dx \quad .$$

One way to find the value of this integral is to start with the equation

$$\frac{d}{dx}\left[x\,\frac{d}{dx}\,J_\nu(\alpha_i x)\right] + \left[\lambda_i x - \frac{\nu^2}{x}\right]\,J_\nu(\alpha_i x) = 0,$$

and multiply both sides by $2x\,\frac{d}{dx}\,J_\nu(\alpha_i x)$ which yields the equation

$$2x\,\frac{d}{dx}\,J_\nu(\alpha_i x)\frac{d}{dx}\left[x\,\frac{d}{dx}\,J_\nu(\alpha_i x)\right]$$

$$+\ 2\left[\lambda_i x^2 - \nu^2\right]J_\nu(\alpha_i x)\,\frac{d}{dx}\,J_\nu(\alpha_i x) = 0.$$

This equation can be written in the more convenient form

$$\frac{d}{dx}\left[x\,\frac{d}{dx}\,J_\nu(\alpha_i x)\right]^2 + \left[\lambda_i x^2 - \nu^2\right]\frac{d}{dx}\,\left[J_\nu(\alpha_i x)\right]^2 = 0. \tag{2.34}$$

Next we integrate equation (2.34) from 0 to c applying integration by parts to the second term in the lefthand side from which we obtain

$$\left[x\,\frac{d}{dx}\,J_\nu(\alpha_i x)\right]^2\Bigg|_0^c + \left[\lambda_i x^2 - \nu^2\right]J_\nu^2(\alpha_i x)\Bigg|_0^c$$

$$- 2\lambda_i\int_0^c xJ_\nu^2(\alpha_i x)dx = 0 \ .$$

Solving for the integral above we find for $0 \le \nu$

$$\int_0^c xJ_\nu^2(\alpha_i x)dx = \frac{1}{2\alpha_i^2}\left\{\left[c \frac{d}{dx} J_\nu(\alpha_i c)\right]^2 + \left[\alpha_i^2 c^2 - \nu^2\right]J_\nu^2(\alpha_i c)\right\}. \tag{2.35}$$

The value of this integral depends on which boundary condition (2.28) is chosen.

Theorem 4 The value of the integral listed below is given for the two boundary conditions

(i) $J_\nu(\alpha c) = 0;$

(ii) $hJ_\nu(\alpha c) + c \frac{d}{dx} J_\nu(\alpha c) = 0.$

$$\int_0^c xJ_\nu^2(\alpha_i x)dx = \begin{cases} \text{(i)} & \frac{c^2}{2}\left[J_{\nu+1}(\alpha_i c)\right]^2 \\ \\ \text{(ii)} & \left[\frac{c^2}{2} + \frac{h^2-\nu^2}{2\alpha_i^2}\right]J_\nu^2(\alpha_i c) \end{cases}.$$

Proof: Substitute conditions (i), (ii) in equation (2.35).

2.8 THE FOURIER BESSEL SERIES

In Section 2.7 we observed that the set of Bessel functions $\{J_\nu(\alpha_i x)\}$ associated with certain boundary conditions (2.28) form an orthogonal set. Because of this fact we are able to show (formally at least) that the function $f(x)$ defined over the interval $[0,c]$ can be expressed as an infinite series of Bessel functions. In other words can we find the value of the coefficients A_i so that

$$f(x) = \sum_{i=1}^{\infty} A_i J_\nu(\alpha_i x)$$

where $J_\nu(x)$ satisfies one of the boundary conditions

(2.28). Such a series is called a *Fourier-Bessel series* while the coefficients A_j are called the *Fourier-Bessel coefficients*.

Theorem 5: The function $f(x)$ can be expressed as a Fourier-Bessel series,

$$f(x) = \sum_{i=1}^{\infty} A_i J_\upsilon(\alpha_i x) \tag{2.36}$$

where

$$A_i = \begin{cases} \dfrac{2}{c^2[J_{\upsilon+1}(\alpha_i c)]^2} \displaystyle\int_0^c x f(x) J_\upsilon(\alpha_i x) dx \text{ when } J_\upsilon(\alpha c)=0 \\ \dfrac{2\alpha_i^2}{(\alpha_i^2 c^2 + h^2 - \upsilon^2) J_\upsilon^2(\alpha_i c)} \displaystyle\int_0^c x f(x) J_\upsilon(\alpha_i x) dx \text{ when} \end{cases}$$

$$hJ_\upsilon(\alpha c) + c \frac{d}{dx} J_\upsilon(\alpha c) = 0.$$

Proof: We shall show the second case only. The other case is similar. We multiply both sides of series (2.36) by $xJ_\upsilon(\alpha_j x)$ which shows

$$x f(x) J_\upsilon(\alpha_j x) = \sum_{i=1}^{\infty} A_i x J_\upsilon(\alpha_i x) J_\upsilon(\alpha_j x).$$

Next we integrate both sides from 0 to c and assuming we can interchange the order of integration and summation on the righthand side we have

$$\int_0^c x f(x) J_\upsilon(\alpha_j x) dx = \sum_{i=1}^{\infty} A_i \int_0^c x J_\upsilon(\alpha_i x) J_\upsilon(\alpha_j x) dx . \tag{2.37}$$

Since we know $\{J_\upsilon(\alpha_i x)\}$ forms an orthogonal set with respect to the boundary condition $hJ_\upsilon(\alpha c) + c \frac{d}{dx} J_\upsilon(\alpha c) = 0$ the righthand side of equation (2.37) reduces to one term

$$A_j \int_0^c xJ_\upsilon^2(\alpha_j x)dx = \int_0^c xf(x)J_\upsilon(\alpha_j x)dx \quad .$$

But from Theorem 4 (ii) the integral on the lefthand side is equal to

$$A_j \left[\frac{c^2}{2} + \frac{h^2-\upsilon^2}{2\alpha_j^2} \right] J_\upsilon^2(\alpha_j c) = \int_0^c xf(x)J_\upsilon(\alpha_j x)dx$$

and solving for A_j we have

$$A_j = \frac{2\alpha_j^2}{(\alpha_j^2 c^2+h^2-\upsilon^2)J_\upsilon^2(\alpha_j c)} \int_0^c xf((x)J_\upsilon(\alpha_j x)dx.$$

Example 12: Find the Fourier-Bessel series so that

$$x = \sum_{i=1}^{\infty} A_i J_1(\alpha_i x)$$

where $\alpha_i c$ are roots of $\frac{d}{dx} J_1(x) = 0$ and $c = 10$.

$$A_i = \frac{2\alpha_i^2}{(100\alpha_i^2-1)J_1^2(10\alpha_i)} \int_0^{10} x^2 J_1(\alpha_i x)dx$$

$$= \frac{2\alpha_i^2}{(100\alpha_i^2-1)J_1^2(10\alpha_i)} \frac{1}{\alpha_i^3} \int_0^{10} (\alpha_i x)^2 J_1(\alpha_i x)d(\alpha_i x)$$

$$= \frac{2\alpha_i^2}{(100\alpha_i^2-1)J_1^2(10\alpha_i)} \frac{1}{\alpha_i^3} \int_0^{10\alpha_i} s^2 J_1(s)ds$$

$$= \frac{2}{\alpha_i(100\alpha_i^2-1)J_1^2(10\alpha_i)} (10\alpha_i)^2 J_2(10\alpha_i)$$

$$= \frac{200\alpha_i}{(100\alpha_i^2-1)} \frac{J_2(10\alpha_i)}{J_1^2(10\alpha_i)}$$

Therefore, we write of Fourier-Bessel series, on the interval [0,c]

$$x = 200 \sum_{i=1}^{\infty} \frac{\alpha_i J_2(10\alpha_i)}{(100\alpha_i^2-1)J_1^2(10\alpha_i)} J_1(\alpha_i x) \quad .$$

EXERCISES FOR BESSEL FUNCTIONS

1. Use Frobenius series to find the general solution of the equation

$$x^2 \frac{d^2y}{dx^2} + x\frac{dy}{dx} + (x^2 - \tfrac{1}{9})y = 0 \quad .$$

2. Prove $J_n(x)$ is an odd function if n is odd.

3. Prove $J_n(x)$ is an even function if n is even.

4. Show $J_0(0) = 1$.

5. Show $J_n(0) = 0$, $n = 1,2,3,\ldots$

6. By using the transformation $u = x^{\frac{1}{2}} y$ in the Bessel equation, show that this equation reduces to

$$u'' + \left[1 + \frac{1/4-n^2}{x^2}\right]u = 0, \quad n = \text{integer}$$

7. By direct substitution in the Bessel's equation show that

(a) $$J_0(x) = \frac{2}{\pi}\int_0^{\frac{\pi}{2}} \cos(x\sin\theta)d\theta$$

(b) $$J_1(x) = \frac{2}{\pi}\int_0^{\frac{\pi}{2}} \sin(x\sin\theta)\sin\theta d\theta$$

8. Show that

(a) $$J_{\frac{1}{2}}(x) = \left[\frac{2}{\pi x}\right]^{1/2} \sin x.$$

Hint: $$J_{\frac{1}{2}}(x) = \sum_{k=0}^{\infty} \frac{(-1)^k}{k!\Gamma(k+3/2)} (x/2)^{3/2+2k} \quad .$$

9. Obtain an explicit formula for $J_{\frac{3}{2}}(x)$.

10. Prove

(a) $\int x^{\nu} J_{\nu-1}(x)dx = x^{\nu} J_{\nu}(x) + C$

(b) $\int x^{-\nu} J_{\nu+1}(x)dx = -x^{-\nu} J_{\nu}(x) + C$

(c) $\int_0^x sJ_0(s)ds = xJ_1(x)$.

11. Find the indefinite integrals of

(a) $\int x^{10} J_9(x)dx$

(b) $\int x^{-3/2} J_{\frac{5}{2}}(x)dx$

(c) $\int x^5 J_2(x)dx$ Hint: Use integration by parts.

(d) $\int x^{2-\nu} J_{\nu+1} dx$

(e) $\int J_1(x)dx$ Hint: Multiply by $x^{-k}x^k = 1$.

12. Evaluate the integral

(a) $\int (J_5(x) - J_7(x))dx$

(b) $\int_0^x s^4 J_1(s)ds$

(c) $\int x^6 J_3(x)dx$

13. Find the indefinite integral $\int J_3(x)dx$ in terms of $J_1(x)$ and $J_2(x)$.

14. Show

$$\int_0^5 x^3 J_2(\alpha_j x)dx = \frac{125}{\alpha_j} J_3(5\alpha_j) \quad .$$

15. Use integration by parts to prove

$$\int_0^x s^3 J_0(s)ds = x^3 J_1(x) - 2x^2 J_2(x) \quad .$$

16. Establish the following recurrence relation

$$J_{\nu}'(x) = J_{\nu-1}(x) - \frac{\nu}{x} J_{\nu}(x) \quad .$$

17. Find $J_1'(2.3)$ if $J_1(2.3) = .5399$ and $J_2(2.3) = .4139$.

18. Prove

$$x^2 J_n''(x) = n(n-1)J_n(x) - (2n+1)xJ_{n+1}(x) + x^2 J_{n+2}(x).$$

19. (a) Show by solving $y'' = 0$ that $y(x)$ crosses the x-axis at most once if the solution is nontrivial.

(b) Show $\int_1^{+\infty} p(x)dx = 0$.

20. (a) Show $y'' + 4y = 0$ has an infinite number of zeros using Theorem 3.

(b) Solve $y'' + 4y = 0$. Show that a particular solution has an infinite number of zeros.

21. Use Theorem 3 to prove solutions of

$$xy'' + y' + e^x y = 0$$

have an infinite number of zeros.

22. Does the modified Bessel's equation

$$(xy')' - \left[x + \frac{\nu^2}{x}\right]y = 0$$

meet the hypothesis of Theorem 3?

Find the Fourier-Bessel series in Problems 23 - 33.

23. $100 = \sum_{j=1}^{\infty} A_j J_0(\alpha_j x) \quad \alpha_j = \frac{x_j}{c}, \quad J_0(x_j) = 0$

24. $x = \sum_{j=1}^{\infty} A_j J_1(\alpha_j x) \quad \alpha_j = \frac{x_j}{5}, \quad c = 5, \; J_1(x_j) = 0$

25. $10 = \sum_{j=1}^{\infty} A_j J_2(\alpha_j x) \quad \alpha_j = \frac{x_j}{c}, \quad J_2(x_j) = 0$

26. $x^2 = \sum_{j=1}^{\infty} A_j J_0(\alpha_j x) \quad \alpha_j = \frac{x_j}{c}, \quad J_0(x_j) = 0$

27. $x^2 = \sum_{j=1}^{\infty} A_j J_3(\alpha_j x), \quad \alpha_j = \frac{x_j}{c}, \quad J_3(x_j) = 0$

28. Leave A_j in integral form

$$f(x) = \begin{cases} 0 & 0 < x < 5 \\ 1 & 5 < x < 10 \end{cases}$$

$$f(x) = \sum_{j=1}^{\infty} A_j J_0(\alpha_j x), \quad \alpha_j = \frac{x_j}{10}, \quad J_0(x_j) = 0$$

29. $3 = \sum_{j=1}^{\infty} A_j J_0(\alpha_j x)$, $\alpha_j = \frac{x_j}{c}$, $J_0(x_j) + J_0'(x_j) = 0$

30. $2x = \sum_{j=1}^{\infty} A_j J_1(\alpha_j x)$, $\alpha_j = \frac{x_j}{10}$,

$c = 10$, $J_1(x_j) - J_1'(x_j) = 0$

31. $50 = \sum_{j=1}^{\infty} A_j J_1(\alpha_j x)$, $\alpha_j = \frac{x_j}{c}$, $J_1'(x_j) = 0$

32. $x^2 = \sum_{j=0}^{\infty} A_j J_0(\alpha_j x)$, $\alpha_j = x_j$, $c = 1$, $x_0 = 0$,

$J_0'(x_j) = 0$

33. $x = \sum_{j=1}^{\infty} A_j J_2(\alpha_j x)$, $\alpha_j = \frac{x_j}{4}$, $c = 4$, $J_2'(x_j) = 0$

34. Expand $f(x) = \sin x$, $0 < x < \pi$ into a Fourier-Bessel series of Bessel functions of the first kind and order zero, assuming $J_0(x_j) = 0$.

35. Given the equation
$r^2R''(r) + rR'(r) + \lambda r^2 R = 0$.
   (a) Make the transformation $s = \sqrt{\lambda}r$, to convert equation above in a Bessel's equation.
   (b) What is general solution of equation shown above?

36. If $x^2 = \sum_{j=1}^{\infty} c_j J_3(\alpha_j x)$ show

$$c_j = \frac{2\alpha_j^2}{(\alpha_j^2-5)[J_3(\alpha_j)]^2}\int_0^1 x^3 J_3(\alpha_j x)dx \quad j = 1,2,\ldots .$$

if $c = 1$ and $\alpha_j = x_j$ are the positive roots of the equation

$$2J_3(\alpha c) + (\alpha c)J_3'(\alpha c) = 0 .$$

3. LEGENDRE POLYNOMIALS

3.1 MOTIVATION

Certain boundary value problems in three dimensions can be simplified considerably by the introduction of special coordinate systems. In general this is due to the fact that in such coordinate systems the solution to the problem depends on two rather than three variables. However, while solving the original partial differential equations in these special coordinate systems various "special" ordinary differential equations appear. In this section we use this approach to motivate the introduction of Legendre polynomials.

Suppose we want to solve Laplace's equation

$$\nabla^2 u = \frac{\partial^2 u}{\partial x^2} + \frac{\partial^2 u}{\partial y^2} + \frac{\partial^2 u}{\partial z^2} = 0 \tag{3.1}$$

in spherical coordinates (r, θ, φ) where

$$x = r\sin\theta\cos\varphi, \quad y = r\sin\theta\sin\varphi, \quad z = r\cos\theta .$$

For the problem at hand it is known that in this coordinate system the solution does not depend on φ, i.e., $u = u(r, \theta)$.

In this coordinate system equation (3.1) takes the form

$$\frac{1}{r^2}\frac{\partial}{\partial r}\left[r^2 \frac{\partial u}{\partial r}\right] + \frac{1}{r^2 \sin\theta}\frac{\partial}{\partial \theta}\left[\sin\theta \frac{\partial u}{\partial \theta}\right] + \frac{1}{r^2\sin^2\theta}\frac{\partial^2 u}{\partial \varphi^2} = 0 \quad . \tag{3.2}$$

where the last term disappears in our case since u is independent of φ.

To solve equation (3.2) we perform separation of variables by seeking solutions in the form

$$u(r,\theta) = R(r)\Theta(\theta) \quad . \tag{3.3}$$

Substituting (3.3) in (3.2) yields

$$\frac{1}{R(r)} \frac{d}{dr}\left[r^2 \frac{dR(r)}{dr}\right] = -\frac{1}{\Theta(\theta)} \left[\frac{1}{\sin\theta} \frac{d}{d\theta}\left[\sin\theta \frac{d\Theta(\theta)}{d\theta}\right]\right] .$$

However, since the left-hand side of this equation depends only on r while the righthand side depends only on θ we infer that each of these sides is equal to the same constant λ i.e.

$$\frac{d}{dr}\left[r^2 \frac{dR}{dr}\right] + \lambda R = 0$$

$$\frac{1}{\sin\theta} \frac{d}{d\theta}\left[\sin\theta \frac{d\Theta}{d\theta}\right] - \lambda\Theta = 0 \quad . \tag{3.4}$$

Equation (3.4) is called the *Legendre equation*. The standard form of this equation (as it appears in the literature) is obtained by making a change of variable to $x = \cos\theta$ and $\Theta = y$ which yields

$$(1-x^2) \frac{d^2y}{dx^2} -2x \frac{dy}{dx} + n(n+1)y = 0 \tag{3.5}$$

where $\lambda = n(n+1)$ and n is not necessarily an integer.

3.2 SOLUTION OF LEGENDRE'S EQUATION.

Equation (3.5) has regular singular points at ±1 and ∞. Writing the desired solution in descending powers of x (i.e., solving "near" the singularity at ∞) and using standard power series methods we obtain the following two independent solutions of (3.5):

$$y_1(x) = x^n - \frac{n(n-1)}{2(2n-1)} x^{n-2} + \frac{n(n-1)(n-2)(n-3)}{2.4(2n-1)(2n-3)} x^{n-4} + \ldots \tag{3.6}$$

$$y_2(x) = x^{-(n+1)} + \frac{(n+1)(n+2)}{2(2n+3)} x^{-(n+3)}$$

$$+ \frac{(n+1)(n+2)(n+3)(n+4)}{2.4.(2n+3)(2n+5)} x^{-(n+5)} + \ldots \quad (3.7)$$

We now observe that when n is a non-negative integer y_1 is a polynomial. (The same polynomials are reproduced by y_2 when n is a non-positive integer. Therefore, it is enough to consider non-negative integers only). These polynomials when normalized by requiring that $y(1) = 1$ are called Legendre polynomials and are denoted by $P_n(x)$. The first six of these polynomials are

$$P_0(x) = 1, \quad P_1(x) = x, \quad P_2(x) = \frac{1}{2}(3x^2-1)$$
$$P_3(x) = \frac{1}{2}(5x^3-3x), \quad P_4(x) = \frac{1}{8}(35x^4-30x^2+3),$$
$$P_5(x) = \frac{1}{8}(63x^5-70x^3+15x)$$

See Figure 1

When n is not an integer the solutions (3.6), (3.7) converge for $|x| > 1$ and are referred to (with proper normalization constants) as Legendre functions of the first and second kind respectively.

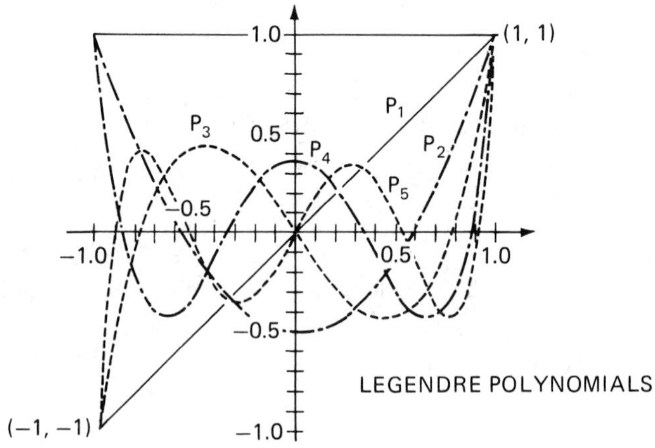

Figure 1

3.3 PROPERTIES OF LEGENDRE POLYNOMIALS

In this subsection we present (usually without proof) various unrelated properties of Legendre polynomials. These properties are useful in the application of these polynomials to electrostatic problems, boundary value problems, numerical calculations as well as other applications.

A. Other Definitions of $P_n(x)$.

Although we introduced Legendre polynomials as solutions of a second order differential equation there are at least three other equivalent ways to introduce these polynomials. These are

a. Rodrigues formula

$$P_n(x) = \frac{1}{2^n n!} \frac{d^n}{dx^n} (x^2-1)^n \quad .$$

b. Laplace integral representation formula

$$P_n(x) = \frac{1}{\pi} \int_0^{\pi} [x+(x^2-1)^{1/2}\cos\theta]^n d\theta \quad .$$

c. Generating function approach

$$\frac{1}{\sqrt{1-2xh+h^2}} = \sum_{n=0}^{\infty} h^n P_n(x) \quad . \tag{3.8}$$

Thus, the polynomials $P_n(x)$ appear as the coefficient of h^n in the Taylor expansion of the generating function $f(h) = (1-2xh+h^2)^{-1/2}$ in terms of h. This formula is particularly useful in solving electrostatic problems.

B. Recurrence Formula for $P_n(x)$.

A recurrence relation for Legendre polynomials relate several of these polynomials and their derivatives. Such relationships are useful in simplifying expressions containing these polynomials. We present here only a few of these identities.

a. $(n+1)P_{n+1}(x) + nP_{n-1}(x) = (2n+1)xP_n(x)$ (3.9)

b. $nP_n(x) = xP_n'(x) - P_{n-1}'(x)$ (3.10)

c. $nxP_n(x) - nP_{n-1}(x) = (x^2-1)P_n'(x)$ (3.11)

C. Orthogonality and Completeness.

Theorem 1: If P_m, P_n are Legendre polynomials, then

$$\int_{-1}^{1} P_m(x)P_n(x)dx = \frac{2}{2n+1}\delta_{m,n} \tag{3.12}$$

where $\delta_{m,n}$ is the Kronecker delta.

Proof (for $m \neq n$): The differential equations satisfied by $P_m(x)$, $P_n(x)$ respectively are

$$\frac{d}{dx}[(1-x^2)P_m'(x)] + m(m+1)P_m(x) = 0$$

$$\frac{d}{dx}[(1-x^2)P_n'(x)] + n(n+1)P_n(x) = 0 \quad .$$

Multiplying these equations by P_n, P_m respectively, integrating over $[-1,1]$ and subtracting the resulting equations we obtain

$$\int_{-1}^{1} P_m(x)[(1-x^2)P_n'(x)]'dx$$

$$- \int_{-1}^{1} P_n(x)[(1-x^2)P_m'(x)]'dx$$

$$+ [n(n+1) - m(m+1)]\int_{-1}^{1} P_m(x)P_n(x)dx = 0 \ . \tag{3.13}$$

Using integration by parts it is easy to show that the first two integrals in (3.13) cancel each other and hence

$$\int_{-1}^{1} P_n(x)P_m(x)dx = 0 \qquad m \neq n \quad .$$

Thus, any two Legendre polynomials of different order are orthogonal to each other on $[-1,1]$.

Another important property of Legendre polynomials is that they form a complete set of functions on $[-1,1]$, i.e. for any piecewise continuous function

$f(x)$ on this interval we can find coefficients c_n so that at the points of continuity of $f(x)$

$$f(x) = \sum_{n=0}^{\infty} c_n P_n(x) \quad . \tag{3.14}$$

To find the coefficients c_m we multiply both sides of (3.14) by $P_m(x)$ and integrate over $[-1,1]$. Using (3.12) it is easy to show that

$$c_m = (m + \tfrac{1}{2}) \int_{-1}^{1} f(x)P_m(x)dx \quad .$$

The series (3.14) is referred to as the "Legendre series" for $f(x)$ and the c_n's as the Fourier coefficients of this series.

D. Miscellaneous Properties.

a. $P_n(x)$ is even when n is even and odd when n is odd hence

$$P_n(-1) = (-1)^n$$

and

$$P_{2k+1}(0) = 0 \quad k = 0,1,\ldots$$

b. The n zeroes of $P_n(x)$ are all real simple and are contained in the interval $(-1,1)$.

3.4 ASSOCIATED LEGENDRE FUNCTIONS AND SPHERICAL HARMONICS

A set of functions which is closely related to the polynomials $P_n(x)$ is given the name the *associated Legendre functions* which are defined as

$$P_{n,m}(x) = (1-x^2)^{m/2} \frac{d^m P_n(x)}{dx^m} \quad .$$

Since the polynomial $P_n(x)$ is of order n, it is obvious from this definition that $P_{n,m}(x)$ are nonzero only when $m \leq n$ and furthermore $P_{n,0}(x) = P_n(x)$.

In a completely analogous fashion to the Legendre polynomials, the functions $P_{n,m}(x)$ satisfy an orthogonality relation

$$\int_{-1}^{1} P_{n,m}(x)P_{k,m}(x)dx = \frac{2}{2n+1}\frac{(n+m)!}{(n-m)!}\delta_{n,k}$$

and some recursion relations similar to (3.9)-(3.11). For example

$$(n+1-m)P_{n+1,m}(x) + (n+m)P_{n-1,m}(x) = (2n+1)xP_{n,m}(x).$$

Moreover, the functions $P_{n,m}(x)$ satisfy the differential equation

$$(1-x^2)y'' - (2m+1)xy' + [n(n+1)-m(m+1)]y = 0$$

which is obtained from (3.2) by separation of variables when $u = u(r,\theta,\varphi)$.

Using the associated Legendre functions we now define the *spherical harmonics* $Y_{n,m}(\theta,\varphi)$ for $0 \leq m \leq n$ as

$$Y_{n,m}(\theta,\varphi) = \frac{(-1)^m}{\sqrt{2\pi}}\left[\frac{(2n+1)(n-m)!}{2(n+m)!}\right]^{1/2} P_{n,m}(\cos\theta)e^{im\varphi}$$

and for $-n < m < 0$ by

$$Y_{n,m}(\theta,\varphi) = (-1)^m Y_{n,-m}(\theta,\varphi) \quad .$$

The spherical harmonics play a central role in the Quantum mechanical description of the atom as well as in many other physical problems. One of the basic properties of these functions is the orthonormality relation

$$\int_0^{\pi}\int_0^{2\pi} Y^*_{n',m'}(\theta,\varphi)Y_{n,m}(\theta,\varphi)\sin\theta d\varphi d\theta = \delta_{n,n'}\delta_{m,m'}.$$

Furthermore the functions $Y_{n,m}(\theta,\varphi)$ are solutions of the angular part of the Laplace equation in spherical coordinates, i.e.

$$\left[\frac{1}{\sin\theta}\frac{\partial}{\partial\theta}(\sin\theta\,\frac{\partial}{\partial\theta}) + \frac{1}{\sin^2\theta}\frac{\partial^2}{\partial\varphi^2} + n(n+1)\right]Y_{n,m}(\theta,\varphi) = 0 .$$

EXERCISE 1

1. Use Rodrigues formula to find $P_n(x)$ for $n = 0,\dots,6$.
2. Use the generating function given by equation (3.8) to prove that
$$P_n(0) = \begin{cases} 0 & n \text{ odd} \\ \dfrac{(-1)^n(2n)!}{2^{2n}(n!)^2} & n \text{ even} \end{cases} .$$
3. Show that
$$\int_{-1}^{x} P_n(t)dt = \frac{1}{2n+1}\left[P_{n+1}(x) - P_{n-1}(x)\right] .$$
4. Find the Legendre series for $P(x) = x^4$.
5. A point electric charge is located at $(0,0,a)$. Evaluate the electrostatic potential at any point in R^3 due to this charge in terms of Legendre polynomials.

   Hint: The electrostatic potential at $\underline{x}$ due to a point charge e given by $\frac{e}{|r'|}$ where $|r'|$ is the distance between $\underline{x}$ and the charge.
6. Use integration by parts to show that
$$\int_a^b f(x)P_n(x)dx = \frac{(-1)^n}{2^n n!}\int_a^b (x^2-1)^n f^{(n)}(x)dx.$$

7. Use Exercise 6 to evaluate

$$\int_{-1}^{1} x^n P_n(x)dx \quad .$$

4. GAMMA FUNCTION

The Gamma function is a useful device when we wish to broaden our ideas concerning factorials. In particular we need the Gamma function if the order ν of a Bessel's equation is neither zero or an integer.

The *Gamma function* is defined as

$$\Gamma(\nu) = \int_0^{\infty} e^{-t}t^{\nu-1}dt \qquad \nu > 0 ,$$

where ν must be positive in order that the improper integral will converge. We can compute some values directly such as

$$\Gamma(1) = \int_0^{\infty} e^{-t}t^0dt = \lim_{b\to\infty} \int_0^b e^{-t}dt$$

$$= \lim_{b\to\infty} -e^{-t}\Big|_0^b = \lim_{b\to+\infty} (1-e^{-b}) = 1 .$$

Using integration by parts we can derive the fundamental identity concerning Gamma functions. If we let

$$dv = e^{-t}dt \qquad u = t^{\nu-1}$$

$$v = -e^{-t} \qquad du = (\nu-1)t^{\nu-2}dt$$

then

$$\Gamma(\nu) = \int_0^{\infty} e^{-t}t^{\nu-1}dt = -t^{\nu-1}e^{-t}\Big|_0^{\infty}$$

$$+ (\nu-1) \int_0^{\infty} e^{-t}t^{\nu-2}dt \qquad (\nu>1) ,$$

or

$$\Gamma(\nu) = (\nu-1)\Gamma(\nu-1) \qquad (\nu>1) . \tag{4.1}$$

If we apply this identity over and over again we arrive at the result,

$$\Gamma(\nu) = (\nu-1)(\nu-2)\cdots\cdots(\nu-r)\Gamma(\nu-r), \quad (\nu>r). \quad (4.2)$$

When $\nu = n$ is a positive integer,

$$\Gamma(n) = (n-1)(n-2)\cdots\cdots 1 = (n-1)!$$

Because of this result $\Gamma(\nu)$ is sometimes known as the *generalized factorial*.

But even with these identities we still have not answered the question of how to find specific values of the Gamma function. True, we can find $\Gamma(n)$ where n is any positive integer but this still leaves us in doubt as to what method we use to find the value of the Gamma function for positive non-integer numbers.

In order to find such values it was necessary for someone to evaluate $\Gamma(\nu)$ by numerical or series methods for a sufficient quantity of real numbers between two integers. The integers chosen are usually one and two. If we observe Figure 1, we notice that the Gamma function is continuous between one and two and the range is between zero and one. Such an interval makes a convenient table.

Therefore, we can find the value of the Gamma function for any real number greater than zero by use of (4.2) along with the appropriate choice from the table.

Example 1: $\Gamma(3) = 2\cdot 1 = 2$

$\Gamma(10) = 9!$

$\Gamma(3.7) = 2.7(2.7) = (2.7)(1.7)\Gamma(1.7) =$
$(2.7)(1.7)(.908) = 4.168$

To find $\Gamma(.7)$ we write (4.1) in the form

$$\Gamma(\nu-1) = \frac{\Gamma(\nu)}{\nu-1} .$$

Therefore,

$$\Gamma(.7) = \frac{\Gamma(1.7)}{.7} = \frac{.908}{.7} = 1.30.$$

What is $\Gamma(\nu)$ when ν is a negative number (not an integer)? We know the integral definition will not work since $\nu < 0$. In its place we require the identity $\Gamma(\nu) = (v-1)\Gamma(v-1)$ hold if $\nu < 0$. It is easier to use this identity if it is written in the form

$$\Gamma(\nu) = \frac{\Gamma(\nu+1)}{\nu} , \quad \nu \neq 0 .$$

Example 2: What is the value of $\Gamma(-1.3)$?

$$\Gamma(-1.3) = \frac{\Gamma(-0.3)}{-1.3} = \frac{\Gamma(.7)}{(-1.3)(-0.3)}$$
$$= \frac{\Gamma(1.7)}{(-1.3)(-0.3)(.7)} = \frac{.908}{(-1.3)(-0.3)(.7)} = 3.33 \quad .$$

Negative integers and zero for ν take special handling. For $\nu = 0$ we define $\Gamma(0)$ as follows:

$$\Gamma(0) = \lim_{\epsilon\to 0} \Gamma(\epsilon) = \lim_{\epsilon\to 0} \frac{\Gamma(1+\epsilon)}{\epsilon} = \pm\,\infty \ .$$

Since every negative integer is related to $\Gamma(0)$,

$$\Gamma(-n) = \pm\,\infty \qquad n = 1,2,3, \ldots \quad .$$

We have now assigned a value to $\Gamma(\nu)$ for every real value ν. See graph, Figure 2.

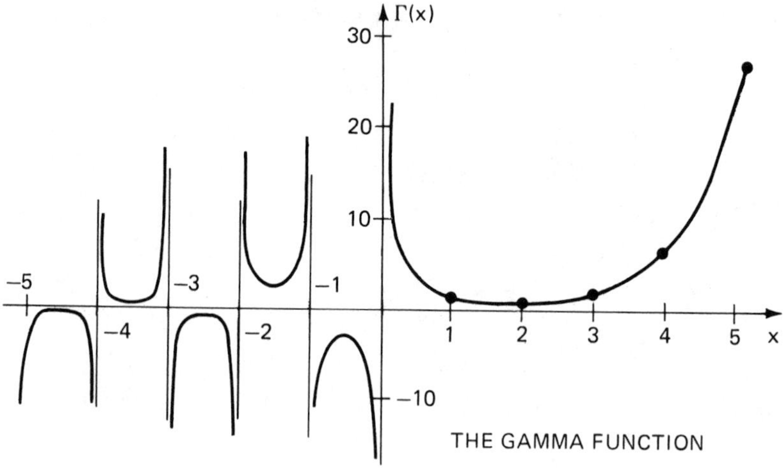

Figure 2

CHAPTER 3. SYSTEMS OF ORDINARY DIFFERENTIAL EQUATIONS

1. INTRODUCTION

In this chapter we discuss systems of first order ordinary differential equations and methods for their solutions. The importance of such systems stems from the following observations:

1. A coupled system of differential equations of any order can always be rewritten as a system of first order equations.

To see how this can be done consider the nth order equation

$$\frac{d^n y}{dx^n} = f\left[x, y, \frac{dy}{dx}, \ldots, \frac{d^{n-1}y}{dx^{n-1}}\right] \tag{1.1}$$

with the initial conditions

$$y(a) = b_0, \; y'(a) = b_1, \ldots, y^{(n-1)}(a) = b_{n-1} \tag{1.2}$$

These are obviously equivalent to the system

$$\frac{dy}{dx} = y_1$$

$$\frac{dy_1}{dx} = y_2 \quad \left[\equiv \frac{d^2y}{dx^2}\right] \tag{1.3}$$

$$\frac{dy_{n-2}}{dx} = y_{n-1} \quad \left[\equiv \frac{d^{n-1}y}{dx^{n-1}}\right]$$

$$\frac{dy_{n-1}}{dx} = f(x,y,y_1,\ldots,y_{n-1})$$

with the initial conditions

$$y(a) = b_0,\ y_1(a) = b_1,\ldots,y_{n-1}(a) = b_{n-1}. \qquad (1.4)$$

Example 1: Rewrite the system

$$x'' + y'' + xy = 0$$

$$y'' + x'y' - x^2y = 0 \qquad (1.5)$$

$$x(0) = 0,\ x'(0) = 1,\ y(0) = 1,\ y'(0) = 2 \qquad (1.6)$$

where primes denote differentiation with respect to t, as a system of first order equations.

Solution: The required system is

$$x' = x_1 \quad , \quad y' = y_1$$

$$x_1' = -y_1 - xy \qquad (1.7)$$

$$y_1' = -x_1y_1 + x^2y$$

$$x(0) = 0,\ x_1(0) = 1,\ y(0) = 1,\ y_1(0) = 2. \qquad (1.8)$$

2. Mathematical models of many problems in science and engineering lead to systems of first order equations.

Example 2: A typical model for an ecological system with a predator and prey, e.g. fish and sharks in a lake is given by;

$$\frac{dF}{dt} = aF - bF^2 - cFS \qquad (1.9)$$

$$\frac{dS}{dt} = -dS + eFS \qquad (1.10)$$

$$a,b,c,d,e > 0.$$

In these equations aF represents the effect of natural birth and death processes on the fish population, $-bF^2$ the effect of their competition for (vegetable) food and $-cFS$ the adverse effect of the predator on the fish population. The terms in equation (1.10) have similar meaning.

Example 3: Derive the equations of motion for the masses m_1, m_2 which are attached to three springs with elastic constants k_1, k_2, k_3 as shown in Figure 1.

Solution: Let x, y be the displacements from equilibrium of m_1, m_2 respectively. From Hooke's law we know that for small displacements from equilibrium the force exerted by the spring is proportional but in the opposite direction to the displacement. Hence if $0 < y < x$ we obtain

$$m_1 \frac{d^2x}{dt^2} = -k_1x - k_2(x-y) \tag{1.11}$$

$$m_2 \frac{d^2y}{dt^2} = -k_3y + k_2(x-y) \quad . \tag{1.12}$$

We leave it to the reader to verify that the same equations hold for other values of x, y.

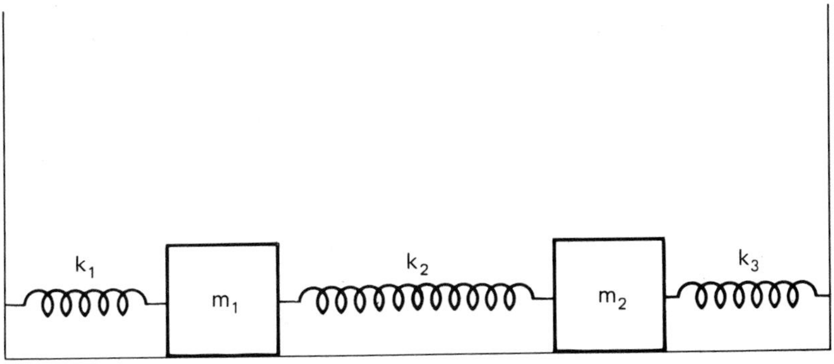

Figure 1

EXERCISE 1

1. Rewrite the following as systems of first order ordinary differential equations.

   a. $x'' + y' - 3xy = t^2$

$$y'' + x' + 4x - 2y = e^t$$

$$x(1) = 1,\ x'(1) = 0,\ y(1) = 2,\ y'(1) = 0$$

b. $$x''' + 2y' + 8xy = 12t^2$$

$$y'' + 2x''y' - 2y = 0$$

$$x(0) = 1,\ x'(0) = 0,\ x''(0) = 6,\ y(0) = 10,$$
$$y'(0) = 1$$

2. Describe the ecological system which is modeled by the following system of equations (all the coefficients are positive)

$$\frac{dF}{dt} = a_1F - a_2F^2 - a_3FS$$

$$\frac{dS}{dt} = -b_1S + b_2FS - b_3SG$$

$$\frac{dG}{dt} = -c_1G + c_2SG$$

3. Derive a mathematical model which describe the vibrations of two masses attached to a spring as shown in Figure 2. Rewrite this model as a system of first order equations.

4. Verify equations (1.11)-(1.12) for all possible combinations of x,y.

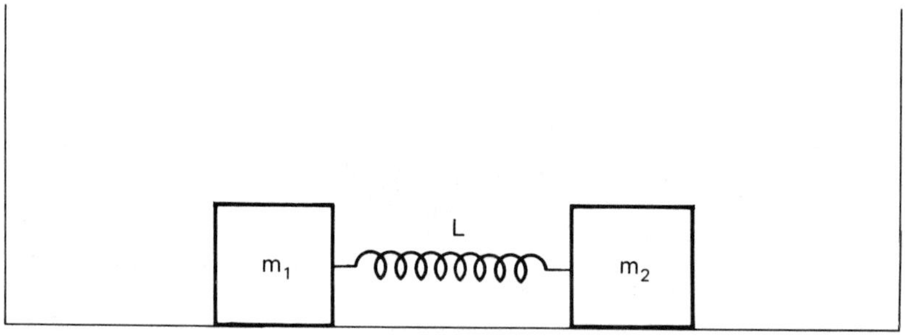

Figure 2

2. METHOD OF ELIMINATION

This technique attempts to solve a given system of equations by reducing it to a system of uncoupled higher order equations (i.e., each equation in the new system contains only one dependent variable). We illustrate this method through examples:

Example 1: Solve the system

$$\frac{dx}{dt} = ax + by$$

$$\frac{dy}{dt} = cx + dy$$

$$x(0) = 1 \quad , \quad y(0) = 2 \tag{2.3}$$

where a,b,c,d are constants.

Solution: First note that if b or c are zero this system can be solved directly. Thus, if $b = 0$ then $x(t) = e^{at}$. Substituting this result in equation (2.2) we obtain

$$\frac{dy}{dt} - dy = ce^{at} \tag{2.4}$$

which can be solved by elementary methods.

When b and c are different from zero we solve this system of equations by elimination. Thus from (2.2) we have

$$x = \frac{1}{c}\frac{dy}{dt} - \frac{d}{c}y \tag{2.5}$$

and hence also

$$\frac{dx}{dt} = \frac{1}{c}\frac{d^2y}{dt^2} - \frac{d}{c}\frac{dy}{dt} \quad . \tag{2.6}$$

Substituting (2.5), (2.6) in (2.1) we obtain

$$\frac{d^2y}{dt^2} - (a+d)\frac{dy}{dt} + (ad-bc)y = 0 \tag{2.7}$$

with the initial conditions (using (2.2))

$$y(0) = 2, \; y'(0) = c + 2d \quad . \tag{2.8}$$

Similar equation can be derived for x.

Example 2: Use the method of elimination to solve

$$x' = x + y \tag{2.9}$$

$$y'' = -x' + 4x + 2y \tag{2.10}$$

$$x(0) = 1, \; y(0) = 1, \; y'(0) = 0 \quad . \tag{2.11}$$

Solution: From (2.9) we obtain

$$y = x' - x \tag{2.12}$$

and hence

$$y'' = x''' - x'' \quad . \tag{2.13}$$

Substituting this in (2.10) yields

$$x''' - x'' - x' - 2x = 0 \tag{2.14}$$

with the initial conditions (using (2.9)-(2.11))

$$x(0) = 1 \quad x'(0) = 2, \quad x''(0) = 2 \quad . \tag{2.15}$$

Equation (2.14) is a homogeneous differential equation with constant coefficients which can be solved by elementary methods. Furthermore, we observe that when $x(t)$ is known $y(t)$ can be computed from equation (2.12).

From these two examples we see that this method is useful only for small systems of equations and becomes rapidly cumbersome in other cases.

EXERCISE 2

1. Find the explicit solution of equation (2.4).
2. Derive a differential equation for x in the system (2.1)-(2.3).
3. Discuss the solutions of equation (2.7) for different values of the parameters.
4. Derive an equation containing x only for the system

   $x' + 3y' = 2xy$

   $3x' - y' = \sin t$

   $x(0) = 1, \; y(0) = 2$
5. Solve the system

   $x' + 2x - 4y = t$

   $x' + 2y' - 2y = t^2$

   $x(0) = 1, \; y(0) = 2.$

3. SOME LINEAR ALGEBRA

In this section we review some topics from linear algebra regarding eigenvalues and eigenvectors, introduce the notion of matrix exponentiation and finally discuss an algorithm to compute it.

A. Eigenvalues and Eigenvectors

Definition 1: Let A be a nxn matrix. We say that a vector $\mathbf{v} \neq \mathbf{0}$ is an eigenvector of A if

$$A\mathbf{v} = \lambda\mathbf{v} \tag{3.1}$$

λ is called the eigenvalue of A which is related to the eigenvector $\mathbf{v}$.

Corollary 1: If $\mathbf{v}$ is an eigenvector of A then $c\mathbf{v}$ for all $c \neq 0$ is also an eigenvector of A (with eigenvalue λ).

Definition 2: The characteristic polynomial of A is

$$p(\lambda) = |A - \lambda I| \tag{3.2}$$

Theorem 1: The eigenvalues of A are the roots of its characteristic polynomial.

We note that to each eigenvalue we have a related eigenvector, however, the number of independent eigenvectors related to a given eigenvalue does not have to equal to its algebraic multiplicity, i.e. the number of times it appears as a root of $p(\lambda)$.

Example 1: Find the eigenvalues and eigenvectors of

$$A = \begin{bmatrix} 1 & 1 & 0 \\ 0 & 2 & 1 \\ 0 & 0 & 2 \end{bmatrix}$$

Solution: The characteristic polynomial of A is

$$p(\lambda) = \begin{bmatrix} 1-\lambda & 1 & 0 \\ 0 & 2-\lambda & 1 \\ 0 & 0 & 2-\lambda \end{bmatrix} = (1-\lambda)(2-\lambda)^2 \tag{3.3}$$

hence $\lambda = 1$ is a simple root (i.e., multiplicity 1) and $\lambda = 2$ is a double root.

To find the corresponding eigenvectors we solve equation (3.1) with each of these eigenvalues

a. $\lambda = 1$

If $\mathbf{v} = \begin{bmatrix} x_1 \\ x_2 \\ x_3 \end{bmatrix}$ we obtain the system

$$\begin{aligned} x_1 + x_2 &= x_1 \\ 2x_2 + x_3 &= x_2 \\ 2x_3 &= x_3 \quad . \end{aligned}$$

Hence up to a multiplicative constant which can be expected in view of corollary 1

$$\mathbf{v} = \begin{bmatrix} 1 \\ 0 \\ 0 \end{bmatrix} \quad .$$

b. $\lambda = 2$

Denoting once again the required eigenvector by $\mathbf{v} = \begin{bmatrix} x_1 \\ x_2 \\ x_2 \end{bmatrix}$ we obtain the system

$$\begin{aligned} x_1 + x_2 &= 2x_1 \\ 2x_2 + x_3 &= 2x_2 \\ 2x_3 &= 2x_3 \quad . \end{aligned}$$

Hence $\mathbf{v} = \begin{bmatrix} 1 \\ 1 \\ 0 \end{bmatrix}$. Thus, there exist only one independent eigenvector related to $\lambda = 2$ although the multiplicity of this eigenvalue is 2.

Theorem 2: If A has n independent eigenvectors $\mathbf{v}_1, \ldots, \mathbf{v}_n$ then

$$M^{-1}AM = \begin{bmatrix} \lambda_1 & & O \\ & \ddots & \\ O & & \lambda_n \end{bmatrix} \tag{3.4}$$

where $M = (\mathbf{v}_1, \ldots, \mathbf{v}_n)$. Such a matrix A is said to be diagonalizable.

Example 2: Verify theorem 2 for

$$A = \begin{bmatrix} 1 & 1 & 0 \\ 0 & 2 & 0 \\ 0 & 0 & 2 \end{bmatrix}$$

Solution: The eigenvalues of A are $\lambda = 1$ and $\lambda = 2$ (with multiplicity 2). However, contrary to example 1 A has three independent eigenvectors

$$\mathbf{v}_1 = \begin{bmatrix} 1 \\ 0 \\ 0 \end{bmatrix}, \quad \mathbf{v}_2 = \begin{bmatrix} 1 \\ 1 \\ 0 \end{bmatrix} \quad \text{and} \quad \mathbf{v}_3 = \begin{bmatrix} 0 \\ 0 \\ 1 \end{bmatrix}. \quad \text{Hence}$$

$$M = \begin{bmatrix} 1 & 1 & 0 \\ 0 & 1 & 0 \\ 0 & 0 & 1 \end{bmatrix}, \quad M^{-1} = \begin{bmatrix} 1 & -1 & 0 \\ 0 & 1 & 0 \\ 0 & 0 & 1 \end{bmatrix}$$

and a direct multiplication shows that

$$M^{-1}AM = \begin{bmatrix} 1 & 0 & 0 \\ 0 & 2 & 0 \\ 0 & 0 & 2 \end{bmatrix} .$$

When a matrix A does not possess n independent eigenvectors it is impossible to find a similarity transformation which diagonalizes it. However, it is possible then to bring it into "Jordan canonical form" by an appropriate similarity transformation.

Theorem 3 (Jordan): Let A be a nxn matrix. There exists a matrix M such that

$$M^{-1}AM = J = \begin{bmatrix} D & & & \bigcirc \\ & J_1 & & \\ & & \ddots & \\ \bigcirc & & & J_m \end{bmatrix}$$

where the submatrices D, J_i are of the form

$$D = \begin{bmatrix} \lambda_1 & & \bigcirc \\ & \ddots & \\ \bigcirc & & \lambda_k \end{bmatrix}, \quad J_i = \begin{bmatrix} \lambda_{k+i} & 1 & & \bigcirc \\ & \ddots & \ddots & \\ & & \ddots & 1 \\ \bigcirc & & & \lambda_{k+i} \end{bmatrix}$$

and $\lambda_1, \ldots, \lambda_{k+m}$ are the eigenvalues of A.

Example 3: For the matrix

$$A = \begin{bmatrix} 2 & 0 & 0 \\ 1 & 2 & 0 \\ 1 & 1 & 2 \end{bmatrix}$$

a similarity transformation with

$$M = \begin{bmatrix} 0 & 0 & 1 \\ 0 & 1 & 0 \\ 1 & -1 & 0 \end{bmatrix}, \qquad M^{-1} = \begin{bmatrix} 0 & 1 & 1 \\ 0 & 1 & 0 \\ 1 & 0 & 0 \end{bmatrix}$$

yields

$$M^{-1}AM = \begin{bmatrix} 2 & 1 & 0 \\ 0 & 2 & 1 \\ 0 & 0 & 2 \end{bmatrix}$$

which is the Jordan canonical form of A. Note that in this case the matrix D is "missing".

We shall not present here an algorithmn to find the appropriate M which transforms a matrix A into its Jordan canonical form but refer the reader to the bibliography at the end of this chapter.

B. Matrix exponentiation

Definition 2: Let A be nxn matrix and $p(x)$ a polynomial of order k

$$p(x) = a_0 + a_1 x + \ldots + a_k x^k \tag{3.5}$$

we define

$$p(A) = a_0 I + a_1 A + \ldots + a_k A^k \tag{3.6}$$

Example 4: If $A = \begin{bmatrix} 0 & 1 \\ -1 & 0 \end{bmatrix}$ and $p(x) = a_0 + a_1 x$

then

$$p(A) = \begin{bmatrix} 1 & 1 \\ -1 & 1 \end{bmatrix} .$$

Motivated by definition 2 and the observation that

$$e^x = 1 + x + \frac{x^2}{2!} + \ldots = \sum_{k=0}^{\infty} \frac{x^n}{n!} \tag{3.7}$$

we now make the following definition .

Definition 3: Let A be nxn matrix. We define

$$e^A = \sum_{k=0}^{\infty} \frac{A^k}{k!} = I + A + \frac{A^2}{2!} + \ldots \tag{3.8}$$

Example 5: If $A = I_n = \begin{bmatrix} 1 & & O \\ & \ddots & \\ O & & 1 \end{bmatrix}$ then

$$e^{\alpha A} = \sum_{k=0} \frac{(\alpha I_n)^k}{k!} = \left[\sum_{k=0}^{\infty} \frac{\alpha^k}{k!}\right] I_n = e^{\alpha} I_n = \begin{bmatrix} e^{\alpha} & & O \\ & \ddots & \\ O & & e^{\alpha} \end{bmatrix}$$

Example 6: If

$$A = \begin{bmatrix} 0 & 1 \\ -1 & 0 \end{bmatrix}$$

then

$$A^2 = -I, \; A^3 = -A, \; A^4 = I, \text{ etc.}$$

Hence,

$$e^{\alpha A} = \sum_{k=0}^{\infty} \frac{(\alpha A)^k}{k!} = I + \frac{\alpha A}{1!} - \frac{\alpha^2 I}{2!} - \frac{\alpha^3 A}{3!} + \frac{\alpha^4 I}{4!} + \ldots =$$

$$= I\left[1 - \frac{\alpha^2}{2!} + \frac{\alpha^4}{4!} - \ldots\right] + A\left[\frac{\alpha}{1!} - \frac{\alpha^3}{3!} + \ldots\right] =$$

$$= I\cos\alpha + A\sin\alpha = \begin{bmatrix} \cos\alpha & \sin\alpha \\ -\sin\alpha & \cos\alpha \end{bmatrix} .$$

We now enumerate some of the properties of e^A.

Theorem 4: For any nxn matrix A (with constant entries) e^A is well defined, i.e. the series (3.8) converges. Moreover,

$$(e^A)^{-1} = e^{-A} \qquad . \qquad (3.9)$$

Thus e^A is always nonsingular matrix (even if A is singular).

Theorem 5: (1) $\frac{d}{dt} e^{At} = Ae^{At}$ (3.10)

(2) $e^{A(t+s)} = e^{At}e^{As}$. (3.11)

The proof of these two properties follows directly from the definition (3.8) and the fact that A^k and A^m commute.

Although the definition (3.8) can be used to compute e^A for some "simple" matrices it is impossible to do so in general. An algorithm towards this end is given by Cayley-Hamilton theorem.

Theorem 6 (Cayley-Hamilton): Let A be nxn matrix and f a continuous and differentiable function. There exists a polynomial r(x) of degree less than n (that depend on A and f) so that

$$f(A) = r(A) \qquad . \qquad (3.12)$$

Moreover, if λ_i, $i = 1 \ldots n$ are the eigenvalues of A then $f(\lambda_i) = r(\lambda_i)$.

Remark: The classical formulation of Cayley-Hamilton theorem states that if $p(\lambda)$ is the characteristic polynomial of A then $p(A) = 0$. Theorem 5 is a consequence of this result.

We now show how to use this theorem to compute e^A.

Example 7: Compute e^{At} if

$$A = \begin{bmatrix} 2 & -1 \\ 3 & -2 \end{bmatrix} .$$

Solution: In this example $f(x) = e^x$ and $n = 2$. From theorem 5 we infer that there exist a polynomial $r(x)$ of degree 1

$$r(x) = a_0 + a_1 x \tag{3.13}$$

so that

$$e^{At} = f(At) = r(At) \tag{3.14}$$

i.e.

$$e^{At} = a_0 I + a_1 (At) \quad . \tag{3.15}$$

To evaluate a_0, a_1 we now use the second part of theorem 5. To this end we note that the eigenvalues of (At) are

$$\lambda_1 = t \quad , \quad \lambda_2 = -t$$

and, therefore,

$$f(t) = r(t) \quad , \quad f(-t) = r(-t) \tag{3.16}$$

i.e.

$$e^t = a_0 + a_1 t \quad , \quad e^{-t} = a_0 - a_1 t \quad .$$

Thus,

$$a_0 = \cosh t \quad , \quad a_1 = \frac{\sinh t}{t}$$

and

$$e^{At} = \cosh t \cdot I + \sinh t \cdot A$$

$$= \begin{bmatrix} \cosh t + 2 \sinh t \ , & -\sinh t \\ 3 \sinh t \ , & \cosh t - 2 \sinh t \end{bmatrix} .$$

It is obvious from theorem 5 that if some of the eigenvalues of A are not simple then the number of equations obtained through this theorem for the coefficients of $r(x)$ will be less than n. To overcome this deficiency we state the following extension of theorem 5:

Theorem 7: Under the same assumptions as in theorem 5 if λ_i is of multiplicity $k > 1$ then

$$f(\lambda_i) = r(\lambda_i) \tag{3.17}$$

and

$$f^{(m)}(\lambda_i) = r^{(m)}(\lambda_i), \ m = 1,\dots,k-1 \tag{3.18}$$

where $r^{(m)}$, $f^{(m)}$ are the mth derivatives of r and f respectively.

Example 8: Compute e^{At} if

$$A = \begin{bmatrix} 2 & 1 & 0 \\ 0 & 2 & -1 \\ 0 & 0 & 2 \end{bmatrix}$$

Solution: Since A is a 3×3 matrix it follows that $r(x)$ is (at most) of degree two i.e.,

$$r(x) = a_0 + a_1x + a_2x^2$$

and

$$e^{At} = f(At) = a_0I + a_1(At) + a_2(At)^2 \ . \tag{3.19}$$

To compute the coefficients of $r(x)$ we note that (At) has one eigenvalue $\lambda = 2t$ with multiplicity three. Hence since $f(x) = e^x$ we infer from theorem 6 that

$$e^{2t} = a_0 + a_1(2t) + a_2(2t)^2$$

$$e^{2t} = a_1 + 2a_2(2t)$$

$$e^{2t} = 2a_2 \ .$$

Hence,

$$a_2 = \frac{e^{2t}}{2}, \quad a_1 = e^{2t}(1-2t)$$

$$a_0 = e^{2t}(1-2t+2t^2)$$

and, therefore, from (3.19)

$$e^{At} = e^{2t}\begin{bmatrix} 1 & t & -\frac{t^2}{2} \\ 0 & 1 & -t \\ 0 & 0 & 1 \end{bmatrix}$$

EXERCISE 3

1. Find the relationship between the eigenvalues of A and those of A^{-1} and A^T (where A^T is the transpose of A).
2. Let $p(x)$ be a polynomial show that the eigenvalues of $p(A)$ are given by $p(\lambda_i)$ where λ_i are the eigenvalues of A.

Hint: Show that if $\mathbf{v}$ is an eigenvector of A then it is also an eigenvector of A^k.

3. Find the eigenvalues and eigenvectors of the following matrices:

   a. $A = \begin{bmatrix} 1 & 2 \\ 4 & 3 \end{bmatrix}$ b. $A = \begin{bmatrix} 1 & 2 \\ 2 & 4 \end{bmatrix}$

   c. $A = \begin{bmatrix} 2 & 1 & 0 \\ 0 & -2 & -1 \\ 0 & 0 & 2 \end{bmatrix}$ d. $A = \begin{bmatrix} 1 & 1 & -1 \\ 0 & 0 & 0 \\ 1 & -2 & 3 \end{bmatrix}$

4. Compute $p(A)$ for each of the matrices in exercise 3 if

   $p(x) = 1 + 2x + 5x^2$

5. If

   $A = \begin{bmatrix} 2 & -1 \\ 3 & -2 \end{bmatrix}$ and

   $p(x) = 1 + x^{300} + 2x^{401}$ compute $p(A)$.

6. Compute e^{At} directly (i.e., using equation 3.8) for the following matrices

a. $A = \begin{bmatrix} 1 & 0 \\ 0 & -1 \end{bmatrix}$ b. $A = \begin{bmatrix} 0 & 1 \\ 1 & 0 \end{bmatrix}$

$A = \begin{bmatrix} 0 & i \\ -i & 0 \end{bmatrix}$

7. If

$$A = \begin{bmatrix} 0 & 1 & 0 \\ 0 & 0 & -1 \\ 0 & 0 & 0 \end{bmatrix}$$

compute e^{At} directly and by Cayley-Hamilton theorem.

8. Show that $(e^{A})^{-1} = e^{-A}$

9. Compute e^{At} using theorems 5 and 6 for the matrices of exercise 3.

10. Show that if A is nonsingular then

$$\int_0^t e^{As}ds = A^{-1}(e^{At}-I)$$

4. LINEAR SYSTEMS WITH CONSTANT COEFFICIENTS

In this section we describe methods to solve the system

$$\frac{d\mathbf{x}}{dt} = A\mathbf{x} + \mathbf{f}(t) \quad , \quad \mathbf{x}(a) = \mathbf{c} \tag{4.1}$$

where A is a matrix with constant entries,

$$\mathbf{x}(t) = \begin{bmatrix} x_1(t) \\ \vdots \\ x_n(t) \end{bmatrix} \quad \text{and} \quad \mathbf{f}(t) = \begin{bmatrix} f_1(t) \\ \vdots \\ f_n(t) \end{bmatrix} .$$

To begin with we observe that if A has n independent eigenvectors then the system (4.1) can be solved through a change of variables.

Thus if

$$M = (\mathbf{v}_1, \ldots, \mathbf{v}_n)$$

where $\mathbf{v}_i$, $i = 1,\ldots,n$ are the eigenvectors of A we define

$$\mathbf{y} = M^{-1}\mathbf{x} \quad . \tag{4.2}$$

Substituting (4.2) in (4.1) we obtain

$$M \frac{d\mathbf{y}}{dt} = AM\mathbf{y} + \mathbf{f}(t)$$

and, therefore,

$$\frac{d\mathbf{y}}{dt} = M^{-1}AM\mathbf{y} + M^{-1}\mathbf{f}(t)$$

$$\mathbf{y}(a) = M^{-1}\mathbf{c} \tag{4.3}$$

but

$$M^{-1}AM = D = \begin{bmatrix} \lambda_1 & & \bigcirc \\ & \ddots & \\ \bigcirc & & \lambda_n \end{bmatrix}$$

and, therefore, the system (4.3) is uncoupled and each equation in this system can be solved independently of the others

Example 1: Solve the system (4.1) if

$$A = \begin{bmatrix} 1 & 1 & 1 \\ 0 & 3 & 3 \\ -2 & 1 & 1 \end{bmatrix} \qquad \mathbf{f} = \begin{bmatrix} 1 \\ 0 \\ 1 \end{bmatrix}$$

and $\mathbf{x}(0) = \begin{bmatrix} 1 \\ -1 \\ 0 \end{bmatrix}$

Solution: The eigenvalues of A are $\lambda_1 = 0$, $\lambda_2 = 2$, $\lambda_3 = 3$ and the corresponding eigenvectors are

$$\mathbf{v}_1 = \begin{bmatrix} 0 \\ -1 \\ 1 \end{bmatrix} , \quad \mathbf{v}_2 = \begin{bmatrix} 2 \\ 3 \\ -1 \end{bmatrix} , \quad \mathbf{v}_3 = \begin{bmatrix} 1 \\ 2 \\ 0 \end{bmatrix} .$$

Hence,

$$M = \begin{bmatrix} 0 & 2 & 1 \\ -1 & 3 & 2 \\ 1 & -1 & 0 \end{bmatrix}, \quad M^{-1} = \begin{bmatrix} 1 & -1/2 & 1/2 \\ 1 & -1/2 & -1/2 \\ -1 & 1 & 1 \end{bmatrix}$$

and from (4.3) we infer that the original system is equivalent to

$$\frac{d}{dt}\begin{bmatrix} y_1 \\ y_2 \\ y_3 \end{bmatrix} = \begin{bmatrix} 0 & 0 & 0 \\ 0 & 2 & 0 \\ 0 & 0 & 3 \end{bmatrix} \begin{bmatrix} y_1 \\ y_2 \\ y_3 \end{bmatrix} + \begin{bmatrix} 3/2 \\ 1/2 \\ 0 \end{bmatrix}, \mathbf{y}(0) = \begin{bmatrix} 3/2 \\ 3/2 \\ -2 \end{bmatrix} .$$

The solution of this system is

$$y_1(t) = \frac{3}{2}(t+1), \quad y_2(t) = \frac{7}{4} e^{2t} - \frac{1}{4}, \quad y_3 = -2e^{3t} .$$

Finally, we express the solution in terms of the original variables by using equation (4.2) which yields

$$x_1 = 2y_2 + y_3, \quad x_2 = -y_1 + 3y_2 + 2y_3$$

$$x_3 = y_1 - y_2 \quad .$$

In more general situations, however, we have the following

Theorem 1: The solution of the system (4.1) is given by

$$\mathbf{x}(t) = e^{A(t-a)}\mathbf{c} + e^{At}\int_a^t e^{As}\mathbf{f}(s)ds \tag{4.4}$$

Remark: We remind the reader that the integral of a vector function

$$g(t) = \begin{bmatrix} g_1(t) \\ \vdots \\ g_n(t) \end{bmatrix}$$ is obtained by integrating each of its components, i.e.

$$\int_a^t \mathbf{g}(s)ds = \begin{bmatrix} \int_a^t g_1(s)ds \\ \vdots \\ \int_a^t g_n(s)ds \end{bmatrix} \tag{4.5}$$

Proof of theorem: First note that at $t = a$ the vector $\mathbf{x}(t)$ defined by equation (4.4) satisfies $\mathbf{x}(a) = \mathbf{c}$ as required. To show that it satisfies the differential equation we differentiate (4.4) with respect to t. We obtain

$$\frac{d\mathbf{x}}{dt} = Ae^{A(t-a)}\mathbf{c} + Ae^{At}\int_a^t e^{-As}\mathbf{f}(s)ds +$$

$$+ e^{At}e^{-At}\mathbf{f}(t) = A\mathbf{x}(t) + \mathbf{f}(t)$$

as required.

Example 2: Solve

$$\frac{d}{dt}\begin{bmatrix} x_1 \\ x_2 \end{bmatrix} = \begin{bmatrix} 2 & -1 \\ 3 & -2 \end{bmatrix}\begin{bmatrix} x_1 \\ x_2 \end{bmatrix} + \begin{bmatrix} 1 \\ 2 \end{bmatrix}$$

$$x_1(0) = 0 \quad , \quad x_2(0) = 2$$

Solution: We already computed e^{At} in example 3.6. Substituting this result in equation (4.4) we obtain

$$\mathbf{x}(t) = \begin{bmatrix} x_1(t) \\ x_2(t) \end{bmatrix} = e^{At}\begin{bmatrix} 0 \\ 2 \end{bmatrix} + e^{At}\int_0^t e^{-As}\begin{bmatrix} 1 \\ 2 \end{bmatrix} ds =$$

$$e^{At}\left\{\begin{bmatrix} 0 \\ 2 \end{bmatrix} + \int_0^t \begin{bmatrix} \cosh s - 2\sinh s , & \sinh s \\ -3\sinh s , & \cosh s + 2\sinh s \end{bmatrix}\begin{bmatrix} 1 \\ 2 \end{bmatrix} ds\right\}$$

$$= e^{At}\left\{\begin{bmatrix} 0 \\ 2 \end{bmatrix} + \int_0^t \begin{bmatrix} \cosh s \\ 2\cosh s + \sinh s \end{bmatrix} ds\right\}$$

$$= \begin{bmatrix} -\sinh t \\ \cosh t - 2\sinh t + 1 \end{bmatrix}$$

EXERCISE 4

Solve the following systems by change of variable and/or matrix exponentiation.

1. $\dot{x}_1 = x_1 + 2x_2 + t$

   $\dot{x}_2 = 4x_1 + 3x_2 + (t + 1)$

   $x_1(0) = 1, \quad x_2(0) = 0$

2. $\ddot{x}_2 + 3\dot{x} + 4x = e^{-t}$

   $x(0) = 1, \quad \dot{x}(0) = 0$

3. $\ddot{x} = 2x + 5y + t$

   $\dot{y} = \dot{x} - 2y + 1$

   $x(0) = 1, \ \dot{x}(0) = -1, \ y(0) = 2$

4. $\ddot{x} + 2\dot{x} + 3x = 2y$

   $\ddot{y} + 2\dot{y} - 4y = x$

   $x(0) = y(0) = 0, \quad \dot{x}(0) = 1 \ , \quad \dot{y}(0) = 2.$

5. Show that if $\mathbf{f}(t) = \mathbf{0}$ and A is diagondizable then the solution of the system (4.1) is of the form

$$\mathbf{x}(t) = \Sigma\, a_i \mathbf{v}_i e^{\lambda_i t}$$

where λ_i, $\mathbf{v}_i$ are the eigenvalues and eigenvectors of A.

5. LINEAR SYSTEMS WITH VARIABLE COEFFICIENTS

In this section we discuss methods to solve systems of the form

$$\dot{\mathbf{x}} = A(t)\mathbf{x}(t) + \mathbf{f}(t)$$

$$\mathbf{x}(a) = \mathbf{c} \tag{5.1}$$

where we assume that all the entries of $A(t)$, $f(t)$ are smooth functions of t. To begin with we observe, once again, that if $A(t)$ has n independent eigenvectors over the time interval under consideration then the solution of the system (5.1) can be obtained through a change of variables identical to the one discussed in the previous section provided that $M^{-1}(t) \cdot \frac{dM(t)}{dt}$ is diagonal.

Example 1: Solve the system

$$\frac{d}{dt}\begin{bmatrix} x_1 \\ x_2 \end{bmatrix} = \begin{bmatrix} 0 & t \\ t & 0 \end{bmatrix}\begin{bmatrix} x_1 \\ x_2 \end{bmatrix} + \begin{bmatrix} t \\ -t \end{bmatrix}$$

$$x_1(0) = 1 \quad , \quad x_2(0) = 0$$

Solution: The eigenvalues of the coefficient matrix one $\lambda = \pm t$ and their corresponding eigenvectors are

$$v_1 = \begin{bmatrix} 1 \\ 1 \end{bmatrix} \quad , \quad v_2 = \begin{bmatrix} 1 \\ -1 \end{bmatrix}$$

(in general, however, the eigenvectors $\mathbf{v}_i$ will depend on t). Hence

$$M = \begin{bmatrix} 1 & 1 \\ 1 & -1 \end{bmatrix} \quad , \quad M^{-1} = \begin{bmatrix} 1/2 & 1/2 \\ 1/2 & -1/2 \end{bmatrix} \quad .$$

By introducing $\mathbf{y} = M^{-1}\mathbf{x}$ we obtain the system

$$\begin{bmatrix} \frac{dy_1}{dt} \\ \frac{dy_2}{dt} \end{bmatrix} = \begin{bmatrix} t & 0 \\ 0 & -t \end{bmatrix} \begin{bmatrix} y_1 \\ y_2 \end{bmatrix} + \begin{bmatrix} 0 \\ t \end{bmatrix}$$

$$y_1(0) = \frac{1}{2} , \quad y_2(0) = \frac{1}{2} .$$

The solutions of this system are

$$y_1 = \frac{1}{2} e^{t^2/2} , \quad y_2 = 1 - \frac{1}{2} e^{-t^2/2}$$

and, therefore,

$$x_1 = 1 + \frac{1}{2}\left[e^{t^2/2} - e^{-t^2/2}\right] = 1 + \sinh(t^2/2)$$

$$x_2 = -1 + \frac{1}{2}\left[e^{t^2/2} + e^{-t^2/2}\right] = -1 + \cosh(t^2/2) .$$

In general, however, a system of the form (5.1) is not easy to solve analytically although we now show that formally an expression similar to equation (4.4) is true.

Definition 1: The transition matrix $\Phi(t,a)$ of the system (5.1) is the solution of the system

$$\frac{d}{dt} \Phi(t,a) = A(t)\Phi(t,a)$$

$$\Phi(a,a) = I \tag{5.2}$$

Note that $\Phi(t,a)$ is an nxn matrix.

The basic property of the transition matrix is given by

Theorem 1: For all $t_1 \in R$ the transition matrix satisfies

$$\Phi(t,t_1)\Phi(t_1,a) = \Phi(t,a) \tag{5.3}$$

Corollary: $[\Phi(t,a)]^{-1} = \Phi(a,t)$ (5.4)

Proof: By theorem 1

$$\Phi(a,t)\Phi(t,a) = \Phi(a,a) = I$$

which proves our statement.

The relationship between the transition matrix and the solution of the system (5.1) is given by

Theorem 2: The solution of the system (5.1) is given by

$$\mathbf{x}(t) = \Phi(t,a)\mathbf{c} + \int_a^t \Phi(t,s)\mathbf{f}(s)ds =$$

$$= \Phi(t,a)\mathbf{c} + \Phi(t,a)\int_a^t \Phi(a,s)\mathbf{f}(s)ds \qquad (5.5)$$

Proof: The proof is obtained by a straightforward differentiation of equation (5.5).

This theorem provides an explanation for the name "transition matrix" for Φ. In fact if $\mathbf{f} = 0$ in equation (5.1) then for any two times t, t_0 we have

$$\mathbf{x}(t) = \Phi(t,t_0)\mathbf{x}(t_0)$$

i.e. Φ acts as a translation operator in time on the solution $\mathbf{x}(t)$. From another point of view it is easy to see from (5.5) that if $\mathbf{x}_i((t), i = 1,\ldots,n$ is a fundamental set of solutions for (5.1) with $\mathbf{f} = 0$ satisfying $x_i(t_0) = e_i = (\delta_{ij})$ then the corresponding fundamental matrix corresponding to these solutions is $\Phi(t,t_0) = (\mathbf{x}_1(t),\ldots,\mathbf{x}_n(t))$. Hence the computation of the transition matrix is equivalent to finding the general solution of the homogeneous part of equation (5.1) and is, therefore, rather difficult in general.

Example 2: Find the transition matrix for the system (5.1) if

$$A = \begin{bmatrix} 1 & t \\ 0 & -1 \end{bmatrix}, \quad a = 0 \qquad (5.6)$$

Solution: The transition matrix satisfies

$$\frac{d\Phi(t,0)}{dt} = A(t)\Phi(t,0) \quad , \quad \Phi(0,0) = I \qquad (5.7)$$

hence, if

$$\Phi(t,0) = \begin{bmatrix} \varphi_1(t) & \varphi_2(t) \\ \varphi_3(t) & \varphi_4(t) \end{bmatrix} \qquad (5.8)$$

then

$$\dot{\varphi}_1 = \varphi_1 + t\varphi_3 \quad , \quad \varphi_1(0) = 1 \qquad (5.9)$$

$$\dot{\varphi}_2 = \varphi_2 + t\varphi_4 \qquad \varphi_2(0) = 0 \tag{5.10}$$

$$\dot{\varphi}_3 = -\varphi_3 \qquad \varphi_3(0) = 0 \tag{5.11}$$

$$\dot{\varphi}_4 = -\varphi_4 \qquad \varphi_4(0) = 1 \quad . \tag{5.12}$$

The system (5.9)-(5.12) is "weakly coupled" since we can solve (5.11), (5.12) individually and then equations (5.9)-(5.10) reduce to linear inhomogeneous equations.

Thus,

$$\varphi_3 = 0 \quad , \quad \varphi_4 = e^{-t}$$

and, therefore,

$$\varphi_1 = e^t$$

$$\varphi_2 = \frac{5}{4} e^t - \frac{e^t}{2} (t + 1/2)$$

Obviously, similar "weakly coupled" systems of equations for the elements of the transition matrix will arise whenever $A(t)$ is triangular, i.e. all the element of $A(t)$ below (or above) the main diagonal are zero. (See exercises.)

EXERCISE 5

1. Show that the transition matrix of the system (5.1) when A has constant entries is given by $\Phi(t,a) = e^{A(t-a)}$
2. Show that if $C(t)$ is a nonsingular differentiable matrix then

   $$\frac{dC^{-1}(t)}{dt} = -C^{-1}(t)\left[\frac{dC(t)}{dt}\right] C^{-1}(t)$$

   Hint: differentiate $C(t)C^{-1}(t) = 1$
3. Show that the solution to the system

   $$\dot{\mathbf{x}}(t) = -A^T(t)\mathbf{x}(t) + \mathbf{f}(t)$$

   $$\mathbf{x}(a) = \mathbf{c}$$

(where T denote the transpose) is

$$\mathbf{x}(t) = \Phi^T(a,t)\mathbf{c} + \int_a^t \Phi^T(s,t)\mathbf{f}(s)ds$$

where Φ is the transition matrix for the system (5.1).

Hint: Show using exercise 2 that

$$\frac{d}{dt}(\Phi^T(t,a))^{-1} = -A^T(t)[\Phi^T(t,a)]^{-1}$$

4. Compute the transition matrix if $a = 0$ and

a. $A(t) = \begin{bmatrix} 1 & t & -t \\ 0 & -1 & t \\ 0 & 0 & 2 \end{bmatrix}$ b. $A(t) = \begin{bmatrix} 2 & 0 & 0 \\ -t & 1 & 0 \\ t & -t & 1 \end{bmatrix}$

5. Show that the solution of

$$\dot{\mathbf{x}}(t) = (A + B(t))\mathbf{x}(t) \quad , \quad \mathbf{x}(0) = \mathbf{c}$$

satisfies

$$\mathbf{x}(t) = e^{At}\mathbf{c} + e^{At}\int_0^t e^{-As}B(s)\mathbf{x}(s)ds$$

6. Show that the system (5.1) can be transformed to a diagonal system

$$\frac{d\mathbf{x}}{dt} = D(t)\mathbf{x} + \mathbf{g}(t), \qquad D = \begin{bmatrix} b_1(t) & & \bigcirc \\ & \ddots & \\ \bigcirc & & b_n(t) \end{bmatrix}$$

if $M^{-1}(t)\dfrac{dM(t)}{dt}$ is diagonal.

6. ELEMENTS OF LINEAR CONTROL THEORY

In this section we introduce some of the basic concepts and theorems of linear control theory. However, we postpone to the next section the discussion of those topics which are related to the Laplace transform (such as the transfer matrix).

To motivate this study we consider first the following example;

Example 1: Control of a satellite in orbit

Consider a satellite in a circular orbit around the earth which is equipped with thrusters to correct deviations in its orbits (these are due to several "secondary effects" e.g. the gravitation of the sun and the moon). To find out how these thrusters should be used to effect the needed corrections we consider the following model.

a. Equations of motion.

The expression for the two-dimensional acceleration of a body in radial coordinates is

$$\mathbf{a} = (\ddot{r}-r\dot{\theta}^2)\mathbf{e}_r + (r\ddot{\theta}+2\dot{r}\dot{\theta})\mathbf{e}_\theta \qquad (6.1)$$

where r,θ are the polar coordinates of the body and $\mathbf{e}_r, \mathbf{e}_\theta$ are radial and tangential unit vectors respectively (see Figure 3). If we assume that the satellite always remain in one plane and that the only force acting on it is the earth's gravitaion we deduce from (6.1) that its equations of motion are

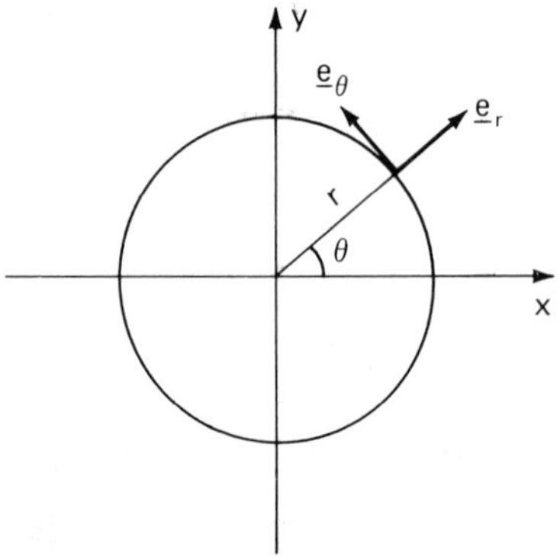

Figure 3

$$\ddot{r} = r\dot{\theta}^2 - \frac{GM}{r^2} + u_1 \qquad (6.2)$$

$$\ddot{\theta} = -\frac{2\dot{r}\dot{\theta}}{r} + \frac{1}{r}u_2 \quad . \qquad (6.3)$$

In these equations G is the gravitational constant, M the earth's mass and u_1, u_2 are the radial and tangential thrusts..

If initially the satellite is in a circular orbit of radius R and (constant) angular velocity then (since in this orbit the centrifugal and gravitational forces must be equal and opposite).

$$mR\omega^2 = \frac{GMm}{R^2} \quad \text{i.e.} \quad R^3\omega^2 = GM. \tag{6.4}$$

and this is also a solution of equations (6.2)-(6.3) with $u_1 = u_2 = 0$.

b. Small deviations.

Since we are interested only in small deviations from the circular orbit described above it is natural to scale R to 1 and introduce the variables

$$x_1 = r - 1 \quad , \quad x_2 = \dot{r}$$

$$x_3 = (\theta - \omega t) \quad , \quad x_4 = (\dot{\theta} - \omega) \quad . \tag{6.5}$$

Substituting these variables in (6.1)-(6.2) and neglecting all the nonlinear terms that appear (remember: the deviations are small.) we obtain the following system:

$$\frac{d\mathbf{x}}{dt} = A\mathbf{x} + B\mathbf{u} \tag{6.6}$$

where $\mathbf{x} = \begin{bmatrix} x_1 \\ x_2 \\ x_3 \\ x_4 \end{bmatrix}$, $\mathbf{u} = \begin{bmatrix} u_1 \\ u_2 \end{bmatrix}$ and

$$A = \begin{bmatrix} 0 & 1 & 0 & 0 \\ 3\omega^2 & 0 & 0 & 2\omega \\ 0 & 0 & 0 & 1 \\ 0 & -2\omega & 0 & 0 \end{bmatrix} , \quad B = \begin{bmatrix} 0 & 0 \\ 1 & 0 \\ 0 & 0 \\ 0 & 1 \end{bmatrix} \tag{6.7}$$

Example 2: Insect control system.

One of the methods used to control the population of an insect f which attacks citrus fruit is to introduce a predator P into the ecosystem. Using the results of example 2 in Section 1 of this chapter we infer that the ecosystem will be modelled by the equations

$$\frac{df}{dt} = af - bf^2 - cfP \tag{6.8}$$

$$\frac{dP}{dt} = -kP + dfP + u \tag{6.9}$$

$$T = F + P \tag{6.10}$$

where u is the rate at which the predator is being introduced into the ecosystem and T is the total population.

These two examples motivate the following definition.

Definition 1: A system that is described by a set of equations of the form

$$\dot{\mathbf{x}}(t) = A(t)\mathbf{x}(t) + B(t)\mathbf{u}(t) \tag{6.11}$$

$$\mathbf{y}(t) = C(t)\mathbf{x}(t) + D(t)\mathbf{u}(t) \tag{6.12}$$

$$\mathbf{x}(0) = \mathbf{c} \tag{6.13}$$

is called a linear control system.

In these equations $\mathbf{x} \in R^n$, $\mathbf{y} \in R^m$ and $\mathbf{u} \in R^k$. We shall refer to $\mathbf{u}(t)$ as the input (or control) $\mathbf{y}(t)$ as the output and $\mathbf{x}(t)$ as the state of the system.

Two major problems that arise naturally in relation to linear control systems are those of Controllability and Observability.

Definition 2: a. Those states of the system (6.11)-(6.12) that can be reached from the zero initial state $\mathbf{x}(0) = \mathbf{0}$ through a proper input are called controllable. On the other hand those states that can not be reached from this initial state for any input are called uncontrollable.

b. A linear system is called completely controllable if any state $\mathbf{x} \in R^n$ is controllable.

When A,B are matrices with constant entries we have the following characterization of completely controllable systems.

Theorem 2: If A,B are matrices with constant entries then the system (6.11)-(6.12) is completely controllable if and only if the matrix

$$M = [B, AB, \ldots, A^{n-1}B] \tag{6.14}$$

is of rank n.

Example 1: Show that the system

$$m\ddot{x} + c\dot{x} + kx = bu(t) \quad , \quad b \neq 0 \tag{6.15}$$

is completely controllable.

Solution: Equation (6.15) can be rewritten as a system of first order equations in the form

$$\frac{d}{dt}\begin{bmatrix} x \\ x_1 \end{bmatrix} = \begin{bmatrix} 0 & 1 \\ \frac{-k}{m} & \frac{-c}{m} \end{bmatrix}\begin{bmatrix} x \\ x_1 \end{bmatrix} + \begin{bmatrix} 0 \\ \frac{b}{m} \end{bmatrix} u(t) \tag{6.16}$$

Hence, the matrix $M = [B, AB]$ is given by

$$M = \begin{bmatrix} 0 & \frac{b}{m} \\ \frac{b}{m} & -\frac{bc}{m^2} \end{bmatrix} . \tag{6.17}$$

The rank of this matrix is 2 if $\det M = -\left[\frac{b}{m}\right]^2 \neq 0$.

It follows then that the system is completely controllable if $b \neq 0$.

To gain some intuitive insight why theorem 1 is true we consider a system with a single input, i.e. $\mathbf{u}(t) = u(t) \in R$. In this case B is a nx1 matrix and, therefore, the solution of equation (6.11) with $\mathbf{x}(0) = 0$ is given by (see equation (4.4)).

$$\mathbf{x}(t) = e^{At}\int_0^t e^{-As}Bu(s)ds =$$

$$= \int_0^t e^{A\sigma}Bu(t-\sigma)d\sigma \tag{6.18}$$

where $\sigma = t-s$. However, from Cayley-Hamilton theorem it follows that

$$e^{At} = \sum_{i=0}^{n-1} a_i(t)A^i \tag{6.19}$$

and, therefore,

$$\mathbf{x}(t) = \sum_{i=0}^{n-1} (A^iB)\int_0^t a_i(\sigma)u(t-\sigma)d\sigma \tag{6.20}$$

since the integrals that appear in equations (6.20) represent scalars we infer that the system is completely controllable, i.e. we can represent any vector $\mathbf{x} \in R^n$ as a linear combination of the vectors $\{B, AB, \ldots, A^{n-1}B\}$, if and only if these vectors are linearly independent or equivalently that the rank of the matrix M is n.

While the controllability of a system relates to those states which can be reached via arbitrary input from $\mathbf{x}(0) = \mathbf{0}$ the "inverse problem" of observability requires us to determine whether the system has nonzero initial states whose corresponding output is $\mathbf{0}$ if $\mathbf{u} \equiv \mathbf{0}$. From a practical point of view the interest in such states stems from the fact that they do not influence the output of the system and hence their existence is "not observable".

Definition 3: a. A state $\mathbf{x}(0) = \mathbf{c}$ of the system (6.11)-(6.12) is unobservable if the corresponding output $y(t) \equiv 0$ when $\mathbf{u}(t) \equiv 0$.

b. If the system has no unobservable state we say that the system is completely observable. Otherwise we say that the system is unobservable.

As for controllability the question of observability is easy to settle when some of the matrices in the system are constant.

Theorem 2: If the matrices A, C in the system (6.11)-(6.12) have constant entries then the system is completely observable if and only if the rank of

$$N = \begin{bmatrix} C \\ CA \\ \vdots \\ CA^{n-1} \end{bmatrix} \tag{6.21}$$

is n.

Example 2: Find for what values of c_1, c_2 the system

$$\begin{aligned} \dot{x}_1 &= x_1 + 2u \\ \dot{x}_2 &= -x_1 + x_2 - u \\ y &= (c_1, c_2)\mathbf{x} \end{aligned} \tag{6.22}$$

is completely observable.

Solution: Since

$$CA = (c_1, c_2) \begin{bmatrix} 1 & 0 \\ -1 & +1 \end{bmatrix} = (c_1 - c_2, c_2) \tag{6.23}$$

it follows that

$$N = \begin{bmatrix} c_1 & c_2 \\ c_1 - c_2 & c_2 \end{bmatrix} \tag{6.24}$$

Hence, $\det N = c_2^2$ and, therefore, the system is completely observable if and only if $c_2 \neq 0$.

Definition 4: A system that is completely controllable and observable is called canonical.

The following theorem is a refinement of theorems 1, 2.

Theorem 3: If the matrices A,B,C in the linear system (6.11)-(6.12) are constant then

a. A state $\mathbf{x}$ is uncontrollable if and only if it is in the kernal of M.

b. A state $\mathbf{x}$ is not observable if and only if it is in the kernal of N.

Remark: The kernal of a matrix G is the set of all vectors which satisfy $G\mathbf{x} = \mathbf{0}$.

EXERCISE 6

1. Discuss the controllability and observability of the following systems

$$\dot{\mathbf{x}} = \begin{bmatrix} \lambda_1 & 0 & 0 \\ 0 & \lambda_2 & 0 \\ 0 & 0 & \lambda_3 \end{bmatrix} \mathbf{x} + \begin{bmatrix} b_1 \\ b_2 \\ b_3 \end{bmatrix} u(t)$$

$$y = (c_1, c_2, c_3)\mathbf{x}$$

Generalize your results to n-dimensions

2. Repeat exercise 1 if A is replaced by

$$A = \begin{bmatrix} \lambda & 1 & 0 \\ 0 & \lambda & 1 \\ 0 & 0 & \lambda \end{bmatrix}$$

3. Repeat exercise 1 for the system

$$m\ddot{x} + u(t)\dot{x} + kx = 0$$

4. Given that $f(t)$ is continuous and

$$0 < m < f(t) < M$$

show that the system

$$\dot{x}(t) = f(t)(Ax(t) + Bu(t))$$

is completely controllable if and only if the rank of

$$M = (B, AB, \ldots, A^{n-1}B)$$

is n.

Hint: Introduce a new time variable.

5. Show that the system

$$\dot{x}(t) = Ax(t) + Bu(t)$$

is controllable if and only if the "dual system"

$$\dot{x}(t) = A^T x(t)$$

$$y(t) = B^T x(t)$$

is completely observable. This shows that any result about controllability can be transcribed in a dual result about observability and vise versa.

7. THE LAPLACE TRANSFORM

In this section we first review the basic properties of the Laplace transform and its use to solve systems of ordinary differential equations and then consider some of its applications to linear control theory.

A. Basic properties.

Definition 1: The Laplace transform $\mathcal{L}(f)$ of a function $f(x)$ is defined as

$$\mathcal{L}(f)(s) = \int_0^{\infty} e^{-st} f(t)dt \quad . \tag{7.1}$$

Obviously, the Laplace transform is a linear operator, i.e.

$$\mathcal{L}(af+bg) = a\mathcal{L}(f) + b\mathcal{L}(g) \quad . \tag{7.2}$$

Example 1: Compute the Laplace transform of x^k, $k > -1$.

Solution:

$$\mathcal{L}(x^k) = \int_0^{\infty} t^k e^{-st} dt = \frac{1}{s^{k+1}} \int_0^{\infty} r^k e^{-r} dr = $$
$$= \frac{\Gamma(k+1)}{s^{k+1}} \tag{7.3}$$

where $r = st$.

Remark: The Gamma function $\Gamma(x)$ can be considered as a generalization of

$$n! = n(n-1) \ldots 2.1$$

to noninteger values of x since it can be shown that $\Gamma(n+1) = n!$

Example 2: Compute the Laplace transform of the Dirac δ - function, i.e. $f(x) = \delta(x-x_0)$, $x_0 > 0$.

Solution: One of the basic properties of the Dirac δ - function is that for any function $g(x)$

$$\int_{-\infty}^{\infty} g(x)\delta(x-x_0)dx = g(x_0) \tag{7.4}$$

(For more details the reader is referred to Section

6.5). It is apparent from (7.4) that $\delta(x-x_0)$ is not a regular function but a "generalized function." Furthermore, from a physical point of view $\delta(x-x_0)$ represents an ideal impulse of zero duration and strength 1 since

$$\int_{-\infty}^{\infty} \delta(x-x_0) = \int_{-\infty}^{\infty} 1 \cdot \delta(x-x_0) = 1 \tag{7.5}$$

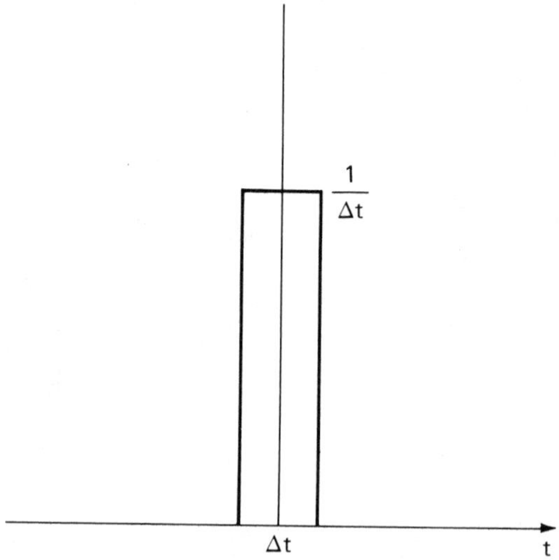

Figure 4: $\delta(x)$ is the idealized limit of a pulse with duration Δt and height $1/\Delta t$ i.e. whose strength (total area) is 1.

Thus many times $\delta(x-x_0)$ is referred to as the "impulse function".

To compute the Laplace transform of this function we now use (7.1), (7.4)

$$\mathcal{L}(\delta(x-x_0)) = \int_0^{\infty} e^{-st}\delta(t-x_0) = e^{-sx_0} \tag{7.6}$$

The most important property of the Laplace transform is the relation between $\mathcal{L}(f)$ and $\mathcal{L}(f')$, $\mathcal{L}(f'')$ etc.

Theorem 1:

$$\mathcal{L}(f') = s\mathcal{L}(f) - f(0) \tag{7.7}$$

$$\mathcal{L}(f'') = s^2\mathcal{L}(f) - sf(0) - f'(0) \tag{7.8}$$

and so on.

Proof: The proof of these formulas requires a repeated integration by parts, e.g.

$$\mathcal{L}(f') = \int_0^\infty e^{-st}f'(t)dt = f(t)e^{-st}\Big|_0^\infty$$

$$+ s\int_0^\infty e^{-st}f(t)dt = s\mathcal{L}(f) - f(0).$$

As the reader might recall from an elementary course in differential equations, the final step in the application of the Laplace transform to the solution of differential equations requires the inversion of the transform. This is usually done with the aid of a table of Laplace transforms and some "factor theorems", e.g.

Theorem 2: If $\mathcal{L}(f)(s) = F(s)$ then

1. $\mathcal{L}(e^{at}f(t)) = F(s-a)$ (7.9)
2. $\mathcal{L}(t^n f(t)) = (-1)^n F^{(n)}(s)$ (7.10)
3. $\mathcal{L}\left[\int_0^t f(\tau)d\tau\right] = \frac{1}{s}F(s)$. (7.11)

Example 3: Compute

$$f(t) = \mathcal{L}^{-1}\left[\ell n\,\frac{s+2}{s-2}\right] .$$

Solution: From equation (7.10) it follows that

$$\mathcal{L}(tf(t)) = -\frac{d}{ds}\left[\ell n\,\frac{s+2}{s-2}\right] = \frac{1}{s-2} - \frac{1}{s+2}$$

Hence, using a table of Laplace transforms we infer that

$$f(t) = \frac{1}{t}\left[e^{2t}-e^{-2t}\right] = 2t^{-1}\sinh(2t) .$$

Another important property of the Laplace transform is related to the convolution of two functions.

Definition 2: The convolution of two functions f, g is defined as

$$f*g = \int_0^t f(\tau)g(t-\tau)d\tau \tag{7.12}$$

Theorem 3: Let $F(s)$, $G(s)$ be the Laplace transform of f, g respectively then

$$\mathcal{L}(f*g) = F(s)G(s) \tag{7.13}$$

B. **Applications to systems of linear equations.**

The Laplace transform can be used to solve systems of differential equations by converting them into a system of algebraic equations. We illustrate the essence of this technique through the following example.

Example 4: Use the Laplace transform to solve

$$\dot{x} = y + 3 \tag{7.14}$$

$$\dot{y} = 4x + 1 \tag{7.15}$$

$$x(0) = 0, \quad y(0) = 1$$

Solution: Applying the Laplace transform to equation (7.14), (7.15) and using theorem 1 we obtain

$$sX(s) = Y(s) + \frac{3}{s} \tag{7.16}$$

$$sY(s) - 1 = 4X(s) + \frac{1}{s} \tag{7.17}$$

where $X(s)$, $Y(s)$ are the Laplace transforms of x, y respectively. Hence

$$X(s) = \frac{4s+1}{s(s^2-4)} = -\frac{1}{4s} - \frac{7}{8}\frac{1}{s+2} + \frac{9}{8} \cdot \frac{1}{s-2} \tag{7.18}$$

$$Y(s) = \frac{s^2+s+12}{s(s^2-4)} = -\frac{3}{s} + \frac{7}{4}\frac{1}{s+2} + \frac{9}{4}\frac{1}{s-2} \quad . \tag{7.19}$$

Using a table of Laplace transforms we now infer that

$$x(t) = \frac{1}{4} - \frac{7}{8}e^{-2t} + \frac{9}{8}e^{2t} \tag{7.20}$$

$$y(t) = -3 + \frac{7}{4}e^{-2t} + \frac{9}{4}e^{2t} \quad . \tag{7.21}$$

C. **Applications to linear control theory.**

Theorem 4: The output $\mathbf{y}(t)$ of the linear control system

$$\dot{\mathbf{x}} = A\mathbf{x} + B\mathbf{u} \tag{7.22}$$
$$\mathbf{y} = C\mathbf{x} + D\mathbf{u} \tag{7.23}$$
$$\mathbf{x}(0) = \mathbf{0}$$

is linearly related to the input $\mathbf{u}(t)$, i.e. if $\mathbf{y}_1, \mathbf{y}_2$ are the outputs which correspond to $\mathbf{u}_1, \mathbf{u}_2$ respectively then the output which correspond to $a\mathbf{u}_1 + b\mathbf{u}_2$ is $a\mathbf{y}_1 + b\mathbf{y}_2$.

Proof: Since equation (7.22) is linear it follows that if $\mathbf{x}_1, \mathbf{x}_2$ are the states which correspond to the inputs $\mathbf{u}_1, \mathbf{u}_2$ respectively then $\mathbf{x}_1 + \mathbf{x}_2$ is the state which correspond to the input $\mathbf{u}_1 + \mathbf{u}_2$. Therefore, the output which correspond to the input $a\mathbf{u}_1 + b\mathbf{u}_2$ is

$$\mathbf{y} = C(a\mathbf{x}_1+b\mathbf{x}_2) + D(a\mathbf{u}_1+b\mathbf{u}_2) = a\mathbf{y}_1 + b\mathbf{y}_2. \tag{7.24}$$

Theorem 5: The Laplace transforms $\mathbf{Y}(s)$, $\mathbf{U}(s)$ of $\mathbf{y}(t)$, $\mathbf{u}(t)$ respectively are linearly related, i.e. there exist a matrix $W(s)$ so that

$$\mathbf{Y}(s) = W(s)\mathbf{U}(s) \tag{7.25}$$

Proof: Applying the Laplace transform to (7.22)-(7.23) (and using the initial condition) we obtain

$$s\mathbf{X}(s) = A\mathbf{X}(s) + B\mathbf{U}(s) \tag{7.26}$$
$$\mathbf{Y}(s) = C\mathbf{X}(s) + D\mathbf{U}(s) \tag{7.27}$$

where $\mathbf{X}(s)$ is the Laplace transform of $\mathbf{x}(t)$. Hence

$$\mathbf{Y}(s) = [C(sI-A)^{-1}B+D]\mathbf{U}(s) \tag{7.28}$$

i.e.

$$W(s) = C(sI-A)^{-1}B + D \quad . \tag{7.29}$$

Definition 3: $W(s)$ is called the transfer function of the system (7.22)-(7.23).

Corollary: The transfer function for the system (7.22)-(7.23) is a matrix whose entries are rational functions of s.

Proof: Since $(sI-A)^{-1}$ consists of rational functions of s the result is obvious from equation (7.29).

An important problem which is related to the transfer function is the realization problem which can be stated as follows: For a given matrix $W(s)$ with rational entries in s determine a linear control system so that

(1) $W(s)$ is the transfer function of the system

(2) The system is canonical, i.e. completely controllable and observable.

Surprisingly this problem has a definite affirmative answer;

Theorem 6: For any transfer function of a system having a finite dimensional realization there exist a canonical model which is unique up to a choice of coordinate system in the state space.

(For more information regarding this theorem see R. Kalman et al, Topic in Mathematical System Theory, McGraw-Hill, NY 69).

EXERCISE 7

1. Prove the properties of the Laplace transform stated in theorem 2.
2. Show that

$$f*g = g*f \tag{7.30}$$

3. Use the Laplace transform to solve the following systems of equations.

a. $$\ddot{x} + \dot{x} + 2y = 0 \tag{7.31}$$

$$\ddot{y} + 2\dot{y} + x = 0 \tag{7.32}$$

$x(0) = y(0) = 0, \quad \dot{x}(0) = 1, \quad \dot{y}(0) = 0$.

b. $$\frac{d^4x}{dt^4} + 2x = 1 \tag{7.33}$$

$x(0) = x'(0) = x''(0) = 0, \; x'''(0) = 1$

4. Use the Laplace transform to solve

$$\ddot{x} + \dot{x} + 2x = 2\delta(t-t_0) \tag{7.34}$$

$x(0) = 0, \quad \dot{x}(0) = 0$

5. Solve equation (7.34) with the boundary conditions

$x(0) = 0, \quad x(a) = 0$

Hint: At first ignore the condition $x(a) = 0$ and use $\dot{x}(0) = c$ to find a solution. Then determine c so that the condition $x(a) = 0$ is satisfied.

6. Find the transfer function for the system. What is the relation between this transfer function and the solution found in exercise 4?

$\ddot{x} + 2\dot{x} + 2x = 2u$.

7. Find the transfer function of the system

$\ddot{x} + b\dot{x} + kx = c_1 u + c_2 \dot{u}$

BIBLIOGRAPHY

1. R. Bronson - Matrix Methods, Academic Press, 1970.

2. R. W. Brockett - Finite Dimensional Linear Systems, J. Wiley, 1970.

3. J. L. Casti - Dynamical Systems and Their Applications, Academic Press, 1977.

4. A. S. Deif - Advanced Matrix Theory for Scientists and Engineers, Abacus Press, 1982.

5. M. R. Spiegel - Applied Differential Equations, 3rd edition, Prentice Hall, 1981.

6. W. J. Palm, III - Modeling, Analysis and Control of Dynamical Systems, J. Wiley, 1983.

7. R. Haberman - Mathematical Models, Prentice Hall, 1977.

CHAPTER 4. APPLICATIONS OF SYMMETRY PRINCIPLES TO DIFFERENTIAL EQUATIONS

1. INTRODUCTION

The application of symmetry principles to differential equations might lead to their simplification or solution in many important cases. Furthermore, it yields insights into the properties of these solutions as well as inter-relations in between them. To see how this is done we present, in this chapter, a short introduction to Lie groups and Lie algebras and then explore the various applications of these algebraic tools to several classes of first and second order equations.

To motivate our discussion we present the following elementary example;

Example 1: Consider Schrödinger equation for the one dimensional harmonic oscillator in Quantum mechanics which is given by

$$\left[\frac{d^2}{dx^2} + a^2x^2\right] \psi(x) = \lambda\psi(x) \tag{1.1}$$

with the boundary conditions

$$\lim_{x\to\pm\infty} \psi(x) = 0 \quad . \tag{1.2}$$

It is easy to see that (1,1), (1,2) remain invariant (i.e. unchanged) under the transformation

$$x_1 = -x \quad . \tag{1.3}$$

Hence if $\psi(x)$ is the solution of (1,1), (1,2) then it follows that $\psi(-x)$ is also a solution of these equations. However, since the solution of (1.1), (1.2) is unique we infer that

$$\psi(-x) = c\psi(x) \tag{1.4}$$

and hence also

$$\psi(x) = c\psi(-x) \quad . \tag{1.5}$$

Thus, from (1,4), (1.5) it follows that

$$\psi(x) = c^2\psi(x) \tag{1.6}$$

i.e. $c = \pm 1$. We conclude then that the solution of (1.1), (1.2) must be either even or odd function of x. We observe that this result was obtained without actually solving equations (1.1), (1.2).

In the following we shall see how such invariance properties of differential equations can be applied systematically to solve and study various properties of these equations.

2. LIE GROUPS.

A group is a set of elements $G = \{a,b,\ldots\}$ with a binary operation $\cdot$ (usually the dot is omitted) which satisfies the following properties;

1. If $a,b \in G$ then ab is well defined (i.e. unique) and $ab \in G$
2. There exists an element $e \in G$ so that
   $$ea = ae = a$$
   for all $a \in G$
3. for all $a \in G$ there exist $b \in G$ so that $ab = e$. We usually denote this b by a^{-1}.
4. The associative law $(ab)c = a(bc)$ is true for all $a,b,c \in G$.

Furthermore, a group is called commutative if for all $a, b \in G$

$$ab = ba.$$

Groups are usually divided into two classes; discrete (both finite and infinite) and continuous, in which case we need a set of continuous parameters to characterize the elements of the group. Although discrete groups play major roles in various applications, our main interest in this chapter is in continuous, or Lie groups which act as transformation groups on R^n. (However, not every continuous group is a Lie group.)

We present some important examples of these groups.

Example 1: The translation group in R^n.

Let $\mathbf{a} \in R^n$ then the transformation

$$\mathbf{x}_1 = \mathbf{x} + \mathbf{a}, \qquad \mathbf{x} \in R^n \tag{2.1}$$

defines a translation of the origin of the coordinate system in R^n to $\mathbf{a}$. Obviously, the set of all these transformations in R^n form a commutative group which we denote by T_n.

Example 2: Magnification (or Dilatation) group in R^n.

This group acts on R^n through the transformations

$$\mathbf{x}_1 = a\mathbf{x} \quad , \quad a > 0 \quad . \tag{2.2}$$

Since $a > 0$ we can rewrite (2.2) also in the form $\mathbf{x}_1 = e^{\alpha}\mathbf{x}$.

Example 3: Rotations in two dimensions.

As is well known a rotation of the coordinate system in R^2 by an angle θ leads to the transformation

$$\begin{aligned} x_1 &= x\cos\theta + y\sin\theta \\ y_1 &= -x\sin\theta + y\cos\theta \end{aligned} \tag{2.3}$$

which can be rewritten in matrix form as

$$\begin{bmatrix} x_1 \\ y_1 \end{bmatrix} = \begin{bmatrix} \cos\theta & \sin\theta \\ -\sin\theta & \cos\theta \end{bmatrix} \begin{bmatrix} x \\ y \end{bmatrix} = O\mathbf{x} . \tag{2.4}$$

From (2.4) it is easy to verify that O is an orthogonal matrix, i.e. $O^TO = I$ (where O^T is the transpose of O) and $\det O = 1$. In fact it can be shown that the group of (pure) rotations in R^2 is isomorphic (or equivalent) to

$$SO(2) = \{O, 2\text{x}2 \text{ matrices} \mid O^TO = I, \det O = 1\} .$$

Similarly it is possible to show that the group of all (pure) rotations in R^n is isomorphic to

$$SO(n) = \{O, n\times n \text{ matrices} \mid O^TO = I, \det O = 1\}$$

We observe that though $SO(2)$ is a commutative group $SO(n)$ for $n > 2$ is not so.

We also note that in some applications it is natural to consider the group of rotations and reflections in R^n. Reflections are transformations of the form

$$\bar{x}_i = \begin{cases} x_i & i \neq k \\ -x_i & i = k \end{cases} \qquad k = 1,\ldots n. \tag{2.5}$$

It is possible to show that this group is isomorphic to $O(n)$ where

$$O(n) = \{O, n\times n \text{ matrix} \mid O^TO = I\}.$$

Example 4: The group of Euclidian transformations in R^n.

This group consists of all rotations and translations in R^n. i.e. each group element is given by a pair $g = (R, \mathbf{a})$ where $R \epsilon O(n)$ is an orthogonal matrix and $\mathbf{a} \in R^n$. An element g of this group acts on R^n through the transformation

$$T_g(\mathbf{x}) = R\mathbf{x} + \mathbf{a} \quad . \tag{2.6}$$

Using elementary geometrical considerations the group multiplication is seen to be given by

$$(R_2, \mathbf{a}_2) \cdot (R_1, \mathbf{a}_1) = (R_2 R_1, a_2 + R_2 \mathbf{a}_1) \tag{2.7}$$

we denote this group acting on R^n by $E(n)$.

For completeness we define here the following groups which appear in various applications.

1. The group of all $n \times n$ unitary matrices viz.

   $U(n) = \{U, n\times n \text{ matrix} | U^{\dagger}U = I\}$

   where $U^{\dagger} = (U^T)^*$, i.e. the $\dagger$ operation denote the combined application of the transpose and complex conjugation.
2. The group of all special unitary matrices

   $SU(n) = \{U | U \in U(n), \det U = 1\}$
3. The general linear group

   $GL(n) = \{A, n\times n \text{ matrix} | A \text{ is nonsingular}\}$
4. The special linear group

   $SL(n) = \{A | A \; GL(n), \det A = 1\}$.

   Finally, if the general element of a Lie group G is specified in terms of n nonredundant parameters then we say that G is a n-parameter group, e.g. $SO(2)$ is a one-parameter group.

EXERCISE 1

1. Show that $\{e, a\}$ with $ea = ae - a$ and $a^2 = e$ is a group.

2. Prove that $O(2)$ is a group.
3. Prove equation (2.7).
4. Show that the transformations

$$x_1 = ax \quad , \quad y_1 = a^2 y \quad , \quad a > 0$$

in R^2 form a group.

5. Show that $x^2 + y^2$ remains invariant under the transformations of $O(2)$. What is the corresponding invariant for the dilatation group in R^2.
6. Generalize the results of exercise 5 to $O(n)$ and the dilatation groups in R^n.
7. An affine transformation in R^2 is given by

$$x_1 = ax + by + \alpha_1$$
$$y_1 = cx + dy + \alpha_2$$

where $a,b,c,d,\alpha_1,\alpha_2$ are constants and $ad - bc \neq 0$. Show that the set of all affine transformations in R^2 is a group. Generalize to R^n.

3. LIE ALGEBRAS.

The definitions introduced in the last section for Lie groups are rather straightforward. However, since these groups are continuous it is desirable to try and find for them, as for vector spaces, a finite set of entities which act as a "basis" in some generalized sense.

To carry this program out we introduce the following definitions:

Definition 1: A differential operator D is called a derivation if

$$D(f_1(x)f_2(x)) = (Df_1(x))f_2(x) + f_1(x)(Df_2(x)) \tag{3.1}$$

$\mathbf{x} \in R^n$.

Example 1: The operators $\frac{\partial}{\partial x_i}$, $x_j \frac{\partial}{\partial x_i}$, $x_i \frac{\partial}{\partial x_j} - x_j \frac{\partial}{\partial x_i}$, $i,j=1,\ldots n$ are all derivations. However, the operator $\frac{\partial^2}{\partial x_j^2}$ is not a derivation.

Definition 2: The commutator of two derivations D_1, D_2 is defined as

$$[D_1,D_2] = D_1D_2 - D_2D_1 \quad . \tag{3.2}$$

From this definition it follows directly that

$$[D,D] = 0 \tag{3.3}$$

$$[D_1,[D_2,D_3]] + [D_2,[D_3D_1]] + [D_3,[D_1,D_2]] = 0 \ . \tag{3.4}$$

Equation (3.4) is called Jacobi identity.

Theorem 1: If D_1, D_2 are derivations and α, β constants then $\alpha D_1 + \beta D_2$ and $[D_1,D_2]$ are derivations.

Thus, although, in general, D_1D_2 is not a derivation the commutator of two derivations is a derivation.

Proof: By direct verification.

Definition 3: A set of derivations A is said to be Lie algebra if

1. It is a vector space over R (or C)
2. If $D_1, D_2 \in A$ then $[D_1,D_2] \in A$

Definition 4: A finite set of derivations $\{D_1,\ldots,D_n\}$ is said to be a basis for the Lie algebra A if $D_i \in A$ and

1. $D_1 \ldots D_n$ form a basis for the vector space A.

2. $$[D_i, D_j] = \sum_{k=1}^{n} c_{ij}^{k} D_k \tag{3.5}$$

The coefficients c_{ij}^{k} are called the structure constants of the algebra (with respect to the given basis).

Example 2: The set $A = \left\{ \sum_{i=1}^{n} a_i \frac{\partial}{\partial x_i},\ a_i \in R \right\}$ is a Lie algebra since for any $D_1, D_2 \in A$ we have $[D_1, D_2] = 0$. A basis for this Lie algebra if given by $\left\{ \frac{\partial}{\partial x_1}, \ldots, \frac{\partial}{\partial x_n} \right\}$. Since

$$\left[\frac{\partial}{\partial x_i}, \frac{\partial}{\partial x_j} \right] = 0 \tag{3.6}$$

it follows that all the structure constants are zero.

Example 3: The set

$$A = \{a_1 D_1 + a_2 D_2 + a_3 D_3,\ a_i \in R\}$$

where

$$D_1 = \frac{\partial}{\partial x},\quad D_2 = \frac{\partial}{\partial y} \tag{3.7}$$

$$D_3 = x \frac{\partial}{\partial y} - y \frac{\partial}{\partial x}$$

is a Lie algebra and $\{D_1, D_2, D_3\}$ is a basis for A since

$$\begin{aligned} &[D_1, D_2] = 0,\ [D_1, D_3] = D_2 \\ &[D_2, D_3] = -D_1 \end{aligned} \tag{3.8}$$

The basic theorem which relates Lie groups and Lie algebras is due to S. Lie.

Theorem 2: For every (connected) Lie group G there exists a Lie algebra $A(G)$ so that for every $g \in G$ there exists $D \in A(G)$ with

$$g = \exp D = \sum_{n=0}^{\infty} \frac{D^n}{n!} \tag{3.9}$$

i.e., the transformations on R^n defined by g and $\exp D$ are the same.

Remark: A connected group G is one in which every element $g \in G$ can be connected to the identity element (of G) by a continuous path in G. If G is disconnected (e.g. O(n)) then the theorem holds for the maximal connected subgroup of G.

Example 4: Let $G = T_n$ then A(G) is the Lie algebra given in example 2. In fact it is easy to see that

$$\left[\exp\left[\sum_{i=1}^{n} a_i \frac{\partial}{\partial x_i}\right]\right] \cdot x_j = x_j + a_j, \quad j = 1,\ldots,n \tag{3.10}$$

i.e. $\exp\left[\Sigma\, a_i \frac{\partial}{\partial x_i}\right]$ correspond to a translation by $\mathbf{a} = (a_1,\ldots,a_n)$.

Example 5: Let G = SO(2) then

$$A(G) = \{aD,\ a \in R\}$$

where $D = -x \frac{\partial}{\partial y} + y \frac{\partial}{\partial x}$. In fact

$$\begin{aligned} \exp(aD)x &= x + \frac{a}{1!} y - \frac{a^2}{2!} x^2 - \frac{a^3}{3!} y + \frac{a^4}{4!} x - \ldots \\ &= x\left[1 - \frac{a2}{2!} + \frac{a^4}{4!} - \ldots\right] + y\left[\frac{a}{1!} - \frac{a^3}{3!} + \ldots\right] \\ &= x\cos a + y\sin a \end{aligned} \tag{3.11}$$

$$\begin{aligned} \exp(aD)y &= y\, \frac{a}{1!} x - \frac{a^2}{2!} y + \frac{a^3}{3!} x + \frac{a^4}{4!} y + \ldots \\ &= -x\left[\frac{a}{1!} - \frac{a^3}{3!} + \ldots\right] + y\left[1 - \frac{a^2}{2!} + \frac{a^4}{4!} - \ldots\right] = \\ &= -x\sin a + y\cos a \end{aligned} \tag{3.12}$$

Similarly it is possible to show that the Lie algebra of SO(n) is generated by

$$J_{ij} = x_j \frac{\partial}{\partial x_i} - x_i \frac{\partial}{\partial x_j} \tag{3.13}$$

and the commutation relations between these generators are

$$\left[J_{ij}, J_{mn}\right] = \left[\delta_{im} J_{jn} + \delta_{jn} J_{im} - \delta_{in} J_{jm} - \delta_{jm} J_{in}\right] \tag{3.14}$$

where δ_{ij} is the Kronecker delta viz.

$$\delta_{ij} = \begin{cases} 1 & i = j \\ 0 & i \neq j \end{cases} \tag{3.15}$$

Example 6: The Lie algebra for the dilatation group in R^n is generated by

$$D = \sum_{i=1}^{n} x_i \frac{\partial}{\partial x_i} \quad . \tag{3.16}$$

In fact it is easy to verify that

$$e^{aD} x_j = e^a x_j, \quad j = 1,\ldots,n. \tag{3.17}$$

Example 7: Since the Euclidian group $E(n)$ consists of rotations and translations in R^n we infer that its Lie algebra is generated by $\{J_{ij}, T_k\}$ where $T_k = \frac{\partial}{\partial x_k}$. The commutation relations between the generators J_{ij} are given by (3.14). As to the other commutation relations we have,

$$[T_k, T_m] = 0 \tag{3.18}$$

$$[T_k, J_{ij}] = \delta_{ki} T_j - \delta_{kj} T_i \quad . \tag{3.19}$$

Finally, to conclude this section we observe that we exhibited explicitly the Lie algebras of some important Lie groups. However, we did not give a constructive method to find the Lie algebra of a given Lie group. Such a technique which is called the "method of infinitesimal transformations" will be discussed in the next section.

EXERCISE 2

1. Verify the commutation relations (3.14) and (3.18).

2. Show that the Lie algebra of the group $x_1 = ax$, $y_1 = a^2 y$ is given by

$$D = x \frac{\partial}{\partial x} + 2y \frac{\partial}{\partial y}$$

3. Show that the operators $\left\{1, x, \frac{\partial}{\partial x}\right\}$ generate a Lie algebra, (i.e., satisfy equation (3.5)). Note that 1,x are not derivations since

$$1 \cdot f = f,\ x \cdot f = xf.$$

4. Find the Lie group which is generated by the operator

$$D = x^2 \frac{\partial}{\partial x} + xy \frac{\partial}{\partial y}$$

i.e., evaluate exp(aD) explicitly.

4. PROLONGATION OF THE ACTION.

So far we defined the action of Lie groups on R^n. We now consider the extension of this action to functions and their derivatives defined on these spaces.

Definition 1: Let G be a Lie group acting on R^n and let $f \in C^\infty(R^n)$(i.e., f is infinitely differentiable). For $g \in G$ we now define

$$T_g f(x) = f(g \cdot x) \quad . \tag{4.1}$$

Example 1: If G is the translation group in R^n then from (4.1) we have

$$T_{\mathbf{a}} f(\mathbf{x}) = f(\mathbf{x} + \mathbf{a}) \tag{4.2}$$

To examine the ramification of this definition we consider the infinitesimal transformations of G on R^n viz. transformations induced by elements of G which are close to the identity element of G.

Example 2: Consider the group SO(2) acting on R^2. The identity element of SO(2) is the unit matrix and the infinitesimal transformations of G are given by $\epsilon = |\theta| << 1$. Under this restriction we can use the Taylor expansion of $\sin\theta$, $\cos\theta$ to rewrite (2.4) in the form

$$x_1 = x + \epsilon y + O(\epsilon^2)$$
$$\tag{4.3}$$
$$y_1 = -\epsilon x + y + O(\epsilon^2)$$

hence for these transformations we have

$$T_g f(x) = f(x_1, y_1) = f(x+\epsilon y,\ y-\epsilon x)$$
$$= f(x,y) + \epsilon\left[\frac{\partial f}{\partial x} - x\,\frac{\partial f}{\partial y}\right] + O(\epsilon^2) \quad . \tag{4.4}$$

In fact if we introduce the operator

$$D = y\,\frac{\partial}{\partial x} - x\,\frac{\partial}{\partial y} \tag{4.5}$$

then it is possible to prove that for any θ

$$T_g f(x) = f(x_1, y_1) = f(x,y) + \theta Df + \frac{1}{2!}\,\theta^2 D^2 f + \ldots$$
$$e^{\theta D} f(x,y) \tag{4.6}$$

i.e.,

$$f(e^{\theta D}(x,y)) = e^{\theta D} f(x,y) \tag{4.7}$$

which is in concert with (4.1). Equation (4.7) demonstrates, however, that the Lie algebra generator of SO(2) can be computed directly by considering the infinitesimal transformations of SO(2) on $C^\infty(R^2)$ and this result is true in general.

Example 3: Find the generator of the Lie algebra for the group of transformations

$$x_1 = e^a x \ , \ y_1 = e^{an} y \quad . \tag{4.8}$$

An infinitesimal transformation of this group which correspond to $|a| << 1$ is given by

$$x_1 = x + ax + O(a^2)$$
$$(4.9)$$
$$y_1 = y + nay + O(a^2) \quad .$$

Hence for $f \in C^\infty(R^2)$

$$f(x_1, y_1) = f(x+ax, y + nay)$$
$$(4.10)$$
$$= f(x,y) + a\left[x \frac{\partial}{\partial x} + ny \frac{\partial}{\partial y}\right]f + O(a^2) \quad .$$

Hence we infer that the Lie algebra generator is given by

$$D = x \frac{\partial}{\partial x} + ny \frac{\partial}{\partial y} \quad . \qquad (4.11)$$

We may summarize this result as follows:

Theorem 1: Let G be a one parameter Lie group. If the infinitesimal transformations of G are given by

$$\mathbf{x}_1 = \mathbf{x} + a\xi(\mathbf{x}) + O(a^2), |a| \ll 1 \qquad (4.12)$$

then the Lie algebra generator of G is given by

$$D = \sum_{i=1}^{n} \xi^i(\mathbf{x}) \frac{\partial}{\partial x_i} \quad . \qquad (4.13)$$

If G has more than one parameter then we have to find the Lie generator for each independent parameter in order to find the Lie algebra of G.

Example 4: Find the Lie algebra for the nonisotropic dilatation group

$$x_1 = e^a x \ , \ y_1 = e^{2a} y \ , \ z_1 = e^b z \quad . \qquad (4.14)$$

Solution: Since this is a two parameter group we first consider infinitesimal transformations with $b = 0$ and $|a| \ll 1$. We then have

$$x_1 = x + ax + O(a^2), \ y_1 = y + 2ay + O(a^2), \ z_1 = z \qquad (4.15)$$

The associated Lie algebra generator is then

$$D_1 = x \frac{\partial}{\partial x} + 2y \frac{\partial}{\partial y} \quad . \qquad (4.16)$$

Similarly when $a = 0$ and $|b| \ll 1$ the corresponding infinitesimal transformations are

$$x_1 = x, \ y_1 = y, \ z_1 = z + bz + O(b^2) \quad . \tag{4.17}$$

Hence we infer that

$$D_2 = z \frac{\partial}{\partial z} \quad . \tag{4.18}$$

The operators D_1, D_2 together generate the Lie algebra of the group (4.14) and we observe that

$$[D_1, D_2] = 0 \tag{4.19}$$

Consider now the situation when $y = y(x)$ and G is a one parameter Lie group of transformations acting on R^2. Obviously, a transformation of x, y will induce a transformation on the derivative $p = \frac{dy}{dx}$. To compute this effect we consider the infinitesimal transformations of G given by

$$\begin{aligned} x_1 &= x + a\xi(x,y) + O(a^2) \\ y_1 &= y + a\eta(x,y) + O(a^2) \quad . \end{aligned} \tag{4.20}$$

The effect on p is then given by

$$p_1 = \frac{dy_1}{dx_1} = \frac{d[y+a\eta(x,y)+O(a^2)]}{d[x+a\xi(x,y)+O(a^2)]}$$

$$= \frac{dy+a\left[\frac{\partial\eta}{\partial x}dx+\frac{\partial\eta}{\partial y}dy\right]+O(a^2)}{dx+a\left[\frac{\partial\xi}{\partial x}dx+\frac{\partial\xi}{\partial y}dy\right]+O(a^2)}$$

$$= \frac{p+a\left[\frac{\partial\eta}{\partial x}+\frac{\partial\eta}{\partial y}dy\right]+O(a^2)}{1+a\left[\frac{\partial\xi}{\partial x}+\frac{\partial\xi}{\partial y}p\right]+O(a^2)} \quad . \tag{4.21}$$

Considering the denominator in the last expression as a function $g(a)$ and expanding around $a = 0$ we obtain;

$$p_1 = p + a\zeta(x,y,p) + O(a^2) \tag{4.22}$$

where

$$\zeta = \frac{\partial\eta}{\partial x} + \left[\frac{\partial\eta}{\partial y} - \frac{\partial\xi}{\partial x}\right] p - \frac{\partial\xi}{\partial y} p^2 \quad . \tag{4.23}$$

We infer then that when an infinitesimal transformation of G acts on a function $f = f(x,y,p)$ we have (similar to (4.1) and (4.4))

$$T_g f = f(x_1,y_1,p_1) = f(x,y,p) + a\left[\xi\frac{\partial}{\partial x} + \eta\frac{\partial}{\partial y} + \zeta\frac{\partial}{\partial p}\right] f + O(a^2) \tag{4.24}$$

or globally

$$e^{aD^{(1)}} f(x,y,p) = f\left[e^{aD^{(1)}}(x,y,p)\right] \tag{4.25}$$

where $D^{(1)} = \xi\frac{\partial}{\partial x} + \eta\frac{\partial}{\partial y} + \zeta\frac{\partial}{\partial p}$

the operator $e^{aD^{(1)}}$ represents then the first prolongation of the action of G to functions of x,y,p.

In a similar fashion we can investigate the effect of G on second order (or higher order) derivatives $q = \frac{d^2y}{dx^2}$ and obtain for infinitesimal transformations

$$q_1 = q + a\mu(x,y,p,q) + O(a^2) \tag{4.27}$$

where

$$\mu = \frac{d\zeta}{dx} - q\frac{d\xi}{dx} = \eta_{xx} + (2\eta_{xy} - \xi_{xx})p + (\eta_y - 2\xi_x)q$$
$$(\eta_{yy} - 2\xi_{xy})p^2 - 3\xi_y pq - \xi_{yy}p^3 \quad . \tag{4.28}$$

This lead to the second prolongation of G given by

$$T_g f(x,y,p,q) = e^{aD^{(2)}} f(x,y,p,q) = f(e^{aD^{(2)}}(x,y,p,q)) \tag{4.29}$$

where

$$D^{(2)} = \xi\frac{\partial}{\partial x} + \eta\frac{\partial}{\partial y} + \zeta\frac{\partial}{\partial p} + \mu\frac{\partial}{\partial q} \tag{4.30}$$

Remark: When G acts on $R^3 = \{(x,y,u)\}$ and $u = u(x,y)$ we can apply the same procedure described above to compute the first prolongation of G to

$f\left[x,y,u,\frac{\partial u}{\partial x},\frac{\partial u}{\partial y}\right]$ by considering the infinitesimal transformations

$$\begin{aligned} x_1 &= x + a\xi(x,y,u) + O(a^2) \\ y_1 &= y + a\eta(x,y,u) + O(a^2) \\ u_1 &= u + a\nu(x,y,u) + O(a^2) \end{aligned} \quad . \qquad (4.31)$$

Obviously, the same procedure can be followed in higher dimensions.

EXERCISE 3

1. Verify equation (4.28).

2. Let $g \in O(2)$ represent a rotation by $\frac{\pi}{4}$. Evaluate $T_g f(x,y,p)$ if

   a. $f = y + xp + x^4p^2$

   b. $f = x^2 + xy$.

3. Repeat exercise 2 if $g \in E(2)$ correspond to a rotation by $\frac{\pi}{3}$ and a translation of the origin to (1,1).

4. Find the finite transformations of the groups whose infinitesimal transformations are

   a. $D = y\frac{\partial}{\partial x} + \frac{\partial}{\partial y}$

   b. $D = \frac{\partial}{\partial x} + y\frac{\partial}{\partial y}$

   Hint: Apply e^{aD} to the coordinate functions $f = x$, $g = y$.

5. Let G act on R^2 as follows;

   $x_1 = x\cosh\theta + y\sinh\theta$

   $y_1 = x\sinh\theta + y\cosh\theta$

   find the generator of the Lie algebra of this group and its effect on

1. $f = x^2 - y^2$

2. $f = x^2 - y^2 + p^2$.

6. Find the first prolongation to (4.31)

5. INVARIANT DIFFERENTIAL EQUATIONS.

In this section we discuss the application of invariance principles to differential equations and the simplifications obtained when they apply.

Definition 1: We say that a function $f(\mathbf{x})$ is invariant with respect to a Lie group G acting on R^n if

$$f(\mathbf{x}_1) = T_g f(\mathbf{x}) = f(\mathbf{x}) \quad , \ \mathbf{x}_1 = g\mathbf{x} \tag{5.1}$$

for all $g \in G$.

Theorem 1: Let G be a one parameter Lie group and let D be the Lie algebra generator of G. The function $f(\mathbf{x})$ is invariant with respect to G if and only if $Df = 0$.

Proof: Since

$$f(\mathbf{x}_1) = e^{aD} f(\mathbf{x}) = f(\mathbf{x}) + aDf(\mathbf{x}) + \frac{a^2}{2!} D^2 f + \ldots \tag{5.2}$$

for all a (especially $|a| \ll 1$) we infer that a necessary and sufficient condition for f to be invariant is that $Df = 0$.

Example 1: Find all functions $f(x,y)$ which are invariant with respect to the group $SO(2)$.

Solution: Since the Lie algebra generator of $SO(2)$ is

$$D = y \frac{\partial}{\partial x} - x \frac{\partial}{\partial y} \tag{5.3}$$

we infer that f is invariant with respect to this group if

$$y \frac{\partial f}{\partial x} - x \frac{\partial f}{\partial y} = 0 \quad . \tag{5.4}$$

To solve this first order partial differential equation (*see appendix* at the end of this section) we consider the simultaneous equation

$$\frac{dx}{y} = -\frac{dy}{x} \tag{5.5}$$

whose general solution is $x^2 + y^2 = c$. Hence the general solution of (5.4) is

$$f = f(x^2 + y^2) \tag{5.6}$$

where f is an arbitrary (differentiable) function.

Example 2: Find all functions which are invariant with respect to the dilatation group in R^2.

Solution: The generator of the Lie algebra of the dilatation group is $x \frac{\partial}{\partial x} + y \frac{\partial}{\partial y}$. Hence we must solve

$$x \frac{\partial f}{\partial x} + y \frac{\partial f}{\partial y} = 0 \tag{5.7}$$

or equivalently the system

$$\frac{dx}{x} = \frac{dy}{y} \quad . \tag{5.8}$$

Since the solution of (5.8) is $\frac{y}{x} = c$ it follows that the general solution of (5.7) is

$$f = f\left[\frac{y}{x}\right] \quad . \tag{5.9}$$

It is natural to extend definition 1 to differential equations as follows:

Definition 2: The differential equation

$$f(x,y,p) = 0 \tag{5.10}$$

is said to be invariant with respect to a one parameter Lie group of transformations acting on R^2 if the first prolongation of G leaves f unchanged.

Similar to theorem 1 it is easy to show that $f(x,y,p)$ is invariant with respect to G if

$$D^{(1)}f = 0 \tag{5.11}$$

where $D^{(1)}$ is the first prolongation of the Lie algebra generator of G.

Remark: The generalization of the definitions and theorems above to higher order equations (or more variables, i.e partial differential equations) should be obvious. Thus, a second order ordinary differential equation

$$f(x,y,p,q) = 0 \tag{5.12}$$

is invariant with respect to a one parameter group of transformations G is and only if

$$D^{(2)}f = 0 \tag{5.13}$$

where $D^{(2)}$ is the second prolongation of the Lie algebra generator of G.

Example 3: Find all first order differential equations which are invariant with respect to the transformations

$$\begin{aligned} x_1 &= x + a \\ y_1 &= y \end{aligned} \tag{5.14}$$

Solution: The Lie algebra generator for the transformations (5.14) is $D = \frac{\partial}{\partial x}$, i.e. $\xi = 1$, $\eta = 0$. Hence using (4.23) we find that $\zeta = 0$. The condition (5.11) is then

$$\frac{\partial f}{\partial x} = 0 \tag{5.15}$$

i.e., f is independent of x. Thus, the most general differential equation invariant under this group is $f(y,p) = 0$ or if one can use the implicit function theorem

$$p = f(y) \tag{5.16}$$

The following theorem due to S. Lie is the key to the solution (or reduction) of invariant equations.

Theorem 2: If $f(x,y,p)$ is invariant with respect to a one parameter group G acting on R^2 then the equation is integrable by quadratures. In particular if the differential equation can be written in the form

$$Pdx + Qdy = 0 \tag{5.17}$$

and $D = \xi \frac{\partial}{\partial x} + \eta \frac{\partial}{\partial y}$ is the Lie algebra generator of G then

$$\frac{1}{P\xi + Q\eta} \tag{5.18}$$

is an integrating factor for (5.17) and, therefore, its solution is given by

$$\Phi(x,y) = \int \frac{Pdx + Qdy}{P\xi + Q\eta} = C = \text{constant}. \tag{5.19}$$

For higher order equations invariance with respect to a one parameter group will reduce the order of the equation by one.

Example 4: Differential equations of the form

$$f(y,y',\dots,y^{(n)}) = 0 \tag{5.20}$$

are obviously invariant with respect to translations in x, i.e. with respect to the one parameter group $x_1 = x + a$, $y_1 = y$ (and its proper prolongations). Hence their order can be reduced by one. In fact if we let y and $p = y'$ be the new independent and dependent variables respectively then

$$y'' = \frac{dp}{dx} = \frac{dp}{dy} \cdot \frac{dy}{dx} = p\frac{dp}{dy} \tag{5.21}$$

$$y''' = p\frac{d}{dy}\left[p\frac{dp}{dy}\right], \text{ etc.} \tag{5.22}$$

and the resulting equation will be of order $n - 1$. In particular a second order equation

$$F(y,y',y'') = 0 \tag{5.23}$$

will be transformed into a first order equation in p

$$F\left[y,p,p\frac{dp}{dy}\right] = 0 \quad . \tag{5.24}$$

Example 5: Find (and solve) all first order equations which are invariant with respect to $SO(2)$.

Solution: The first prolongation of the Lie algebra of SO(2) is given by

$$D^{(1)} = x \frac{\partial}{\partial y} - y \frac{\partial}{\partial x} + (1 + p^2)\frac{\partial}{\partial p} \quad . \tag{5.25}$$

Hence $f(x,y,p)$ is invariant with respect to SO(2) if

$$x \frac{\partial f}{\partial y} - y \frac{\partial f}{\partial x} + (1 + p^2)\frac{\partial f}{\partial p} = 0 \quad . \tag{5.26}$$

To solve this equation we must find two independent solutions of the system

$$\frac{dx}{-y} = \frac{dy}{x} = \frac{dp}{1+p^2} \quad . \tag{5.27}$$

From the first equation we readily obtain

$$x^2 + y^2 = c^2 \quad . \tag{5.28}$$

Substituting for x in the second term of (5.27) from (5.28) leads to

$$\frac{dy}{\sqrt{c^2-y^2}} = \frac{dp}{1+p^2} \tag{5.29}$$

whose solution is

$$\arcsin \frac{y}{c} - \arctan p = c_2 \tag{5.30}$$

substituting for c we have

$$\arcsin \frac{y}{c} = \arcsin \frac{y}{\sqrt{x^2+y^2}} = \arctan \frac{y}{x} \quad . \tag{5.31}$$

Hence (5.30) can be rewritten as

$$\arctan \frac{y}{x} - \arctan p = c_2 \quad . \tag{5.32}$$

Letting $\upsilon = \arctan \frac{y}{x}$ and $\mu = \arctan p$ we see that equation (5.32) is equivalent to $\tan(\upsilon - \mu) = \tan c_2$. Using the formula for the tangent of $(\alpha - \beta)$ we obtain

$$\frac{\frac{y}{x} - p}{1 + \frac{y}{x}p} = \text{constant} \ .$$

We infer then that the general solution of (5.26) is given by

$$\frac{xp - y}{yp + x} = g(x^2 + y^2)$$

where g is an arbitrary function. Rewriting this equation in the form

$$(y + xg)dx + (yg - x)dy = 0$$

we conclude from theorem 2 that it admits an integrating factor of $1/(x^2 + y^2)$.

Appendix

First Order Partial Differential Equations

A first order partial differential equation

$$f_1(\mathbf{x}) \frac{\partial u}{\partial x_1} + \dots + f_n(\mathbf{x}) \frac{\partial u}{\partial x_n} = 0, \quad \mathbf{x} \in R^n \tag{5.33}$$

can be rewritten as

$$\mathbf{f} \cdot \text{grad}u = 0 \;, \quad \mathbf{f} = \mathbf{f} = \begin{bmatrix} f_1 \\ \vdots \\ f_n \end{bmatrix} \quad . \tag{5.34}$$

Hence, if $u = u(x_1,\dots,x_n)$ is a solution to equation (5.33) then the normal to the surface $u(\mathbf{x}) = \text{const.}$ is orthogonal to $f(\mathbf{x})$ which is equivalent to saying that $f(\mathbf{x})$ is tangent to any such surface. It follows then that there exists a vector $\mathbf{dx}$ on $u(\mathbf{x}) = \text{const.}$ so that

$$\mathbf{dx} = \mathbf{f}dt$$

or equivalently

$$\frac{dx_1}{f_1} = \dots = \frac{dx_n}{f_n} \; (= dt) \quad . \tag{5.35}$$

We conclude then that in order to find a solution for (5.33) we can use (5.35) to find the surfaces $u(\mathbf{x}) = \text{const.}$ of such a solution.

Theorem 3: If $u_1(\mathbf{x}) = c_1, \ldots, u_{n-1}(\mathbf{x}) = c_{n-1}$ (where c_i, $i = 1, \ldots, n-1$ are constants) are $(n-1)$ indpendent solutions of (5.35) then the general solution of (5.33) is given by

$$u = F(u_1, \ldots, u_{n-1}) \tag{5.36}$$

where F is any C^1 function.

Proof: (We only show that (5.36) is a solution of (5.33)). Since

$$\frac{\partial u}{\partial x_i} = \sum_{j=1}^{n-1} \frac{\partial F}{\partial u_j} \frac{\partial u_j}{\partial x_i}$$

we conclude that

$$\sum_{i=1}^{n} f_i \frac{\partial u}{\partial x_i} = \sum_{j=1}^{n-1} \frac{\partial F}{\partial u_j} \left[\sum_{i=1}^{n} f_i \frac{\partial u_j}{\partial x_i} \right] = 0$$

i.e. u is a solution of (5.33).

Although there exists no standard technique to solve the system (5.35) when $\mathbf{f}$ is nonlinear we demonstrate the "usual" technique used to solve such equation through the following example (see also Chapter 3).

Example 6: Find the general solution of

$$x(y + z) \frac{\partial u}{\partial x} + y \frac{\partial u}{\partial y} - z \frac{\partial u}{\partial z} = 0 \tag{5.37}$$

Solution: For equation (5.37) the system (5.35) is

$$\frac{dx}{dt} = x(y + z), \frac{dy}{dt} = y, \frac{dz}{dt} = -z \quad . \tag{5.38}$$

This can be rewritten as

$$\frac{dx}{x(y+z)} = \frac{dy}{y} = \frac{dz}{-z} \; (= dt) \tag{5.39}$$

One solution of this system is easy to compute by considering the second and third terms in (5.39). We obtain

$$u_1 = yz = c_1 \quad . \tag{5.40}$$

To find a second independent solution we have to use the first term in (5.39) in conjunction with the second or third term. This, however, leads to "noseparable" system due to the presence of a third variable. To overcome this difficulty we use (5.40) to eliminate z (or y) from the first equation in the system. Thus, $z = c_1/y$ and hence

$$\frac{dx}{x(y+\frac{c_1}{y})} = \frac{dy}{y} \tag{5.41}$$

which lead to

$$u_2 = \ln x - y + \frac{c_1}{y} = c_2 \tag{5.42}$$

substituting for c_1 from (5.40) leads to

$$u_2 = \ln x - y + z = c_2 \quad .$$

Thus, the general solution of (5.37) is

$$u = F(yz,\ \ln x - y + z) \quad . \tag{5.43}$$

EXERCISE 4

In exercises 1-5 find and solve (by quadratures) the first order differential equations which are invariant with respect to the Lie groups whose Lie algebra generators are:

1. $D = a\frac{\partial}{\partial x} + b\frac{\partial}{\partial y}$ a,b constants
2. $D = ax\frac{\partial}{\partial x} + by\frac{\partial}{\partial y}$
3. $D = ax\frac{\partial}{\partial y} - by\frac{\partial}{\partial x}$
4. $D = g(x)\frac{\partial}{\partial y}$
5. $D = g(y)\frac{\partial}{\partial x}$

Remark: In exercises 1-3 pay special attention to the cases $a = 0$ or $b = 0$.

6. Find the second order differential equations which are invariant with respect to the generators of exercises 1,2.

6. THE FACTORIZATION METHOD

The same program of classification and reduction which was carried for first order equations can be carried out, in principle, for second order equations. However, for these equations the most important application of Lie groups is related to a method of solution called the Factorization method. This method which is applicable to many of the differential equations which define the special functions of Mathematical physics has a strong Lie algebraic content. This content can be used advantageously to solve these equations and obtain insights about the properties of the solutions and recursion relations between them.

To begin with we observe that any second order equation

$$p_2(\theta)\frac{d^2P}{d\theta^2} + p_1(\theta)\frac{dP}{d\theta} + p_0(\theta)P = 0 \tag{6.1}$$

on the interval $[\alpha,\beta]$ with $p_2 \neq 0$ can be brought into the (self adjoint) form

$$\frac{d}{d\theta}\left[q_2(\theta)\frac{dP}{d\theta}\right] + q_1(\theta)P + \lambda q_o(\theta)P = 0 \tag{6.2}$$

if multiplied by the factor

$$\frac{1}{p_2}\exp\left[\int \frac{p_1}{p_2}\,dx\right]$$

(where λ is a constant). Furthermore, if $q_2, q_0 > 0$ we can transform equation (6.2) into a "standard form"

$$y'' + r(x,m)y + \lambda y = 0,\ x \in [a,b] \tag{6.3}$$

(where primes denote differentiation with respect to x and m is a constant) by the transformation

$$y = (q_2q_0)^{1/4}P,\ dx = \left[\frac{q_o}{q_2}\right]^{1/2} d\theta \quad . \tag{6.4}$$

We note that equations of the form (6.3) are obtained repeatedly as a result of applying the method of separation of variables (in various coordinate systems) to the Laplacian operator in two and three dimensions.

The basic idea of the Factorization method is to solve equations of the form (6.3) (for some $r(x,m)$) subject to the conditions

1. $y(a) = y(b) = 0$

$$\tag{6.5}$$

2. $\int_a^b y^2(x)dx = 1$ (normalization condition) by reducing these equations to first order equations for particular values of λ and m. Other solutions of these equations for different λ, m can be obtained then by repeated application of the "raising" and "lowering" operators on the known solutions.

Remark: We observe that if one solution of (6.3) is known then a second independent solution can be obtained by quadratures.

Definition 1: We say that equation (6.3) is factorizable if it can be replaced by each of the following two equations

$$H^-_{m+1}H^+_m y(\lambda,m) = [\lambda - L(m+1)]y(\lambda,m) \tag{6.6}$$

$$H^+_{m-1}H^-_m y(\lambda,m) = [\lambda - L(m)]y(\lambda,m) \tag{6.7}$$

where $H^{\pm}_m$ are the linear operators

$$H^+_m = k(x,m+1) - \frac{d}{dx} \tag{6.8}$$

$$H^-_m = k(x,m) + \frac{d}{dx} \tag{6.9}$$

In these equations $k(x,m), L(m)$ are some functions of x,m (to be determined) and $y(\lambda,m) = y_{\lambda,m}(x)$.

When (6.3) is factorizable we have the following important theorem:

Theorem 1: If equation (6.3) is factorizable and $y(\lambda,m)$ is a solution of this equation then

$$y(\lambda, m+1) = H_m^+ \, y(\lambda, m) \tag{6.10}$$

$$y(\lambda, m-1) = H_m^- \, y(\lambda, m) \tag{6.11}$$

are also solutions of (6.3) with $m + 1$, $m - 1$ respectively.

We see then that by repeated applications of H^+ and H^- on a known solution of (6.3) we obtain a "ladder" of solutions with the same λ but with different m's. It is natural, therefore, to refer to these operators as raising and lowering operators respectively. However, this "ladder" operation can terminate itself as follows:

Theorem 2: a. If $\lambda = L(\ell+1)$ then $H_\ell^+ \, y(\lambda, \ell) = 0$.

b. If $\lambda = L(\ell)$ then $H_\ell^- \, y(\lambda, \ell) = 0$.

Thus, at these points of λ and m the second order differential equation has been reduced effectively to a first order equation (since $H_\ell^\pm$ are first order differential operators) and other solutions of (6.3) can be computed from these solutions by repeated application of $H^\pm$.. However, we must still find out to what extent the solutions obtained by this process satisfy the conditions (6.5). The following theorem answers this question.

Theorem 3: If $y(\lambda, m)$ satisfy the conditions (6.5) and $L(m)$ is an increasing function of m then H_m^- applied to $y(\lambda, m)$ produces a solution which satisfy the conditions (6.5) up to normalization (i.e. $\int_a^b y^2(\lambda, m-1)dx < \infty$). Similarly if $L(m)$ is monotonically decreasing then the application of H_m^+ to $y(\lambda, m)$ produces a solution of (6.3) which satisfies (6.5) up to normalization.

From a practical point of view the following two situations can happen:

1. $\lambda = L(\ell+1)$ and $L(m)$ is monotonically increasing.

Under these circumstances we find $y(\lambda,\ell)$ from $H^+_\ell\, y(\lambda,\ell) = 0$ and other solutions (with the same λ) by repeated application of H^- to $y(\lambda,m)$ we obtain

$y(\lambda,\ell-1)$, $y(\lambda,\ell-2)$ ...

This process can terminate if there exist $m = \ell$, such that $\lambda = L(\ell_1)$ or continue indefinitely.

2. $\lambda = L(\ell)$ and $L(m)$ monotonically decreasing.

In this case we find $y(\lambda,\ell)$ from $H^-_\ell\, y(\lambda,\ell) = 0$ and other solutions by repeated application of H^+ to this solution. Once again the process can terminate after a finite number of steps on continue indefinitely.

Finally we show how to determine the class of factorizable equations. To this end we substitute (6.8), (6.9) in (6.6), (6.7) and subtract the result from (6.3). We obtain

$$k^2(x,m) - k'(x,m) + L(m) = -r(x,m) \tag{6.12}$$

$$k^2(x,m+1) + k'(x,m+1) + L(m+1) = -r(x,m). \tag{6.13}$$

By subtracting (6.12) from (6.13) and introducing the ansatz

$$k(x,m) = \frac{k_{-1}(x)}{m} + k_0(x) + mk_1(x)$$

we obtain first order differential equations for k_{-1}, k_0, k_1 whose solution then yield $L(m)$ and (from (6.12) or (6.13)) $r(x,m)$. The class of factorizable

TABLE 1: Factorization Types

Type	r(x,m)	k(x,m)	L(m)
A	$\dfrac{[a^2(m+\gamma)(m+\gamma+1)+d^2 + 2ad(m+ +\frac{1}{2})\cos a(x+\alpha)]}{\sin^2 a(x+\alpha)}$	$(m+\gamma)a \cot a(x+\alpha) + \dfrac{d}{\sin a(x+\alpha)}$	$a^2(m+\gamma)^2$
B	$-d^2\exp(2ax)+2ad(m+\gamma+\frac{1}{2})\exp(ax)$	$d\exp(ax)-m-\gamma$	$-a^2(m+\gamma)^2$
C	$\dfrac{(m+\gamma)(m+\gamma+1)}{x^2} - \dfrac{b^2x^2}{4} + b(m-\gamma)$	$\dfrac{m+\gamma}{x} + \dfrac{bx}{2}$	$-2bm+b/2$
D	$-(bx+d)^2 + b(2m+1)$	$bx+d$	$-2bm$
E	$-\dfrac{m(m+1)a^2}{\sin^2(x+\alpha)} - 2ap \cot a(x+\alpha)$	$ma \cot a(x+\alpha)+\dfrac{p}{m^2}$	$a^2m^2 - \dfrac{p^2}{m^2}$
F	$-\dfrac{2p}{x} - \dfrac{m(m+1)}{x^2}$	$\dfrac{m}{x} + \dfrac{p}{m}$	$-\dfrac{p^2}{m^2}$

In this table a,b,d,p,α,γ are arbitrary constants to be adjusted to the equation under consideration.

equations is thus seen to consist of six interrelated types (see Table 1).

By proper adjustments of the various parameters that appear in the formulas describing these types it can be shown that a large class of special functions are defined by factorizable equations.

Some important examples of factorizable equations and their relation to various Lie algebras are considered in the next section.

EXERCISE 5

Show (using Table 1) that the following equations are factorizable and write down the corresponding raising and lowering operators

1. $$y'' + 2c\cot\theta y' - \frac{m^2 + 2mc - m}{\sin^2\theta} y + \lambda y = 0$$

2. $$y'' + \left[\frac{a^2\alpha(\alpha-1)}{\sin^2 a\theta} + \frac{a^2\beta(\beta-1)}{\cos^2 a\theta}\right] y + \lambda y = 0$$

3. $$y'' + \frac{2}{r} y - \frac{(\ell+c)(\ell+c+1)}{r^2} y - \frac{1}{(n+c)^2} y = 0$$

7. EXAMPLES OF FACTORIZABLE EQUATIONS

In this section we treat four types of special functions using the factorization method and examine the underlying Lie algebraic structure which is related to the raising and lowering operators.

Example 1: Associated spherical harmonics.

The differential equation which defines these functions is

$$\frac{1}{\sin\theta}\frac{d}{d\theta}\left[\sin\theta\frac{dP}{d\theta}\right] - \frac{m^2}{\sin^2\theta} P + \lambda P = 0 \tag{7.1}$$

we transform this equation into standard form by introducing

$$y = \sin^{1/2}\theta P \tag{7.2}$$

which leads to

$$y'' - \frac{m^2 - \frac{1}{4}}{\sin^2\theta} y + \left[\lambda + \frac{1}{4}\right] y = 0 \quad . \tag{7.3}$$

Using Table 1 we infer that this equation is factorizable (of type A) with

$$a = 1, \ \alpha = -\frac{1}{2}, \ d = \gamma = 0, \ x = \theta, \ \bar{\lambda} = \lambda + \frac{1}{4} \ .$$

The factorization is given then by the functions

$$k(\theta, m) = \left[m - \frac{1}{2}\right]\cot\theta, \ L(m) = \left[m - \frac{1}{2}\right]^2 \quad . \tag{7.5}$$

Since $L(m)$ is an increasing functions of m we must have at the top of the ladder

$$\lambda + \frac{1}{4} = L(\ell+1), \ \text{i.e.} \ \lambda = \ell(\ell+1) \quad . \tag{7.6}$$

Solving $H_\ell^+ y_{\ell,\ell}(\theta) = 0$ (where the first subscript reprsent λ and the second m) we obtain the normalized solution

$$y_{\ell,\ell}(\theta) = \left[\frac{1.3.5. \ \ldots \ (2\ell+1)}{2.2.4. \ \ldots \ 2\ell}\right]^{1/2} \sin^{\ell+1/2}\theta \tag{7.7}$$

which satisfies the conditions (6.5). To obtain the other solutions of (7.3) for this λ we apply H^- to this solution and obtain the following recursion relations for the normalized solutions

$$\left[(m-\tfrac{1}{2})\cot\theta + \frac{d}{d\theta}\right] y_{\ell,m} = \left[(\ell+m)(\ell+1-m)\right]^{1/2} y_{\ell,m-1} \tag{7.8}$$

$$\left[(m+\tfrac{1}{2})\cot\theta - \frac{d}{d\theta}\right] y_{\ell,m} = \left[(\ell+m+1)(\ell-m)\right]^{1/2} y_{\ell,m+1} \tag{7.9}$$

We observe, however, that besides (7.6)

$$\lambda + \frac{1}{4} = L(-\ell)$$

and, therefore, the ladder solutions which are obtained by repeated applications of H^- will terminate when $m = -\ell$. Thus, for $\lambda = \ell(\ell+1)$ equation (7.3) has solutions which satisfy (6.5) only when $-\ell \le m \le \ell$.

We note, however, that from a practical point of view it is useful to find raising and lowering operators for $P_{\ell,m}$ rather than $y_{\ell,m}$. This can be easily accomplished using (7.2), (7.9). A little algebra then yields

$$\left[m\cot\theta + \frac{d}{d\theta}\right]P_{\ell,m} = \left[(\ell+m)(\ell+1-m)\right]^{1/2} P_{\ell,m-1} \quad (7.10)$$

$$\left[m\cot\theta - \frac{d}{d\theta}\right]P_{\ell,m} = \left[(\ell+m+1)(\ell-m)\right]^{1/2} P_{\ell,m+1} \quad (7.11)$$

To consider the algebraic aspects of this factorization we introduce the extraneous variable φ which can be interpreted as the azimuthal angle in R^3 and the spherical harmonic functions

$$Y_{\ell,m}(\theta,\varphi) = e^{im\varphi}P_{\ell,m}(\theta) \quad . \quad (7.12)$$

For these functions the m dependence of the raising and lowering operators (7.10), (7.11) can be replaced by $-i\frac{\partial}{\partial\theta}$ and we obtain

$$J_+ = e^{i\varphi}\left[-i\cot\theta\frac{\partial}{\partial\varphi} - \frac{\partial}{\partial\theta}\right] \quad (7.13)$$

$$J_- = e^{i\varphi}\left[-i\cot\theta\frac{\partial}{\partial\varphi} + \frac{\partial}{\partial\theta}\right] \quad (7.14)$$

as the raising and lowering operators respectively. Moreover, if we define the operator

$$J_3 = i\frac{\partial}{\partial\varphi} \quad (7.15)$$

and compute the commutation relations between the operators J_+, J_- and J_3 we find

$$[J_+, J_-] = 2J_3, \ [J_3, J_+] = J_+, [J_3, J_-] = -J_-. \quad (7.16)$$

These commutation relations characterize the Lie algebra of SO(3).

Remark: The relationship between these operators and the real operators $J_{\mu\upsilon}$ which were introduced in section 3 is as follows:

$$J_+ = iJ_{23} + J_{31}, J_- = iJ_{23} - J_{31}, J_3 = iJ_{12}. \quad (7.17)$$

Another factorization of equation (7.1) is possible if we interchange the roles of λ and m in this equation and rewrite it in the form

$$\frac{1}{\sin\theta}\frac{d}{d\theta}\left[\sin\theta \frac{dP}{d\theta}\right] + \ell(\ell+1)P - \frac{m^2}{\sin^2\theta} P = 0 \quad (7.18)$$

viz. seek raising and lowering operators in ℓ (for fixed m). To accomplish this goal we transform equation (7.18) into standard form by the transformation

$$z = \ell n \tan\frac{\theta}{2}, \ y(z) = P(\theta) \quad (7.19)$$

which leads to

$$y'' + \frac{\ell(\ell+1)}{\cosh^2 z} y - m^2 y = 0 \quad (7.20)$$

Once again this is type A equation with the parameters (of table 1)

$$a = i, \ \gamma = 0, \ \alpha = i\frac{\pi}{2}, \ d = 0$$

and the identifications

$$x = z, \ m = \ell, \ \lambda = -m^2 \quad .$$

Hence we infer that

$$k(z,\ell) = \ell \tanh z, \ L(\ell) = -\ell^2 \quad . \quad (7.21)$$

Since $L(\ell)$ is a decreasing function of ℓ we must solve for $\lambda = L(\ell)$, i.e. $\ell = m$ and obtain other solutions in the ladder by repeated applications of H^+. Thus, for fixed m we obtain the infinite sequence

$$y(m,m),\ y(m+1,m),\ldots \qquad .$$

Solving $H_m^- y(m,m) = 0$ and normalizing we obtain

$$y(m,m) = \left[\frac{1.3.5.\ \ldots\ (2m-1)}{2.2.4.\ \ldots\ (2m-2)}\right]^{1/2} \cosh^{-m} z \tag{7.22}$$

and the recursion relation (for the normalized solutions).

$$\left[(\ell+1)\tanh z - \frac{d}{dz}\right] y(\ell,m) = \left[(\ell+1+m)(\ell+1-m)\right]^{1/2} y(\ell+1,m) \tag{7.23}$$

Rewriting the ladder operators in terms of θ leads to

$$\left[-(\ell+1)\cos\theta - \sin\theta\frac{d}{d\theta}\right] P_{\ell,m} = \left[(\ell+1-m)(\ell+1+m)\right]^{1/2} P_{\ell+1,m} \tag{7.24}$$

$$\left[-\ell\cos\theta + \sin\theta\frac{d}{d\theta}\right] P_{\ell,m} = \left[(\ell-m)(\ell+m)\right]^{1/2} P_{\ell-1,m} \tag{7.25}$$

where $\ell \geq m$.

Once again we can obtain ladder operators which are independent from the explicit dependence on ℓ by introducing the extraneous variable ψ (which has no physical meaning in R^3) and consider

$$\tilde{Y}_{\ell,m}(\theta,\psi) = e^{i\ell\psi} P_{\ell,m}(\theta) \tag{7.26}$$

The raising and lowering operators (7.24), (7.25) take the form

$$k_+ = e^{i\psi}\left(i\cos\theta\frac{\partial}{\partial\psi} - \sin\theta\frac{\partial}{\partial\theta}\right) \tag{7.27}$$

$$k_- = e^{-i\psi}\left(i\cos\theta\frac{\partial}{\partial\psi} - \sin\theta\frac{\partial}{\partial\theta}\right) \quad . \tag{7.28}$$

Furthermore, if we introduce

$$k_3 = \left(\frac{1}{2} - i\frac{\partial}{\partial\psi}\right) \tag{7.29}$$

we obtain the following commutation relations

$$[k_+,k_-] = -2k_3,\quad [k_3,k_+] = k_+,\quad [k_3,k_-] = -k_3 \quad . \tag{7.30}$$

These commutation relations characterize the Lie algebra of SO(2,1).

Remarks: 1. The pseudo-rotation groups O(p,q) are defined geometrically as the groups of transformations which leave the pseudo spheres

$$\sum_{i=1}^{p} x_i^2 - \sum_{j=1}^{q} x_{p+j}^2 = 1 \tag{7.31}$$

invariant. The commutation relations for the generators of the Lie algebras of these groups are given by

$$[J_{\mu\upsilon}, J_{\rho\pi}] = g_{\mu\rho}J_{\upsilon\pi} + g_{\upsilon\pi}J_{\upsilon\rho} - g_{\mu\pi}J_{\upsilon\rho} - g_{\upsilon\rho}J_{\mu\pi} \tag{7.32}$$

$\mu, \upsilon, \rho, \pi + 1, \ldots, p + q$ and where

$$g = (g_{\mu\upsilon}) = \left[\begin{array}{c|c} -I_p & 0 \\ \hline 0 & I_q \end{array}\right] \tag{7.33}$$

I_m being the unit matrix of dimension m

The relationship between the operators J_{12}, J_{23}, J_{31} of O(2.1) and k_+, k_-, k_3 is the same as for O(3) (equation (7.17)).

2. The operators $\{k_+, k_-, k_3\}$ do not commute with $\{J_+, J_-, J_3\}$ but at the same time they do not generate directly (by repeated commutation) a finite dimensional Lie algebra.

Example 2: Bessel Functions

Bessel functions are usually defined as solutions of the equation

$$r^2y'' + ry' + (r^2 - m^2)y = 0, \quad y = y(r) \quad . \tag{7.34}$$

The standard form of this equation is obtained by the transformation .

$$y = r^{-1/2}u \tag{7.35}$$

which leads to

$$u'' - \frac{m^2 - \frac{1}{4}}{r^2} u + u = 0 \tag{7.36}$$

(i.e. $\lambda = 1$). Equation (7.34) is type C equation and the factorization is given by

$$k(r,m) = \frac{m-1/2}{r}, \quad L(m) = 0 \quad . \tag{7.37}$$

The ladder operators lead to the following recursion relations

$$\left[\frac{m+1/2}{r} - \frac{d}{dr}\right] u_m(r) = u_{m+1}(r) \tag{7.38}$$

$$\left[\frac{m-1/2}{r} + \frac{d}{dr}\right] u_m(r) = u_{m-1}(r) \quad . \tag{7.39}$$

In standard notation we have

$$u_m(r) = r^{1/2} J_m(r) \quad . \tag{7.40}$$

Once again it is possible to obtain ladder operators acting directly on $J_m(r)$ from (7.35)-(7.37) which lead to

$$\left(\frac{m}{r} - \frac{d}{dr}\right) J_m = J_{m+1} \tag{7.41}$$

$$\left(\frac{m}{r} + \frac{d}{dr}\right) J_m = J_{m-1} \quad . \tag{7.42}$$

By introducing

$$J_m(r,\varphi) = e^{im\varphi} J_m(r) \tag{7.43}$$

we can replace the m dependence of the ladder operators by $-i\frac{\partial}{\partial\varphi}$ and obtain the operators

$$J_+ = \left(-\frac{i}{r}\frac{\partial}{\partial\varphi} - \frac{\partial}{\partial r}\right) \tag{7.44}$$

$$J_- = \left(-\frac{i}{r}\frac{\partial}{\partial\varphi} + \frac{\partial}{\partial r}\right) \quad . \tag{7.45}$$

If we further introduce the operator

$$J_3 = -i\frac{\partial}{\partial\varphi} \tag{7.46}$$

then these three operators satisfy the commutation relations

$$[J_+, J_-] = 0, \ [J_3, J_+] = J_+, \ [J_3, J_-] = J_- \tag{7.47}$$

which characterize the Lie algebra of $E(2)$.

Example 3: Hermite functions

These functions are defined by the differential equation

$$h''(x) - x^2 h(x) + \lambda h(x) = 0 \tag{7.48}$$

which represents also the Schrödinger equation for a one dimensional harmonic oscillator in Quantum Mechanics.

We bring this equation into a factorizable form by introducing m artificially as follows:

$$h''(x) - m^2 x^2 h(x) + \lambda h(x) = 0 \tag{7.49}$$

and seek raising and lowering operators in λ. Performing the standard factorization procedure on (7.49) which is an equation of type D) we obtain when we substitute $m = 1$

$$H_{\pm} = x \pm \frac{d}{dx}, \quad \lambda = 2n + 1, \quad n = 0, 1, \ldots \tag{7.50}$$

and
$$H_+ h_n(x) = (2n + 2)^{1/2} h_{n+1}(x) \tag{7.51}$$

$$H_- h_n(x) = (2n)^{1/2} h_{n-1}(x) \quad . \tag{7.52}$$

Since the ladder has a "bottom" for $n = 0$ we solve

$$H_- h_0(x) = 0 \tag{7.53}$$

which yields

$$h_0(x) = \pi^{-1/4} \exp(-x^2/2) \quad . \tag{7.54}$$

Other functions $h_n(x)$ can be obtained now by repeated application of H_+.

To find the Lie algebraic structure generated by $H_{\pm}$ we introduce the functions

$$H_n(x,t) = e^t h_n(x) \tag{7.55}$$

For these functions the raising and lowering operators can be rewritten as

$$R = x \frac{\partial}{\partial t} - \frac{\partial}{\partial x}, \quad L = x \frac{\partial}{\partial t} + \frac{\partial}{\partial x} \tag{7.56}$$

and we have the commutation relations

$$[R,L] = -\frac{\partial}{\partial t} = -2M$$
$$[R,M] = [L,M] = 0 \qquad . \qquad (7.57)$$

Thus, the Lie algebra generated by these operators is nilpotent.

Remark: Nilpotent Lie algebras are defined as follows: Let L be a Lie algebra and let

$$L^2 = \{L,\ L = [L_1,L_2] \quad \text{for some} \quad L_1,L_2 \in L\}$$

and in general

$$L^{k+1} = \{L,\ L = [L_1,L_2] \quad \text{where} \quad L_1 \in L,\ L_2 \in L^{k-1}\}$$

L is nilpotent if $L^k = 0$ for some finite k.

Example 4: Radial functions for the Hydrogen atom in Quantum Mechanics.

In Quantum Mechanics the stationary states of a particle moving in a spherically symmetric field are given by the solutions of Schrödinger equation which in proper units is

$$(-\nabla^2 + U(r))\psi(r,\theta,\varphi) = \lambda\psi(r,\theta,\varphi) \qquad (7.58)$$

For the Hydrogen atom $U(r) = -\frac{1}{r^2}$ and after the proper separation of variables we obtain the following differential equation for the radial function $R(r)$ of ψ:

$$\frac{d^2R}{dr^2} + \left[\frac{2}{r} - \frac{\ell(\ell+1)}{r^2}\right]R + \lambda R = 0,\ \ell \geq 0 \qquad . \qquad (7.59)$$

Equation (7.59) is to type F with $q = -1$ and m replaced by ℓ. We find that the boundary conditions on $[-\infty,\infty]$ are satisfied when

$$\lambda = -\frac{1}{n^2},\ \ell + 1 < n = 1,2,\ldots$$

and the ladder operators are given by

$$\left[\left[\frac{\ell^2}{r} - 1\right] + \ell \frac{d}{dr}\right]R_{n,\ell}(r)$$
$$= \frac{1}{n}\left[n^2 - \ell^2\right]^{1/2} R_{n,\ell-1}(r) \tag{7.60}$$
$$\left\{\left[\frac{(\ell+1)^2}{r} - 1\right] - (\ell+1)\frac{d}{dr} R_{n,\ell}(r)\right.$$
$$= \frac{1}{n}(n^2 - (\ell+1)^2)^{1/2}R_{n,\ell+1} \tag{7.61}$$

at the top of the ladder we have $H^+_{n,n-1}R_{n,n-1} = 0$ which yields

$$R_{n,n-1}(r) = \left[\frac{2}{n}\right]^{n+\frac{1}{2}} \left[2n)!\right]^{-\frac{1}{2}} r^n\exp(-r/n) \tag{7.62}$$

Other solutions of (7.59) can be obtained from the solution (7.62) by recursion.

Finally we refer the reader to the bibliography for a detailed discussion of the consequences of the Lie algebraic structure which is related to the factorization method.

EXERCISE 6

1. Find the commutation relations for the ladder operators of the radial functions of the Hydrogen atom.
2. The Hypergeometric functions are defined as solutions of $x(1-x)F'' + [c - (a+b+1)x]F' - abF = 0$ (7.63)

   a. Use the transformation

   $$x = \sin^2\theta, \quad F = \sin^{-c+1/2}\theta\cos^{c-a-b-1/2}\theta V$$

   To bring this equation into the form

   $$V'' - \left[\frac{(c-\frac{3}{2})(c-\frac{1}{2})}{\sin^2\theta} - \frac{(a+b-c-\frac{1}{2})(a+b-c+\frac{1}{2})}{\cos^2\theta}\right]V + (a-b)^2V = 0 \tag{7.64}$$

b. Introduce m artificially in the following two forms into equation (7.6) and discuss the corresponding factorizations (when $m = 0$).

$$\text{I} \quad V'' - \left[\frac{(m+c-\frac{3}{2})(m+c-\frac{1}{2})}{\sin^2\theta} - \frac{(m+a+b-c-\frac{1}{2})(m+a+b-c+\frac{1}{2})}{\cos^2\theta}\right]V + (a-b)^2V = 0$$

$$\text{II} \quad V'' - \left[\frac{(m+c-\frac{3}{2})(m+c-\frac{1}{2})}{\sin^2\theta} - \frac{(m+c-a-b-\frac{1}{2})(m+c-a-b+\frac{1}{2})}{\cos^2\theta}\right]V + (a-b)^2V = 0$$

c. Find similar artificial factorizations which raise and lower a only.

BIBLIOGRAPHY

1. I. M. Gelfand et al - Representations of the rotation and Lorentz groups, Pergamon Press, 1963.
2. G. W. Bluman & J. D. Cole - Similarity methods for differential equations, Spring Verlag, N.Y., 1974.
3. W. Miller, Jr. - Lie theory and special functions, Academic Press, 1968.
4. N. J. Vilenkin - Special functions and the theory of group representqations, Amer. Math. Soc., Providence, RI, 1968.
5. P. M. Cohn - Lie groups, Cambridge Univ. Press, 1968.
6. R. Seshadri Y T. Y. Na - Group invariance in engineering
and boundary value problems, Springer Verlag, N.Y., 1959.
7. E. P. Wigner - Group theory, Academic Press, N. Y., 1959.
8. N. H. Ibragimov - Transformation groups applied to mathematical physics, Reidel, Boston, 1984.

CHAPTER 5. EQUATIONS WITH PERIODIC COEFFICIENTS

1. INTRODUCTION

In this chapter we discuss differential equations whose coefficients are periodic and the properties of their solutions. Such equations appear in various fields of science, e.g., solid state physics, celestial mechanics and others. Our objective is then to investiage the implications of periodicity on these systems properties and behavior. From another point of view, many physical systems are invariant with respect to certain transformations of the independent and dependent variables. Accordingly, the corresponding differential equations which model these systems are invariant under the same transformations. However, surprisingly not all solutions to these equations are invariant with respect to these same transformations. We shall illustrate this phenomena (and its partial resolution) within the context of periodic equations. We start with a few examples:

Example 1.1: Consider the equation

$$y'' + k^2 y = 0 \quad , \quad y' = \frac{dy}{dt} \quad . \tag{1.1}$$

Obviously, this equation is invariant with respect to translations in t, i.e. the equation remains unchanged under the transformations

$$\bar{t} = t + a \quad , \quad a \in R \quad . \tag{1.2}$$

However, the general solution of (1.1)

$$y = A \cos kt + B \sin kt \ , \ A,B \text{ constants} \tag{1.3}$$

is invariant with respect to (1.2) only if $a = n\pi$. Thus, none of the nontrivial solutions of equation (1.1) has the same invariance properties as the equation itself. Similarly the equation

$$y'' - k^2 y = 0 \tag{1.4}$$

is also invariant with respect to the transformations (1.2). However, none of the nontrivial solutions of this equation

$$y = A \cosh kt + B \sinh kt \tag{1.5}$$

is invariant with respect to any translations in t.

Example 1.2: The equation of motion of a particle in a central gravitational field (Kepler's problem) such as that of the planets around the sun can be reduced in polar coordinates to

$$\frac{d^2u}{d\theta^2} + u = c \tag{1.6}$$

where c is a constant and $u = \frac{1}{r}$. Obviously, equation (1.6) is invariant under rotations $\bar{\theta} = \theta + \alpha$. However, it is well known that equation (1.6) admits solutions which are not invariant under these transformations, e.g ellipses, etc.

Example 1.3: Energy Bands in Crystaline Solids.

In Quantum mechanics the basic equation describing the state of an electron in a potential well is Schrodinger equation. In one dimension and with proper units we can write this equation as

$$\frac{d^2\psi}{dx^2} + (E - V(x))\psi(x) = 0 \tag{1.7}$$

where E is the electron's energy and $V(x)$ is the potential.

In crystaline solids the potential $V(x)$ is periodic. In particular for (fictitious) one dimensional lattices $V(x)$ can be idealized to be in the form shown in Figure 1.

To investigate the implications of lattice and potential periodicity on the possible electronic states and energies we shall solve equation (1.7) on $[-a,b-a]$ subject to periodic boundary conditions. However, we recall that in Quantum mechanics only $\psi\psi^*$ (where ψ^* is the complex conjugate of ψ) is observable. Hence $\psi(-a)$ might be different from $\psi(b-a)$ by a phase factor, which we denote by e^{-ikb} (k real), without affecting any observable physical quantity.

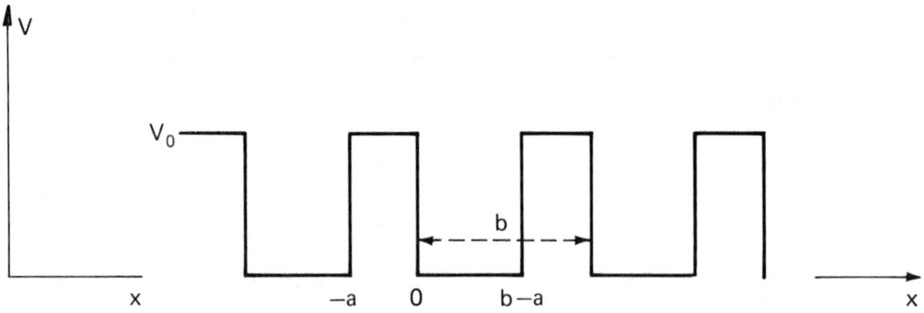

Figure 1
Idealized periodic potential in one dimensional lattice.

Denoting the solution of (1.7) on $[-a,0]$ by ψ_1 and on $[0,b-a]$ by ψ_2 it follows that the periodic boundary conditions imply

$$\begin{aligned}\psi_1(-a) &= e^{-ikb}\psi_2(b-a)\\ \psi_1'(-a) &= e^{-ikb}\psi_2'(b-a)\end{aligned} \qquad . \qquad (1.8)$$

Furthermore, at the juncture point $x = 0$ between the solutions ψ_1 and ψ_2 we must require continuity in the solution and its first order derivative, i.e.

$$\psi_1(0) = \psi_2(0) \ , \ \psi_1'(0) = \psi_2'(0) \quad . \tag{1.9}$$

The general solution of (1.7) on $[-a,0]$ where $V = V_0$ and $E < V_0$ (bound electronic states) is

$$\psi_1 = Ae^{\alpha x} + Be^{-\alpha x} \quad , \quad \alpha = (V_0-E)^{1/2} \tag{1.10}$$

On the other hand on $[0,b-a]$ where $V = 0$ the general solution is

$$\psi_2 = Ce^{i\beta x} + De^{-i\beta x} \ , \ \beta = E^{1/2} \quad . \tag{1.11}$$

The imposition of the periodic and continuity conditions (1.8)-(1.9) leads then to four linear equations in four unknowns A,B,C,D. A nontrivial solution for these unknowns exists only if the determinant of the coefficients of this system is zero. A short algebra shows that this happens when

$$\begin{aligned} \cos kb &= \cosh \alpha a \cos\beta(b-a) \\ &\quad + [(\alpha^2-\beta^2)/2\alpha\beta]\sinh\alpha a \sin\beta(b-a) \ . \end{aligned} \tag{1.12}$$

Since the left hand side of this equation can take values only between -1, $+1$ the right hand side must also be restricted to this range of values (for nontrivial solution for A,B,C,D). It follows then that when the value of the right hand of (1.12) is between $-1,+1$ we obtain allowable electronic states (energy bands) but for values outside this range we obtain forbidden energy values.

Thus, the periodicity of the lattice imposes restrictions on the possible electronic states and predicts the existence of allowable and forbidden energy bands. These facts have important physical implications.

In view of these examples it is natural to ask what is the relationship between the symmetries of a given differential equation and those of its solutions. In particular under what conditions there exist at least one nontrivial solution of the equation which is invariant with respect to the same transformations as the equation itself.

In order to make a meaningful progress towards the resolution of the questions posed above we restrict our attention in this chapter to periodic linear differential equations or systems thereof and investigate the properties of their solutions.

Definition 1.1: A linear differential equation with (piecewise) continuous coefficients

$$a_n(t)y^{(n)}(t) + \ldots + a_1y'(t) + a_0y(t) = 0,$$
$$a_n(t) \neq 0 \qquad (1.13)$$

is said to be periodic if there exists a constant $p \neq 0$ so that

$$a_i(t+p) = a_i(t) \ , \ i = 0,\ldots,n \qquad (1.14)$$

for all t. Similarly a system of equations

$$\mathbf{y}'(t) = A(t)\mathbf{y}(t) \ , \quad \mathbf{y} \in R^n \qquad (1.15)$$

is said to be periodic if there exist $p \neq 0$ so that

$$A(t+p) = A(t) \ . \qquad (1.16)$$

for all t and the entries of A are (piecewise) continuous.

From a historical perspective we note that early results regarding the solutions of periodic equations were derived by G. Floquet in 1883. Hence, much of the theory regarding the solutions of these equations is referred to as "Floquet Theory".

EXERCISE 1

1. Drive equation (1.10) from the conditions (1.8)-(1.9)

2. Plot the right hand side of (1.10) versus E if $a = 2$, $b = 1$ and $V_0 = 10$ (remember bound electronic states are obtained only if $E \leq V_0$).

2. FLOQUET THEORY FOR PERIODIC EQUATIONS.

In this section we study the nature of the solutions to periodic differential equations and systems thereof. However, to simplify the proofs we carry this discussion within the context of periodic second order equations and only state the proper generalization of these results to other cases.

Lemma 2.1: If $u_1(t)$, $u_2(t)$ are linearly independent solutions of the periodic equation

$$a_2(t)y''(t) + a_1(t)y' + a_0(t)y = 0 \qquad (2.1)$$

where $a_i(t+p) = a_i(t)$, $i = 0,1,2$ for all t then $w_i(t) = u_1(t+p)$ and $w_2(t) = u_2(t+p)$ are also linearly independent solutions of (2.1).

Proof: First it is clear that w_1, w_2 are solutions of (2.1) since this equation is invariant under the translation $t = t + p$. Furthermore, the Wronskian of $w_1(t)$, $w_2(t)$ is equal to the Wronskian of u_1, u_2 at $t + p$ which is nonzero since u_1, u_2 are linearly independent. This proves this lemma.

Theorem 2.1: There exists a nontrivial solution $u(t)$ of (2.1) so that

$$u(t+p) = \mu u(t) \quad , \quad \text{for all } t \qquad (2.2)$$

where $\mu \neq 0$ is a constant.

Proof: Let u_1, u_2 be a fundamental set of solutions for (2.1). Any solution $u(t)$ of this equation can be written then as

$$u(t) = c_1u_1(t) + c_2u_2(t) \quad . \tag{2.3}$$

Therefore,

$$u(t+p) = c_1u_1(t+p) + c_2u_2(t+p) \quad . \tag{2.4}$$

But from the fact that u_1, u_2 is a fundamental set of solution for (2.1) and the previous lemma we infer that there exists a nonsingular matrix A with constant entries so that

$$\begin{bmatrix} u_1(t+p) \\ u_2(t+p) \end{bmatrix} = \begin{bmatrix} a_{11} & a_{12} \\ a_{21} & a_{22} \end{bmatrix} \begin{bmatrix} u_1(t) \\ u_2(t) \end{bmatrix} \tag{2.5}$$

Hence, if $u(t)$ is to satisfy equation (2.2) we must have

$$\begin{aligned} \mu u(t) &= \mu(c_1u_1(t) + c_2u_2(t)) \\ &= u(t+p) = c_1(a_{11}u_1(t) + a_{12}u_2(t)) \\ &+ c_2(a_{21}u_1(t) + a_{22}u_2(t)) \end{aligned} \tag{2.6}$$

i.e.

$$\begin{aligned} &[(a_{11}-\mu)c_1 + a_{21}c_2]u_1(t) \\ &\qquad + [a_{12}c_1 + (a_{22}-\mu)c_2]u_2(t) = 0 \end{aligned} \tag{2.7}$$

Since u_1, u_2 are linearly independent this implies that

$$(A^T - \mu I)\mathbf{c} = 0 \tag{2.8}$$

where $\mathbf{c} = \begin{bmatrix} c_1 \\ c_2 \end{bmatrix}$ and A^T is the transpose of A. The system (2.8) has a nontrivial solution for c if and only if μ is an eigenvalue of A. (Remember: the eigenvalues of A and A^T are the same). Since A is nonsingular its eigenvalues are nonzero. Hence, there exists a $\mu \neq 0$ for which the system (2.8) has a nontrivial solution c_1, c_2 and the corresponding solution (2.3) satisifes equation (2.2).

Example 2.1: The equation

$$y'' - \lambda^2 y = 0 \tag{2.9}$$

is periodic equation for any fixed p. Hence, (for any such p) there must exist a nontrivial solution $u(t)$ so that $u(t+p) = \mu u(t)$ for all t. In fact $u(t) = e^{\lambda t}$ is a solution of (2.9) which satisfies

$$u(t+p) = e^{\lambda(t+p)} = \mu u(t) \quad , \tag{2.10}$$

where $\mu = e^{\lambda p}$.

We now state the generalization of theorem 2.1 to systems of periodic equations.

Theorem 2.2: If the system

$$\mathbf{y}'(t) = A(t)\mathbf{y}(t), \; \mathbf{y} \in R^n \tag{2.11}$$

is periodic with period p, i.e. $A(t+p) = A(t)$ then there exists a nontrivial solution $\mathbf{u}(t)$ of (2.11) so that

$$\mathbf{u}(t+p) = \mu \mathbf{u}(t) \tag{2.12}$$

Corollary: Since any linear periodic equation of order n is equivalent to a periodic system of equations with the same period we infer that any such equation has a solution which satisfies equation (2.2).

The following theorem gives a more detailed information regarding the solutions of a periodic equation.

Theorem 2.3: Equation (2.1) has two linearly independent solutions $w_1(t)$, $w_2(t)$ so that either

$$w_1(t) = e^{\lambda_1 t} f_1(t) \quad , \quad w_2(t) = e^{\lambda_2 t} f_2(t) \tag{2.13}$$

or

$$w_1(t) = e^{\lambda t} f_1(t) \quad , \quad w_2(t) = e^{\lambda t}[f_2(t) + t f_1(t)] \; . \tag{2.14}$$

In these equations $\lambda, \lambda_1, \lambda_2$ are constants, called the characteristic or Floquet exponents of (2.1) and $f_1(t)$, $f_2(t)$ are periodic functions with period p.

Proof: (We prove only the first part, i.e. equation (2.13)). If A in equation (2.5) has two distinct eigenvalues λ_1, λ_2 then we infer that there exists two indepedent solutions to equation (2.8). (However, this can happen also when A has only one eigenvalue of multiplicity two). Hence, there exist, under these circumstances, two independent solutions $w_1(t)$, $w_2(t)$ so that

$$w_i(t+p) = \mu_i w_i(t), \quad i = 1,2 \tag{2.15}$$

(μ_1 can be equal to μ_2). Since $\mu_i \neq 0$ (A is nonsingular) there exist λ_i so that $\mu_i = e^{p\lambda_i}$. Now introduce

$$f_i(t) = e^{-\lambda_i t} w_i(t), \quad i = 1,2 \tag{2.16}$$

then

$$w_i(t) = e^{\lambda_i t} f_i(t) \tag{2.17}$$

and

$$\begin{aligned} w_i(t+p) &= e^{\lambda_i (t+p)} f_i(t+p) = \mu_i w_i(t) \\ &= \mu_i e^{\lambda_i t} f_i(t). \end{aligned} \tag{2.18}$$

Hence,

$$f_i(t+p) = f_i(t) \tag{2.19}$$

as required.

To generalize this theorem to systems of periodic equations we recall that a solution of (2.11) subject to the initial condition $\mathbf{y}(t_0) = \mathbf{c}$ (in the following we let $t_0 = 0$) is given by

$$\mathbf{y}(t) = \Phi(t,0)\mathbf{c} = \Phi(t)\mathbf{c} \tag{2.20}$$

where $\Phi(t)$ is the transition matrix of the system (2.11). Hence the structure of the transition matrix determines the properties of the solutions for such systems.

Theorem 2.4: The transition matrix of the system (2.11) can be factored as

$$\Phi(t) = P(t)e^{tB} \tag{2.21}$$

where the matrix function $P(t)$ is periodic of period p and B is some $n\times n$ matrix with constant entries.

Corollary: $\Phi(t+p) = \Phi(t)e^{pB}$ (2.22)

In fact

$$\Phi(t+p) = P(t+p)e^{(t+p)B} = P(t)e^{tB}e^{pB} = \Phi(t)e^{pB} .$$

However, It is easy to see that the matrix B in the factorization (2.21) is not unique. Thus,

$$\Phi(t) = P(t)e^{tB} = [e^{2\pi i t/p}P(t)][e^{tB}e^{-2\pi i t/pI}]$$

$$= \tilde{P}(t)e^{t\tilde{B}} \tag{2.23}$$

where $\tilde{B} = B - \frac{2\pi i}{p} I$

Remark: To obtain the last equality in equation (2.23) we used the fact that if two matrices C, D commute then $e^C e^D = e^{C+D}$.

Nevertheless, in spite of this arbitrariness in the definition of B it can be shown that the eigenvalues of e^{pB} are unique. If we denote these eigenvalues by μ_i, $i = 1,\ldots n$ then there exist (since e^{pB} is nonsingular) $\lambda_1,\ldots,\lambda_n$ so that

$$\mu_i = e^{p\lambda_i} \quad . \tag{2.24}$$

The numbers $\lambda_1,\ldots,\lambda_n$ are called the characteristic or Floquet exponents of equation (2.11) while $\mu_1,\ldots,\mu_n$ are called the characteristic multipliers of this equation.

To characterize the behavior of the solutions of (2.11) more directly we state the following theorem;

Theorem 2.5: (1) For the system (2.11) if

$|\mu_i| < 1$, $i = 1,\ldots,n$ then all the solutions of this system satisfy

$$\lim_{t\to\infty} y(t) = 0$$

(2) The system (2.11) has nontrivial periodic solutions (of period p) if and only if some of the eigenvalues μ_i are equal to 1.

Example 2.2: Find the Floquet exponents for the system

$$\frac{d}{dt}\begin{bmatrix} y_1 \\ y_2 \end{bmatrix} = \begin{bmatrix} 0 & 1 \\ \cos t & \sin t \end{bmatrix}\begin{bmatrix} y_1 \\ y_2 \end{bmatrix} \tag{2.25}$$

and determine whether it has a periodic solution of period 2π.

Solution: The system (2.25) is equivalent to the second order equation

$$u'' - \sin t u' - \cos t u = 0 \quad . \tag{2.26}$$

The solution of this equation with $u(0) = 1$, $u'(0) = 0$ is given by

$$u_1(t) = e^{1-\cos t} \quad . \tag{2.27}$$

Hence, using standard reduction method we find that a second independent solution of (2.26) with $u(0) = 0$, $u'(0) = 1$ is

$$u_2(t) = e^{(1-\cos t)}\int_0^t e^{(\cos\tau-1)}d\tau \quad . \tag{2.28}$$

We deduce then that the general solution of the system (2.25) is

$$y_1 = c_1u_1 + c_2u_2$$

$$y_2 = y_1' \quad . \tag{2.29}$$

Since we know that the general solution of (2.25) can be written as

$$\mathbf{y} = \Phi(t)\mathbf{c}$$

we can use (2.29) to compute the transition matrix. We obtain

$$\Phi(t) = \begin{bmatrix} u_1 & u_2 \\ u_1 \sin t & 1+u_2 \sin t \end{bmatrix} .$$

Since $\Phi(t+2\pi) = \Phi(t)e^{2\pi B}$ and $\Phi(0) = I$ we infer that $(p=2\pi)$

$$e^{2\pi B} = \Phi(2\pi) = \begin{bmatrix} 1 & \int_0^{2\pi} e^{(\cos\tau-1)} d\tau \\ 0 & 1 \end{bmatrix} .$$

Thus, both μ_1 and μ_2 are equal to 1, i.e. $\lambda_1 = \lambda_2 = 0$ and the system must have a periodic solution. In fact it is easy to see that $u_1(t)$ is such a solution.

EXERCISE 2

1. Find the system $y' = Ay$ if the factorization of the fundamental matrix for the system is given by

   (a) $P(t) = \begin{bmatrix} \cos t & -\sin t \\ \sin t & \cos t \end{bmatrix}$, $B = \begin{bmatrix} 1 & 0 \\ 0 & -1 \end{bmatrix}$

   (b) $P(t) = \begin{bmatrix} \cos t & -\sin t \\ \sin t & \cos t \end{bmatrix}$, $B = \begin{bmatrix} 0 & 1 \\ 1 & 0 \end{bmatrix}$

   (c) $P(t) = \begin{bmatrix} e^{it} & e^{it}-e^{-it} \\ e^{it}-e^{-it} & e^{-it} \end{bmatrix}$ $B = \begin{bmatrix} 0 & 1 \\ -i & 0 \end{bmatrix}$

Hint: The fundamental matrix of the system satisfies $\Phi' = A\Phi$

2. Find the Floquet exponents and determine whether the following systems admit a periodic solution.

   a. $\dfrac{d\mathbf{y}}{dt} = \begin{bmatrix} 0 & 1 \\ -\cos t & -(1+\sin t) \end{bmatrix} \mathbf{y}$

Hint: A solution of this system is $y_1 = e^{t+\cos t}$, $y_2 = y_1'$

b. $$\frac{d\mathbf{y}}{dt} = \begin{bmatrix} 0 & 1 \\ -\sin t & 1+\cos t \end{bmatrix} \mathbf{y}$$

Hint: A solution of this system is $y_1 = e^{t+\sin t}$, $y_2 = y_1'$

3. Show that if

$$\mathbf{y}' = A(t)\mathbf{y} + \mathbf{f}(t)$$

is periodic of period p, i.e.

$A(t+p) = A(t)$ and $\mathbf{f}(t+p) = \mathbf{f}(t)$

then a solution $\mathbf{y}(t)$ of the system is periodic if and only if $\mathbf{y}(p) = \mathbf{y}(0)$.

Hint: Consider $\mathbf{x}(t) = \mathbf{y}(t+p) - \mathbf{y}(t)$ and show that $\mathbf{x}'(t) = A\mathbf{x}(t)$.

*3. HILL'S AND MATHIEU EQUATIONS

In general an equation of the form

$$(q_1(t)y'(t))' + q_2((t)y((t) = 0 \tag{3.1}$$

is called a Hill's equation if the functions $q_1(t)$, $q_2(t)$ are periodic of period p and $q_1(t)$ is continuous and nowhere zero. These equations are named after C. W. Hill who first investigated their properties in his work (in 1877) on the rotation of the apogee of the moon. A special case of Hill's equations is Mathieu equation

$$y'' + (a-2b \cos 2t)y = 0 \tag{3.2}$$

which appears in various application in Mechanics. In the following we discuss Mahtieu equation and its solutions since the treatment of this special case is somewhat easier (though similar) than the general case.

Theorem 3.1: The characteristic multipliers μ_1, μ_2 for a periodic equation

$$y'' + q(t)y = 0 \tag{3.3}$$

must satisfy $\mu_1 \cdot \mu_2 = 1$.

Proof: Let u_1, u_2 be a fundamental set of solutions of (3.3) satisfying the initial conditions

$$u_1(0) = 1, \qquad u_1'(0) = 0,$$
$$u_2(0) = 0, \qquad u_2'(0) = 1 \quad . \tag{3.4}$$

For these solutions it is easy to verify that the entries of the matrix A^T in equation (2.5) are

$$a_{11} = u_1(p), \qquad a_{12} = u_1'(p),$$
$$a_{21} = u_2(p), \qquad a_{22} = u_2'(p) \quad . \tag{3.5}$$

Hence A^T is the Wronskian of the solutions u_1, u_2 at p and

$$\det A = W(u_1, u_2)(p) = \exp\left(-\int_0^p \frac{a_1(t)}{a_2(t)}\, dt\right) = 1 \tag{3.6}$$

(since $a_1(t) = 0$ in (3.3)). It follows then that the characteristic equation for the μ_i's (eq. (2.8)) is

$$\mu^2 - (u_1(p) + u_2'(p))\mu + 1 = 0 \tag{3.7}$$

and, therefore, the roots of this equation (which are the characteristic multipliers of (3.3)) must satisfy $\mu_1 \cdot \mu_2 = 1$.

We now state without proof the following:

Theorem 3.2: For certain values of a in Mathieu equation $\mu_1 = \mu_2 = 1$.

From theorems 2.3 and (3.2) we deduce that for these values of a one of the solutions of eq. (3.2) must be periodic. However, the second solution of this equation is always not periodic. In fact from (2.14) we infer that when one of the solutions of (3.2) is periodic we can find two independent solutions of this equation in the form

$$w_1(t) = f_1(t), \quad w_2(t) = f_2(t) + tf_1(t) \tag{3.8}$$

where f_1, f_2 are periodic.

To find for which values of a equation (3.2) has a periodic solution and the nature of this solution we observe that:

Theorem 3.3: There exist nontrivial solutions of equation (3.2) which are either even or odd. Furthermore, equation (3.2) does not possess two independent solutions which are both even or odd.

Proof: First observe that if $u(t)$ is a solution of (3.2) which is neither even or odd then $u(-t)$ is also a solution of (3.2) which is independent of $u(t)$. Hence, $u(t) \pm u(-t)$ are even and odd solutions of (3.2).

Furthermore, if (3.2) possessed two independent solutions which are both even then the general solution of (3.2) will be even and it will be impossible to find a solution of this equation which satisfies the initial conditions $u(0) = 0$, $u'(0) = 1$. Similarly if (3.2) possessed two independent odd solutions it will be impossible then to find a solution with the initial conditions $u(0) = 1$, $u'(0) = 0$.

To obtain an explicit representation of these even and odd (periodic) solutions we write these proposed solutions in terms of an appropriate Fourier series with period π and 2π. viz.

$$\begin{matrix} C_e \\ S_e \end{matrix} = \sum_{k=0}^{\infty} c_k \begin{matrix} \cos 2kt \\ \sin 2kt \end{matrix} \quad \text{(period } \pi) \tag{3.9}$$

and

$$\begin{matrix} C_o \\ S_o \end{matrix} = \sum_{k=0}^{\infty} c_k \begin{matrix} \cos(2k+1)t \\ \sin(2k+1)t \end{matrix} \quad \text{(period } 2\pi) . \tag{3.10}$$

In the following we treat explicitly only the solutions in the form $C_0(t)$, however, similar treatment apply to the other types of solutions.

By substituting the expansion for $C_o(t)$ in (3.2) we obtain recursive relations for the coefficients c_k as follows:

$$(a-b-1)c_0 - bc_1 = 0$$

$$[(2k+1)^2-a]c_k + b(c_{k+1}+c_{k-1})$$
$$= 0, \quad k = 1,2,\ldots \quad . \tag{3.11}$$

A nontrivial solution of this (infinite) system of equations exists only if the determinant of the coefficients is zero, i.e.

$$\Delta(a,b) = \begin{vmatrix} a-1, & -b, & 0, & 0 & \ldots \\ -b & a-9 & -b & 0 & \ldots \\ 0 & -b & a-25 & -b & \\ & \cdot & \cdot \cdot \cdot & \cdot & \end{vmatrix} = 0 \tag{3.12}$$

We observe that when $b = 0$ $\Delta(a,0) = 0$ for $a = 1, 9, 25$, etc. Hence, for other values of b (e.g. $|b| << 1$) it is natural to look for solutions of this equation near these values.

Theorem 3.4: Approximate values of a for which nontrivial periodic solution in the form $C_o(t)$ exist are

$$a_1 = 1 + b + 0(b^2)$$

$$a_3 = 9 + \frac{b^2}{16} + 0(b^2)$$

$$a_{2k+1} = (2k+1)^2 + \frac{b^2}{8k(k+1)} + 0(b^2) \quad k = 2,3,\ldots \quad .$$

As to the nature of the second solution of (3.2) for these values of a we have

Theorem 3.5: If equation (3.2) admits a periodic solution $C_o(t)$ then the second solution of this equation can be written as

$$u_2(t) = tC_o(t) + \tilde{S}_0(t)$$

where $\tilde{S}_0((t) = \sum_{k=0}^{\infty} d_k \sin(2k+1)t.$

Finally, we note that the solutions of (3.2)(for appropriate a's) in the form $C_o(t)$, $S_0(t)$, $S_e(t)$, $S_e(t)$ and with appropriate normalization factors are referred to as Mathieu functions.

EXERCISE 3

1. Find the recursive equations for the coefficients of the solutions $S_0(t)$, $S_e(t)$, $S_e(t)$ of Mathieu equation. For which values of a these recursive equations have a nontrivial solution.
2. Discuss the nature of the solutions for equations in the form (3.3) in view of the fact that $\mu_1 \cdot \mu_2 = 1$.
3. Let $C_0(t) = \Sigma\, c_k \cos(2k+1)t$ be the periodic solution that corresponds to the approximate value of $a = 1 + b(|b| << 1)$. Choose $c_0 = 1$ and find the leading term in the perturbation expansion of $c_1, c_2 \dots$ in terms of b, i.e.

$$c_k = \alpha_{k1} b + \alpha_{k2} b^2 + \dots$$

4. Repeat exercise 3 for $a = 9 + \frac{b^2}{16}$.

BIBLIOGRAPHY

1. Whittaker and Watson - Modern Analysis, Cambridge Univ. Press.
2. N. W. McLachlen - Theory and application of Mathieu functions, Oxford Univ. Press 1947.
3. W. Magnus and S. Winkler - Hill's equation, Interscience, N.Y. 1966.
4. F. M. Arscott - Periodic differential equations, Pergammon Press, London, 1964.
5. M. S. P. Eastham - Theory of ordinary differential equations, Van Nostrand Reinhold, London, 1970.
6. E. L. Ince - Ordinary differential equations, Dover, NY, 1956.

CHAPTER 6. GREENS'S FUNCTIONS

1. INTRODUCTION

In many important applications it is desirable to find an integral representation for the solution to a boundary value problem. To achieve this goal we first discuss in this chapter "nonsmooth" solutions to such problems and then show how the existence of such solutions enable us to solve our original problem.

We motivate our treatment by a simple example.

Example 1: It is obvious that the boundary value problem

$$y'' + m^2 y = 0 \qquad m = 1,2,... \tag{1.1}$$

$$y(0) = y(2\pi) = 0 \tag{1.2}$$

has the solution $y(x) = \sin mx$. However, if the boundary conditions are changed to

$$y(0) = y(1) = 0 \tag{1.3}$$

then it is easy to see that the problem has no smooth nonzero solution. In fact the general solution of (1.1) is

$$y(x) = A \cos mx + B \sin mx \tag{1.4}$$

and the boundary condition $y(0) = 0$ implies that $A = 0$. It follows then that to satisfy the second boundary condition with $B \neq 0$ we must have $\sin m = 0$ which can not be satisfied. (If $\sin m = 0$ then $m = k\pi$ where k is an integer. This would have had implied that π is a rational number.) To obtain a solution of our problem we relax, therefore, the smoothness condition and look for a solution which is smooth everywhere on $[0,1]$ except at a point $\xi \in (0,1)$ where its first order derivative has a jump discontinuity of $+1$ at this point. (We shall interpret and motivate these conditions from an applied point of view in Section 3).

Since (1.4) is a fundamental solution of (1.1) it follows that for $x < \xi$ and $x > \xi$ the solution $G(x,\xi)$ we are seeking must satisfy

$$G(x,\xi) = \begin{cases} A_1 \cos mx + A_2 \sin mx, & x < \xi \\ B_1 \cos mx + B_2 \sin mx, & x > \xi \end{cases}. \quad (1.5)$$

The continuity and jump conditions on this solution at $x = \xi$ then yield

$$(A_1-B_1) \cos m\xi + (A_2-B_2) \sin m\xi = 0 \quad (1.6)$$

$$-m(A_1-B_1) \sin m\xi + m(A_2-B_2) \cos m\xi = -1 . \quad (1.7)$$

Introducing $C_1 = A_1 - B_1$, $C_2 = A_2 - B_2$ we see that the system (1.6)-(1.7) can be rewritten as

$$\begin{bmatrix} \cos m\xi & \sin m\xi \\ -\sin m\xi & \cos m\xi \end{bmatrix} \begin{bmatrix} C_1 \\ C_2 \end{bmatrix} = \begin{bmatrix} 0 \\ -\frac{1}{m} \end{bmatrix} \quad (1.8)$$

The determinant of the coefficients of this system (which is the Wronskian of the fundamental solution) is nonzero and thus the system (1.8) has a unique solution

$$\begin{bmatrix} C_1 \\ C_2 \end{bmatrix} = \begin{bmatrix} \cos m\xi & -\sin m\xi \\ \sin m\xi & \cos m\xi \end{bmatrix} \begin{bmatrix} 0 \\ -\frac{1}{m} \end{bmatrix} \quad (1.9)$$

i.e.

$$C_1 = \frac{\sin m\xi}{m}, \quad C_2 = -\frac{\cos m\xi}{m} . \quad (1.10)$$

Since this solution determines only the difference $A_1 - B_1$, $A_2 - B_2$ we now apply the boundary conditions (1.3) on (1.5) and obtain

$$y(0) = A_1 = 0$$

$$y(1) = B_1 \cos m + B_2 \sin m = 0 \quad . \tag{1.11}$$

Hence, we infer that

$$B_1 = -C_1 = \frac{-\sin m\xi}{m} \tag{1.12}$$

$$B_2 = \frac{\sin m\xi \cot m}{m} \tag{1.13}$$

and finally (from $C_2 = A_2 - B_2$)

$$A_2 = \frac{\sin m(\xi - 1)}{m \sin m} \quad .$$

Hence,

$$G(x,\xi) = \begin{cases} \dfrac{\sin m(\xi-1)}{m \sin m} \sin mx & , x < \xi \\ \dfrac{-\sin m\xi \cos mx}{m} + \dfrac{\sin m\xi \cot m}{m} \sin mx & , x > \xi \end{cases} \tag{1.14}$$

$G(x,\xi)$ is called the Green's function for the boundary value problem (1.1), (1.3).

In the following sections we shall discuss the definition and construction of solutions with similar properties to (1.14) in general setting and discuss their application to the solution of nonhomogeneous boundary value problems.

EXERCISE 1

For the following boundary value problems show that

a. There exists no smooth nonzero solution of the differential equation and boundary conditions on the indicated interval.

b. Find $G(x,\xi)$ which is smooth except at $x = \xi$ where its first order derivative has a jump discontinuity of $+1$.

1. $y'' - m^2y = 0$ $[0,1]$, $m = 1,2,\ldots$
$y(0) = y(1) = 0$.
2. $y'' + m^2y = 0$ on $[-1,1]$, $m = 1,2,\ldots$
$y(1) - y(-1) = 0$, $y'(1) - y'(-1) = 0$.
3. Repeat exercise 2 with the boundary conditions
$y(1) = y(-1) = 0$.
4. $x^2y'' - 2xy' + 2y = 0$ (Euler's equation) on
$[-1,1]$
$y(-1) = y'(1) = 0$

2. GENERAL DEFINITION OF GREEN'S FUNCTION.

Consider the linear homogeneous boundary value problem.

$$L(y) = p_n(x)y^{(n)} + \ldots + p_1(x)y' + p_0(x)y$$
$$= 0,\ p_n(x) \neq 0 \qquad (2.1)$$

$$D_i(y) = a_{i,0}y(a) + a_{i,1}y'(a)$$
$$+\ldots + a_{i,n-1}y^{(n-1)}(a) + b_{i,0}y(b)$$
$$+ b_{i,1}y'(b) + \ldots + b_{i,n-1}y^{(n-1)}(b)$$
$$= 0,\quad i = 1,\ldots,n \qquad (2.2)$$

on the interval $[a,b]$ where a_{ij}, b_{ij} are constants, $p_i(x)$ are smooth functions and the boundary conditions $D_i(y)$ are independent. Furthermore, assume that this system has no smooth nonzero solution on $[a,b]$.

DEFINITION 1: The Green's function $G(x,\xi)$ for the incompatible linear homogeneous system (1.1)-(1.2) is a function with the following properties;

a. $G(x,\xi)$ and its first $(n-2)$ derivatives are continuous on $[a,b]$
b. $G(x,\xi)$ satisfies equations (2.1), (2.2) for all $x \neq \xi$

c. The (n-1) derivative of $G(x,\xi)$ has a jump discontinuity of $\frac{1}{p_n(\xi)}$ at $x=\xi$, i.e.

$$\lim_{\epsilon\to 0}\left[G^{(n-1)}(\xi+\epsilon,\xi) - G^{(n-1)}(\xi-\epsilon,\xi)\right] = \frac{1}{p_n(\xi)} \quad (2.3)$$

We now demonstrate the existence of $G(x,\xi)$ and at the same time show how to compute it if a fundamental set of solutions $y_1(x) \ldots y_n(x)$ for (2.1) is known.

Since $G(x,\xi)$ is a solution of (2.1) for $x < \xi$ and $x > \xi$ it follows that there must exist constants A_i, B_i, $i = 1,\ldots,n$ so that

$$G(x,\xi) = \begin{cases} A_1y_1(x) + \ldots + A_ny_n(x) & x < \xi \\ B_1y_1(x) = \ldots + B_ny_n(x) & x > \xi \end{cases}. \quad (2.4)$$

From the conditions (a), (c) on $G(x,\xi)$ applied at $x = \xi$ we obtain the following system of equations

$$\sum_{i=1}^{n} (A_i-B_i)y_i^{(k)}(\xi) = 0 \quad , \quad k = 0,\ldots,n-2 \quad (2.5)$$

$$\sum_{i=1}^{n} (A_i-B_i)y_i^{(n-1)}(\xi) = \frac{-1}{p_n(\xi)} \quad . \quad (2.6)$$

Introducing $C_i = A_i - B_i$, $i = 1,\ldots,n$ we can rewrite this system in matrix form as

$$\begin{bmatrix} y_1(\xi) & ,\ldots, & y_n(\xi) \\ y_1'(\xi) & ,\ldots, & y'(\xi) \\ y_1^{(n-1)}(\xi) & ,\ldots, & y_n^{(n-1)}(\xi) \end{bmatrix} \begin{bmatrix} C_1 \\ \vdots \\ C_n \end{bmatrix} = \begin{bmatrix} 0 \\ \vdots \\ \frac{-1}{p_n(\xi)} \end{bmatrix}. \quad (2.7)$$

We observe that the coefficient matrix in (2.7) is the Wronskian of the fundamental set of solutions to (2.1) and hence, nonsingular viz. the system (2.7) has a unique solution. Assuming, therefore, that the C_i's are known we now compute the A_i, B_i using the boundary conditions (2.2). We obtain

$$D_i(G) = \sum_{i=0}^{n-1} a_{i,j} G^{(i)}(a,\xi) + \sum_{j=0}^{n-1} b_{i,j} G^{(i)}(b,\xi) = 0. \tag{2.8}$$

Using (2.4) this yields

$$\sum_{k=1}^{n} A_k \left[\sum_{j=0}^{n-1} a_{i,j} y_k^{(j)}(a) \right] + \sum_{k=1}^{n} B_k \left[\sum_{i=0}^{n-1} b_{i,j} y_k^{(j)}(b) \right] = 0. \tag{2.9}$$

adding and subtracting

$$\sum_{k=1}^{n} B_k \left[\sum_{j=0}^{n-1} a_{i,j} y_k^{(j)}(a) \right]$$

to (2.9) leads to

$$\sum_{k=1}^{n} C_k \left[\sum_{j=0}^{n-1} a_{i,j} y_k^{(j)}(a) \right] + \sum_{k=1}^{n} B_k D_i(y_k) = 0, \qquad i = 1,\ldots n \quad . \tag{2.10}$$

This is a system of equations for the B_k whose coefficient matrix is

$$D = (D_i(y_k))$$

However, since $y_1,\ldots,y_n$ is a fundamental set of solutions to (2.1) and the system (2.1)-(2.2) is incompatible we infer that D is nonsingular and hence, the solution of (2.10) is unique. We, thus, proved the follwoing:

Theorem 1: If the system (2.1)-(2.2) is incompatible then the Green's function for the system exists and is unique.

Furthermore if the system (2.1)-(2.2) is self-adjoint it is possible to show that $G(x,\xi)$ is symmetric on $[a,b] \times [a,b]$, i.e.

$$G(x,\xi) = G(\xi,x) \tag{2.11}$$

Example 1: For the system (1.1), (1.3) we showed that the Green's function is given by (1.14). Since (1.1), (1.3) is self-adjoint $G(x,\xi)$ must be symmetric. In fact it is easy to verify that

$$G(\xi,x) = \begin{cases} \dfrac{\sin m(x-1)}{m \sin m} \sin m\xi & , \ \xi < x \\ \dfrac{-\sin mx \cos m}{m} + \dfrac{\sin mx \cot m}{m} \sin m\xi, & \xi > x \end{cases} \tag{2.12}$$

satisfies (2.11).

Example 2: The Green's function for the second order equation

$$p_2(x)y'' + p(x)y' + p_0(x)y = 0, \quad p_2(x) \neq 0 \tag{2.13}$$

on $[a,b]$ with appropriate boundary conditions can be constructed easily if a fundamental set of solutions $y_1(x)$, $y_2(x)$ for (2.13) is known. To construct $G(x,\xi)$ we first introduce the function

$$g(x,\xi) = Ay_1(x) + By_2(x) \pm f(x,\xi) \tag{2.14}$$

where the plus sign is taken for $x < \xi$ and the minus sign for $x > \xi$ and

$$f(x,\xi) = \frac{y_1(x)y_2(\xi)-y_2(x)y_1(\xi)}{2p_2(\xi)[y_1(\xi)y_2'(\xi)-y_2(\xi)y_1'(\xi)]} . \tag{2.15}$$

Obviously, for any fixed $\xi, g(x,\xi)$ is continuous and is a solution of (2.13) for $x \neq \xi$. Moreover, it is easy to verify that $g'(x,\xi)$ has a jump discontinuity of $\dfrac{1}{p_2(\xi)}$ at $x = \xi$. Thus $g(x,\xi)$ satisfies all the requirement for being a Green's function of (2.13) except the boundary condition. However, we can take care of these conditions by appropriate choice of the constants A,B which yields then the desired result.

EXERCISE 2

1. Show that the systems in exercises 1-4 of the

previous section are self-adjoint and verify that the Green's functions are symmetric.

2. Compute the Greenn's function for the system

$y''' + 2y'' - y' - 27 = 0$

$y(0) = y'(0) = 0, \quad y(1) = 0$

3. Repeat exercise 2 for the system

$y''' - y'' - y' + y = 0$

$y(0) = y'(0) = 0, \quad y(1) = y'(1)$

4. Find the Green's function for

$y'' = 0, \quad y(0) = y(a) = 0$

5. Explain in detail why there exist a unique non-zero solution for $B_1, \ldots, B_n$ in equation (2.10).

Hint: Write the first term in (2.10) as $A \cdot W \cdot C$, where $A = (a_{ij})$, W the Wronskian of $y_1, \ldots, y_n$ at a and $C^T = (C_1, \ldots, C_n)$.

3. THE INTERPRETATION OF GREEN'S FUNCTIONS

In many physical theories the relationship between cause and effect is a linear one. Thus, if a certain source is causing a certain effect then by doubling the source strength we double the effect. Similarly if we have several sources then the total effect due to these sources is equal to the sum of the effects due to the individual sources. In general this principle is referred to as the principle of superposition.

Example 1: Newton's second law states that $F = ma$, i.e. the acceleration of a mass m is proportional to the force acting on m. Therefore, if we double the force ($\equiv$ cause) we shall double the effect ($\equiv$ acceleration).

Example 2: By Newton's law of Gravitation the attractive force between a point mass M at $\boldsymbol{\xi}$ and a point mass m at $\mathbf{x}$ is given by

$$\mathbf{F} = \frac{-GMm}{r^3}\,\mathbf{r} \qquad (3.1)$$

where $\mathbf{r} = \mathbf{x} - \boldsymbol{\xi}$. If, however, we consider several point masses M_i, $i = 1,\dots,n$ then the total effect on m will be given by sum

$$\mathbf{F} = -\sum_{i=1}^{n} \frac{GM_i m}{r_i^3}\,\mathbf{r}_i \qquad (3.2)$$

where $\mathbf{r}_i = x - \boldsymbol{\xi}_i$. Thus, the total effect ($\equiv$ force) is the sum of the effects from each of the sources ($\equiv$ masses).

Example 3: Consider a taut metal string which is rigidly attached at $x = 0, a$. By applying a small force F on the string at $x = \xi$ it will be deflected (see Figure 1) and we have the relation

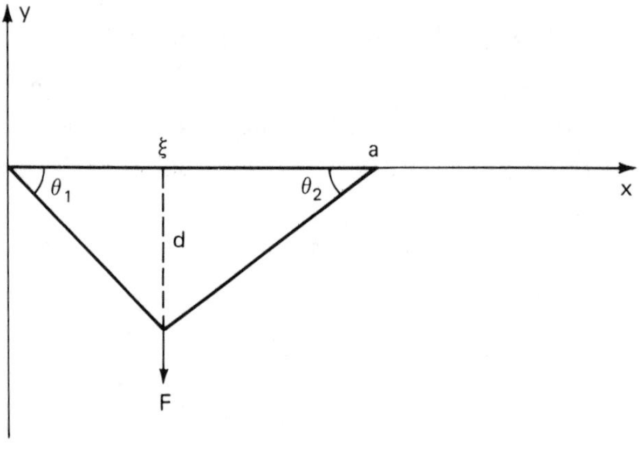

Figure 1

$$T\sin\theta_1 + T\sin\theta_2 = F \qquad (3.3)$$

where T is the tension in the string.
(We assume that the deflection is so small that every

point on the string is constrained to move only in the vertical direction). Since $|\theta_1|, |\theta_2| << 1$ we can apply the approximation $\sin\theta \sim \tan\theta$ and deduce from figure 1 that

$$\sin\theta_1 = \frac{d}{\xi}, \quad \sin\theta_2 = \frac{d}{a-\xi} \quad . \tag{3.4}$$

Hence, from (3.3) we obtain that

$$d = \frac{F}{Ta}\,\xi(a-\xi) \quad . \tag{3.5}$$

Thus, the effect (≡ deflection) on the string due to the force F is given by $y(x) = FG(x,\xi)$ where

$$G(x,\xi) = \begin{cases} \frac{x}{Ta}\,(a-\xi), & 0 < x < \xi \\ \frac{\xi}{Ta}\,(a-x), & \xi < x < a \end{cases} \quad . \tag{3.6}$$

Therefore, we can interpret $G(x,\xi)$ as the effect of a unit force acting at ξ on the string. By the superposition principle we now deduce that the effect caused by several such forces acting at points ξ_i is given by

$$y(x) = \sum_{i=1}^{n} F_i G(x,\xi_i) \quad . \tag{3.7}$$

Moreover, if the loading on the string is continuous, i.e. the force per unit length at x is given by (a continuous function) $F(x)$ then the total effect is given approximately by

$$y(x) \simeq \Sigma(F(\xi)\Delta\xi)G(x,\xi) \quad . \tag{3.8}$$

(To obtain this result we divide $[0,a]$ into subintervals of length $\Delta\xi$. The loading on each of these subintervals is given approximately by $F(\xi)\Delta\xi$.). Letting $\Delta\xi \to 0$ we obtain in the limit

$$y(x) = \int_0^a F(\xi)G(x,\xi)d\xi \quad . \tag{3.9}$$

Thus, whenever the total effect due to a continuous distribution of sources $F(x)$ is given by a formula of the form (3.9) we can interpret $G(x,\xi)$ as the effect at x due to a unit source located at ξ.

We now show that $G(x,\xi)$ is proportional to the Green's function for the homogeneous part of the differential equation which governs the system above.

In fact if we differentiate (3.9) with respect to x and use (3.6) we obtain

$$\frac{dy}{dx} = \int_0^x \frac{F(\xi)(a-\xi)}{Ta}\, d\xi + \int_x^a F(\xi)\left(\frac{-\xi}{Ta}\right) d\xi \ . \tag{3.10}$$

By differentiating (3.10) we infer that $y(x)$ satisfies the differential equation.

$$\frac{d^2y}{dx^2} = \frac{F(x)}{T}\,, \quad y(0) = y(a) = 0 \ . \tag{3.11}$$

A short computation then shows that $G(x,\xi)$ is proportional to the Green's function for

$$y'' = 0, \quad y(0) = y(a) = 0 \tag{3.12}$$

In more general setting we now prove the following theorem;

Theorem 1: Let $G(x,\xi)$ be the Green's function for the incompatible boundary value problem (2.1)-(2.2). The solution of

$$L(y) = f(x), \quad D_i(y) = 0, \quad i = 1,\ldots,n \tag{3.13}$$

is given by

$$y(x) = \int_a^b G(x,\xi) f(\xi)\, d\xi \ . \tag{3.14}$$

Proof: Differentiating (3.14) with respect to x we obtain

$$y^{(k)}(x) = \int_a^b \frac{\partial^k G(x,\xi)}{\partial x^k} f(\xi) d\xi, \quad k = 1,\ldots,n-2. \tag{3.15}$$

Furthermore, since $G^{(n-2)}(x,\xi)$ is uniformly continuous on $[a,b]$ (and its $(n-1)$ derivative is discontinuous only at one point) it follows that

$$y^{(n-1)}(x) = \int_a^b \frac{\partial^{(n-1)} G(x,\xi)}{\partial x^{n-1}} f(\xi) d\xi \ . \tag{3.16}$$

Finally, we obtain $y^{(n)}(x)$ as follows:

$$y^{(n)}(x) = \frac{\partial}{\partial x}\int_a^b \frac{\partial^{(n-1)}G(x,\xi)}{\partial x^{n-1}} f(\xi)d\xi =$$

$$= \frac{\partial}{\partial x}\left[\int_a^x \frac{\partial^{n-1}G(x,\xi)}{\partial x^{n-1}} f(\xi)d\xi + \int_x^b \frac{\partial^{n-1}G(x,\xi)}{\partial x^{n-1}} f(\xi)d\xi\right]$$

$$= \lim_{\epsilon\to 0}\left[\int_a^{x-\epsilon} \frac{\partial^{n}G(x,\xi)}{\partial x^{n}} f(\xi)d\xi + \int_{x+\epsilon}^b \frac{\partial^{n}G(x,\xi)}{\partial x^{n}} f(\xi)d\xi\right]$$

$$+ \frac{\partial^{n-1}G(x,\xi)}{\partial x^{n-1}} f(\xi)\Bigg|_{x-\epsilon}^{x+\epsilon} =$$

$$= \int_a^b \frac{\partial^{n}G(x,\xi)}{\partial x^{n}} f(\xi)d\xi + \frac{f(x)}{p_n(x)} \quad . \tag{3.17}$$

Collecting (3.15), (3.16), (3.17) we infer that

$$L(y) = \int_a^b L(G)f(\xi)d\xi + f(x) = f(x) \ . \tag{3.18}$$

Since G satisfies (2.1) for all x except $x = \xi$.

Similarly since each of the boundary conditions contains derivatives up to order $(n-1)$ only we deduce from (3.15), (3.16) and (2.2) that

$$D_i(y) = \int_a^b D_i(G)f(\xi)d\xi = 0 \quad . \tag{3.19}$$

Since $D_i(G) = 0$. This completes the proof of the theorem.

We can generalize this result to nonhomogeneous boundary value problems as follows:

Theorem 2: Let $G(x,\xi)$ be the Green's function of (2.1), (2.2) and let y_k be the solution of (2.1) with the boundary conditions

$$D_i(y) = \begin{cases} 0 & i \neq k \\ 1 & i = k \end{cases} \quad . \tag{3.20}$$

The solution of

$$L(y) = f(x), \quad D_i(y) = c_i = \text{constant} \tag{3.21}$$

is given by

$$y(x) = \int_a^b G(x,\xi)f(\xi)d\xi + c_1y_1(x) + \ldots + c_ny_n(x). \tag{3.22}$$

Proof: Obvious from theorem 1.

EXERCISE 3

1. The superposition principle for solutions of differential equations states that if y_1, y_2 are solutions of a given differential equation then $c_1y_1 + c_2y_2$ where c_1, c_2 are constants is also a solution of this differential equation. Does this principle hold for

   $$y'' + yy' + 2xy = 0 \quad .$$

   Explain.

2. The superposition principle does not hold for the Ricatti equation

   $$y' + p(x)y^2 + q(x)y + g(x) = 0$$

   (Why?). However, show that if y_1, y_2, y_3 are three independent solutions of this equation then $y(x)$ which is defined by

   $$\frac{y-y_1}{y-y_2} = c\,\frac{y_3-y_1}{y_3-y_2}$$

   (c is a constant) is also a solution of Ricatti equation. This shows that some nonlinear differential equations admit a "nonlinear" superposition principle.

4. Use the Green's function method to solve

   a. $y'' + 4y = \sin 5x$

   $y(0) = y(1) = 0$

b. $y'' - 4y = \cosh 2x$, $y(0) = y(1) = 0$

c. $x^2y'' + xy' + 4y = 0$, $y(1) = y(2) = 0$

5. Repeat exercise 4 with the boundary conditions

$$y(0) = y(1) = 1$$

for a,b and

$$y(1) = y(2) = 1$$

for part c.

4. GENERALIZED FUNCTIONS.

The study of generalized functions in general and the Dirac δ function in particular is intimately connected to Green's functions. In this section we define the Dirac δ function (and some related generalized functions) while in the next section we study its relation to Green's functions.

Consider the following sequence of functions

$$g_n(x) = \begin{cases} \frac{n}{2} & \text{on } [-\frac{1}{n}, \frac{1}{n}] \\ 0 & \text{elsewhere} \end{cases} \tag{4.1}$$

$n = 1,2,\ldots$. Then it is easy to verify that

$$\int_{-\infty}^{\infty} g_n(x)dx = 1, \quad n = 1,2,\ldots$$

Furthermore, if $f(x)$ is continuous then by the mean value theorem for integrals we obtain

$$\int_{-\infty}^{\infty} f(x)g_n(x)dx = \frac{n}{2}\int_{-1/n}^{1/n} f(x)dx = f(\xi_n) \tag{4.2}$$

where $\xi_n \in [-\frac{1}{n}, \frac{1}{n}]$.

Remark: The mean value theorem for integrals states that under proper assumptions on $f(x)$ there exists $c \in (a,b)$ so that

$$\int_a^b f(x)dx = (b-a)f(c) \quad . \tag{4.3}$$

Taking the limit of equation (4.2) as $n \to \infty$ we infer that

$$\lim_{n\to\infty} \int_{-\infty}^{\infty} f(x)g_n(x)dx = f(0) \quad . \tag{4.4}$$

Thus, although the sequence $\{g_n(x)\}$ has no classical limit, since $\lim_{n\to\infty} g_n(0) = \infty$ and $\lim_{n\to\infty} g_n(x) = 0$ for $x \neq 0$, the limit of the integrals in (4.2) is well defined. It is "natural", therefore, to introduce a limit function $\delta(x)$ which would have had existed if the interchange of the integral and limit operations in equation (4.4) were permissible viz.

$$\int_{-\infty}^{\infty} \delta(x)f(x)dx = f(0) \tag{4.5}$$

where $\delta(x)$ is called the Dirac δ-function.

We observe, however, that one can construct many other sequences of functions which "converge" to $\delta(x)$. In particular although the functions $\{g_n(x)\}$ are not smooth there exist other sequences of smooth functions which yield the same result.

Example 1: Consider the sequence of normal distributions

$$F_n(x) = \frac{n}{\sqrt{2\pi}} e^{-(nx)^2/2} \quad , \quad n = 1,2,\ldots \quad . \tag{4.6}$$

then for all $n = 1,2,\ldots$

$$\int_{-\infty}^{\infty} F_n(x)dx = 1 \tag{4.7}$$

and yet for all $x \neq 0$ we have $\lim_{n\to\infty} F_n(x) = 0$. Hence, the same procedure which we applied to the sequence $\{g_n(x)\}$ yields

$$\lim_{n\to\infty} \int_{-\infty}^{\infty} F_n(x)f(x)dx = f(0) \quad . \tag{4.8}$$

Thus, to make the definition of the generalized function (or distribution) $\delta(x)$ sequence independent we introduce the following definition;

1. $\delta(x) = 0$ for $x \neq 0$
2. $\delta(x)$ has a singularity at $x = 0$
3. $\int_{-\infty}^{\infty} \delta(x)dx = 1$.

From these three properties it then follows that

$$\int_{-\infty}^{\infty} \delta(x)f(x)dx = \int_{-1/n}^{1/n} \delta(x)f(x)dx$$

and since $f(x)$ is continuous we can approximate it by $f(\xi_n)$, $\xi_n \in [-\frac{1}{n}, \frac{1}{n}]$.

Hence,

$$\int_{-\infty}^{\infty} f(x)\delta(x)dx = \lim_{n\to\infty} f(\xi_n) \int_{-1/n}^{1/n} \delta(x)dx = f(0) \tag{4.9}$$

We enumerate some of the properties of the δ function in the following:

Theorem 1:

1. $\int_{-\infty}^{\infty} f(x)\delta(x-x_0)dx = f(x_0)$ (4.10)
2. $\int_{-\infty}^{\infty} f(x)\delta(-x)dx = f(0)$ (4.11)
3. $\int_{-\infty}^{\infty} f(x)\delta(ax)dx = \frac{1}{a}f(0) \quad a > 0$ (4.12)

All these properties can be proved by a simple change of variable e.g. to prove (4.9) introduce $\bar{x} = x - x_0$ and substitute

$$\int_{-\infty}^{\infty} f(x)\delta(x-x_0)dx = \int_{-\infty}^{\infty} f(\bar{x}+x_0)\delta(\bar{x})d\bar{x} = f(x_0) \quad . \tag{4.13}$$

Since $\delta(x)$ is not continuous in the classical sense it is not differentiable either. However, one can define the generalized derivatives of $\delta(x)$ to any order by repeated (formal) application of the formula for integration by parts;

$$\int_{-\infty}^{\infty} f(x)\delta'(x)dx = f(x)\delta(x)\Big|_{-\infty}^{\infty}$$

$$- \int_{-\infty}^{\infty} f'(x)\xi(x)dx = -f'(0). \quad (4.14)$$

Similarly it is easy to show that;

$$\int_{-\infty}^{\infty} f(x)\delta^{(n)}(x)dx = (-1)^n f^{(n)}(0) \quad . \quad (4.15)$$

We can define, therefore, $\delta^{(n)}(x)$, $n = 1,2,...$ as generalized functions which are singular at $x = 0$, zero for $x \neq 0$ and satisfy equation (4.15).

We now relate $\delta(x)$ to the generalized derivative of a "classical" function.

Theorem 2: The generalized derivative of the Heaviside function

$$H(x) = \begin{cases} 1 & x \geq 0 \\ 0 & x < 0 \end{cases} \quad (4.16)$$

is $\delta(x)$.

Proof: Using integration by parts we formally obtain

$$\int_{-\infty}^{\infty} f(x)H'(x)dx = f(x)H(x)\Big|_{-\infty}^{\infty} - \int_{-\infty}^{\infty} f'(x)H(x)dx =$$

$$= - \int_{0}^{\infty} f'(x)dx = f(x)\Big|_{0}^{\infty} = f(0) \quad , \quad (4.17)$$

where we used the fact that $f(x)$ is assumed to be integrable over $[-\infty,\infty]$ and hence $\lim_{x\to\infty} f(x) = 0$.

Since (4.17) must hold for arbitrary $f(x)$ we conclude then that $H'(x) = \delta(x)$.

Remark: The n-dimensional δ-function is defined as

$$\delta(x-a) = \delta(x_1-a_1)\cdot...\cdot\delta(x_n-a_n) \quad (4.18)$$

where $x = (x_1,...,x_n)$, $a = (a_1,...,a_n)$. It is obvious from this definition that

$$\int_{R_n} f(x)\delta(x-a)dx = f(a) \quad . \quad (4.19)$$

Finally, to conclude this section we review the convolution operation and its properties.

Definition 1: The convolution of two functions f,g on $[-\infty,\infty]$ is defined as

$$(f*g)(x) = \int_0^x f(\xi)g(x-\xi)d\xi \quad . \tag{4.20}$$

The basic properties of this operation are

1. $f * g = g * f$ (commutativity) (4.21)
2. $(af_1+bf_2) * g = a(f_1*g) + b(f_2*g)$ (linearity) (a,b are constants) (4.22)
3. $\delta * f = f$ (4.23)

To verify the first property we introduce $\bar{\xi} = x - \xi$ and obtain;

$$f * g = \int_0^x f(\xi)g(x-\xi)d\xi = \int_x^0 f(x-\bar{\xi})g(\bar{\xi})(-d\bar{\xi})$$

$$= \int_0^x f(x-\bar{\xi})g(\bar{\xi})d\bar{\xi} = g * f \quad . \tag{4.24}$$

The other properties can be verified in a similar manner.

EXERCISE 4

1. Prove formulas (4.11), (4.12).
2. Show that the sequence
$$f_n(x) = \frac{n}{\pi(1+n^2x^2)} \quad , \quad n = 1,2, \ldots$$
converge to $\delta(x)$.
3. Express the generalized derivative of
$$f_n(x) = \begin{cases} 0 & x < 0 \\ 2 & 1 \le x < 3 \\ 1 & 3 \le x < \infty \end{cases}$$
in terms of Dirac δ-functions. Generalize to a function with n jump discontinuities.
4. Find a function $G(x)$ whose generalized derivatives is $H(x)$.

5. Show that

$$x^m\delta^{(n)}(x) = \begin{cases} 0 & n < m \\ \dfrac{(-1)^m m!\,\delta^{(n-m)}(x)}{(m-n)!} & m \le n \end{cases}$$

(where $\delta^{(0)}(x) = \delta(x)$).

6. Show that the derivatives of the functions $f_n(x)$ in exercise 2 converge to $\delta'(x)$.

7. Verify the properties of the convolution given by equations (4.21), (4.22), and (4.23)

5. ELEMENTARY SOLUTIONS AND GREEN'S FUNCTIONS

In this section we study the connection between generalized functions and Green's functions.

Definition 1: Let L be a linear differential operator

$$L = p_n(x)\frac{d^n}{dx^n} + \ldots + p_1(x)\frac{d}{dx} + p_0(x). \tag{5.1}$$

with $p_n(0) \neq 0$. An elementary solution for L is a solution of

$$Lu = \delta(x) \quad . \tag{5.2}$$

Since no boundary conditions were imposed on the elementary solutions it is obvious that if u is such a solution and v is an arbitrary solution of $Lv = 0$ then $u + v$ is an elementary solution.

To see how one constructs elementary solutions let $u_1,\ldots,u_n$ be a fundamental set of solutions for $Lu = 0$ then since $\delta(x) = 0$ for $x \neq 0$ it follows that

$$u = \begin{cases} \alpha_1u_1 + \ldots + \alpha_nu_n & x > 0 \\ \beta_1u_1 + \ldots + \beta_nu_n & x < 0 \end{cases} \quad . \tag{5.3}$$

To satisfy the requirement $L(u) = \delta(x)$ let $u,\ldots,u^{(n-1)}$ be continuous at 0 while $u^{(n-1)}$ to have a jump discontinuity of $\frac{1}{p_n(0)}$ there. Then since $H'(x) = \delta(x)$ we infer that the term $p_n(x)\frac{d^n}{dx^n}$ will be equivalent to $\delta(x)$ at $x = 0$. Imposing these conditions on (5.3) yields

$$\begin{aligned} &\gamma_1 u_1(0) + \dots + \gamma_n u_n(0) = 0 \\ &\quad \cdots \\ &\gamma_1 u_1^{(n-2)}(0) + \dots + \gamma_n u_n^{(n-2)}(0) = 0 \\ &\gamma_1 u^{(n-1)}(0) + \dots + \gamma_n u^{(n-1)}(0) = -\frac{1}{p_n(0)} \end{aligned} \tag{5.4}$$

where $\gamma_i = \alpha_i - \beta_i$. Since the system (5.4) is the same as the one derived in Section 2 for the construction of Green's function with $\xi = 0$ (and without the imposition of the boundary conditions) we conclude that;

Theorem 1: The Green's function for the system (2.1)-(2.2) is given by the elementary solution of

$$Lu = \delta(x-\xi) \tag{5.5}$$

which satisfies the boundary conditions (2.2).

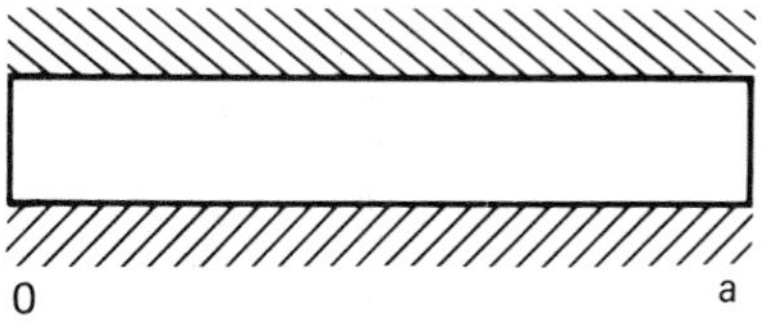

Figure 2
Heat conducting rod with lateral insulation

Example 1: Find the Green's function for the steady state heat conduction in a laterally insulated rod with variable conductivity whose ends are kept at constant temperature. (see Figure 2).

Solution: A mathematical model for the heat conduction problem described above if given by

$$\rho c \frac{\partial u}{\partial t} = \frac{\partial}{\partial x} (k(x) \frac{\partial u}{\partial x}) + r(x) \tag{5.6}$$

$$u(0,t) = 0, \quad u(a,t) = T \tag{5.7}$$

where $u = u(x,t)$ is the temperature in the rod, $k(x)$ is the conductivity, $r(x)$ is the rate of heat production per unit length, ρ is the mass density and c is the specific heat of the rod material.

Remark: Since the ends of the rod are at constant temperature we can always let one of these to be zero by a translation of the temperature scale.

In the steady state $\partial u/\partial t = 0$, i.e. $u = u(x)$ and the partial differential equation (5.6) reduces to

$$(k(x)u')' = -r(x) \quad . \tag{5.8}$$

The general solution for the homogeneous part of this equation is

$$u(x) = au_1(x) + bu_2(x) \tag{5.9}$$

where

$$u_1(x) = \int_0^1 1/k(s)\, ds, \quad u_2(x) = 1 \ . \tag{5.10}$$

To find the Green's function for (5.7)-(5.8) we must solve

$$(k(x)G'(x,\xi))' = \delta(x-\xi) \tag{5.11}$$

subject to the boundry conditions

$$u(0) = 0, \quad u(a) = T \quad . \tag{5.12}$$

To do so we integrate (5.11)(remembering that $H'(x-\xi) = \delta(x-\xi)$) to obtain

$$k(\xi)[G'(\xi+,\xi) - G'(\xi-,\xi)] = 1 \quad . \tag{5.13}$$

Furthermore, since $G(x,\xi)$ is continuous we must have

$$G(\xi+,\xi) - G(\xi-,\xi) = 0 \quad . \tag{5.14}$$

Writing $G(x,\xi)$ in the form

$$G(x,\xi) = \begin{cases} A_1u_1(x) + A_2u_2(x) & x < \xi \\ \\ B_1u_1(x) + B_2u_2(x) & x > \xi \end{cases} \tag{5.15}$$

we can solve for A_i, B_i, $i = 1,2$ using (5.12), (5.13), (5.14). This yields;

$$B_1 = \frac{T+\int_0^\xi k^{-1}(s)ds}{\int_0^a k^{-1}(s)ds}, \quad B_2 = -\frac{1}{T}\int_0^\xi k^{-1}(s)ds$$

$$A_1 = B_1 - 1, \quad A_2 = 0 \tag{5.16}$$

From theorem (3.1) we now infer that the solution of (5.7)-(5.8) is

$$u(x) = -\int_0^a G(x,\xi)r(\xi)d\xi \quad . \tag{5.17}$$

Thus, $-G(x,\alpha)$ represents the steady state temperature in the rod due to a unit source of heat located at $x = \alpha$.

EXERCISE 5

1. If

$$L = \frac{d^2}{dx^2} + 2\frac{d}{dx} - 3$$

find an elementary solution of $Lu = \delta(x)$ and a particular solution of $Lu = \sin x$.

2. Repeat exercise 1 with

$$L = x^2\frac{d^2}{dx^2} + 3x\frac{d}{dx} - 3$$

3. Use the Laplace transform to solve for

$$y'' + 5y' + 4y = \delta(x-1)$$
$$y(0) = 1, \ y'(0) = 0$$

6. EIGENFUNCTION REPRESENTATION OF GREEN'S FUNCTIONS

It is useful in some applications to represent the Green's function of the system (2.1)-(2.2) in terms of an infinite series which consists of the eigenfunctions of a related differential operator. In this section we discuss this technique for second order differential operators but the generalization of these ideas to higher order differential operators is straightforward (at least in principle).

Suppose we want to compute the Green's function for a second order (self-adjoint) operator

$$L = \frac{d}{dx}\left(p(x)\frac{d}{dx}\right) + (q(x)+k) = L_0 + k \tag{6.1}$$

subject to some homogeneous boundary conditions on $[a,b]$. Moreover assume that the eigenvalues and eigenfunctions λ_n, ψ_n, $n = 1,2,\ldots$ of

$$(L_0 + \lambda)\psi(x) = 0 \tag{6.2}$$

(we assume $k \neq \lambda_n$ for all n) subject to the same boundary conditions as L are known and satisfy the following;

1. $\psi_m(x)$ are orthogonal to each other on $[a,b]$ with respect to a weight function $\rho(x)$, i.e.

$$\int_a^b \psi_m(x)\psi_n(x)\rho(x)dx = c_n^2\delta_{mn} \tag{6.3}$$

where c_n^2 is a constant and δ_{mn} is the Kronecker delta

$$\delta_{mn} = \begin{cases} 1 & m = n \\ 0 & m \neq n \end{cases} \tag{6.4}$$

2. $\{\psi_m\}$ form a complete set of functions on $[a,b]$, i.e. for any well behaved function $f(x)$ satisfying the same boundary conditions as $\psi_m(x)$ we can find a_m so that

$$f(x) = \sum_m a_m \psi_m(x) \tag{6.5}$$

on $[a,b]$. To compute the coefficients a_m in (6.5) we multiply both sides of equation (6.4) by $\rho(x)\psi_n(x)$ (where n is arbitrary but fixed) and integrate over $[a,b]$. From (6.3) we then obtain

$$\int_a^b f(x)\psi_n(x)\rho(x)dx = a_n \int_a^b \psi_n^2(x)\rho(x)dx \tag{6.6}$$

i.e.

$$a_n = \frac{1}{c_n^2} \int_a^b f(x)\psi_n(x)\rho(x)dx \quad . \tag{6.7}$$

Returning to our original problem of computing the Green's function $G(x,\xi)$ for L on $[a,b]$ with the given boundary conditions we now expand $G(x,\xi)$ in terms of $\psi_m(x)$

$$G(x,\xi) = \sum_m a_m \psi_m(x) \tag{6.8}$$

where for the moment we assume that ξ is fixed but arbitrary, i.e $a_m = a_m(\xi)$. Substituting (6.8) in

$$LG(x,\xi) = \delta(x-\xi) \tag{6.9}$$

and using (6.2) we obtain

$$\sum_m (k-\lambda_m) a_m \psi_m(x) = \delta(x-\xi) \quad . \tag{6.10}$$

Multiplying both sides of (6.9) by $\rho(x)\psi_n(x)$ (where n is arbitrary but fixed) and integrating over $[a,b]$ we obtain

$$(k-\lambda_n) a_n c_n^2 = \rho(\xi)\psi_n(\xi) \tag{6.11}$$

i.e.

$$a_n = \frac{\rho(\xi)\psi_n(\xi)}{(k-\lambda_m) c_n^2} \quad . \tag{6.12}$$

Hence

$$G(x,\xi) = \sum_m \frac{\rho(\xi)\psi_m(\xi)\psi_m(x)}{(k-\lambda_m)c_m^2} \quad . \tag{6.13}$$

This function satisfies equation (6.9) and the required (homogeneous) boundary conditions since each $\psi_m(x)$ satisfies these conditions hence it is the required Green's function for the problem. It follows then (from the discussion of section 3) that the solution of

$$Ly = f \tag{6.14}$$

subject to the same boundary conditions as above is given by

$$y(x) = \int_a^b G(x,\xi)f(\xi)d\xi =$$

$$= \sum_m \frac{\psi_m(x)}{(k-\lambda_m)c_m^2} \int_a^b \psi_m(\xi)f(\xi)\rho(\xi)d\xi \quad . \tag{6.15}$$

Example 1: Find the Green's function for

$$y'' + k^2 y = 0 \qquad k \neq n\pi \tag{6.16}$$

$$y(0) = y(1) = 0 \tag{6.17}$$

Solution: The eigenvalues and eigenfunctions of

$$y'' + \lambda y = 0 \tag{6.18}$$

with the boundary conditions (6.17) are

$$\lambda_n = (n\pi)^2 \quad , \quad \psi_n(x) = \sin(n\pi x), \; n = 1,2,\ldots \quad . \tag{6.19}$$

Moreover, these eigenfunctions are orthogonal to each other on $[0,1]$ with $\rho = 1$ and $c_n^2 = \frac{1}{2}$. Hence,

$$G(x,\xi) = 2 \sum_m \frac{\sin(m\pi x)\sin(m\pi\xi)}{(k^2-\lambda_m)} \tag{6.20}$$

The procedure described above can be extended to singular Sturm-Liouville systems where the boundary conditions imposed on the solutions are those of boundedness.

Example 2: Find the Green's function for (Legendre equation)

$$(1-x^2)y'' - 2xy' + ky = 0 \quad , \quad k \neq n(n+1)$$ where n is interger which is bounded on $[-1,1]$

Solution: The eigenvalues and eigenfunctions of this singular Sturm-Liouville system are given by

$$\lambda_n = n(n+1) \quad , \quad \psi_n(x) = P_n(x) \tag{6.22}$$

where $P_n(x)$ are the Legendre polynomials. Moreover,

$$P_n(1) = 1 \quad , \quad P_n(-1) = (-1)^n \tag{6.23}$$

and

$$\int_{-1}^{1} P_n(x)P_m(x)dx = \frac{2}{2m+1}\,\delta_{mn} \quad . \tag{6.24}$$

Hence $\rho(x) = 1$ and

$$G(x,\xi) = \sum_m \frac{(2m+1)}{2}\,\frac{P_m(\xi)P_m(x)}{(k-\lambda_m)} \tag{6.25}$$

EXERCISE 6

1. Find the eigenfunction representation for the Green's function of

$$y'' - 2xy' + ky = 0 \tag{6.26}$$

on $[-\infty,\infty]$, $k \neq 2n$ with the boundary conditions

$$\lim_{x\to\pm\infty} y(x) = 0$$

Hint: Equation (6.26) is Hermite equation. If eigenvalues and eigenfunctions are

$$\lambda_n = 2n \quad , \quad \psi_n = e^{-x^2/2}H_n(x) \qquad n = 1,2,\ldots$$

where $H_n(x)$ are Hermite polynomials.

2. Find the eigenfunction representation of the Green's function for the following regular. Sturm-Liouville systems;

   a. $u'' - 2u'\ Lu = 0 \quad , \quad u'(0) = u'(\pi) = 0$
   b. $x^2u + xu' + \lambda u = 0 \quad , \quad u(1) = u(e) = 0$

3. Solve using Green's functions the following equations

   a. $y'' + \lambda y = \sin x \cos 2x$, y is bounded on $[-\infty, \infty]$

   b. $y'' - y' + \lambda y = xe^{x/2}$ $y(0) = y(1) = 0$

   c. $x^2 y'' + xy' + \lambda y = \frac{1}{x}$, $y(1) = y(e) = 0$

7. INTEGRAL EQUATIONS

Integral equations relates an unknown function with its integral. The interest in such equations stems from the fact that in some instances physical problems are conceptually easier to model in terms of integral rather than differential equations (e.g scattering problems).

In this section we discuss the relationship between Green's functions and integral equations and describe some methods for the solution of these equations.

Theorem 1: If the system (2.1)-(2.2) is incompatible then the solution of

$$L(y) = q(x)y + f(x)$$
$$D_i(y) = c_i \quad , \quad i = 1, \ldots, n \tag{7.1}$$

where the c_i's are constants is given by

$$y(x) = c_1 y_1 + \ldots + c_n y_n + \int_a^b G(x,\xi)[q(\xi)y(\xi) + f(\xi)]d\xi = \int_a^b K(x,\xi)y(\xi)d\xi + F(x) \quad . \tag{7.2}$$

In this equation $G(x,\xi)$ is the Green's function for the system (2.1)-(2.2) and $y_1(x), \ldots, y_n(x)$ are the soltuions fo equation (2.1) with the boundary conditions (3.20). We observe that equation (7.2) is an integral equation for $y(x)$ since it appears both inside and outside the integral sign. The proof of this theorem follows the same steps as theorem 1 in section 3.

Corollary 1: The solution of

$$L(y) = \lambda y + f(x) \quad , \quad D_i(y) = 0, \; i = 1 \ldots n \tag{7.3}$$

is given by

$$y(x) = F(x) + \lambda \int_a^b G(x,\xi)y(\xi)d\xi \tag{7.4}$$

where

$$F(x) = \int_a^b G(x,\xi)f(x)d\xi \quad . \tag{7.5}$$

Definition 1: An integral equation in the form

$$p(x)y(x) = F(x) + \lambda \int_a^x K(x,z)y(z)dz \tag{7.6}$$

is called a Fredholm type integral equation. Similarly an equation of the form

$$p(x)y(x) = F(x) + \lambda \int_a^x K(x,z)y(z)dz \tag{7.7}$$

is called a Volterra (type) equation. In these equations $y(x)$ is the unknown function while $p(x)$, $F(x)$ and $K(x,z)$ are known and λ is a parameter. The function $K(x,z)$ is called the Kernel of the integral equation and if $K(x,z) = K(z,x)$ we say that the integral equation is symmetric. Furthermore, we say that equations (7.6), (7.7) are of the first kind if $p(x) = 0$ and of the second kind if $p(x) = 1$.

Corollary 2: It is apparent from the definitions that equation (7.4) is a Fredholm equation of the second kind.

Definition 2: Equations (7.6), (7.7) are said to be homogeneous if $F(x) = 0$.

Solution of Integral Equations:

A. Solution by iterations.

Integral equations of the second kind (regardless of type) under proper restrictions on $K(x,z)$ and λ

can be solved by iterations. In the following we describe this method explicitly for Fredholm equations of the second kind (equation (7.6) with $p = 1$).

Starting with $y_0(x) = F(x)$ we compute $y_1(x)$ as

$$y_1(x) = F(x) + \lambda \int_a^b K(x,z)y_0(z)dz \quad . \tag{7.8}$$

Repeating the process with $y_1(x)$ we now define $y_2(x)$, $y_3(x)$, etc. or in general

$$y_n(x) = F(x) + \lambda \int_a^b K(x,z)y_{n-1}(z)dz \tag{7.9}$$

Theorem 2: If the sequence of functions $y_n(x)$ converges then the limit function $u(x)$ is a solution of the integral equation (7.6).

Proof: To prove this theorem we introduce the operator T as

$$Tf(x) = \int_a^b K(x,z)f(z)dz \quad . \tag{7.10}$$

Equation (7.6) then takes the form

$$y(x) = F(x) + \lambda Ty(x) \tag{7.11}$$

and for $y_n(x)$ we obtain

$$y_n(x) = F(x) + \lambda Ty_{n-1} = F(x) + \lambda T(F(x) + \lambda Ty_{n-2})$$

$$= \ldots = F(x) + \sum_{k=1}^{n} \lambda^n T^n F(x) \tag{7.12}$$

and as $n \to \infty$ this yields;

$$u(x) = \lim_{n\to\infty} y_n(x) = F(x) + \sum_{n=1}^{\infty} \lambda^n T^n F(x) \quad . \tag{7.13}$$

To show that this function satisfies equation (7.6) we observe that

$$F(x) + \lambda Tu(x) = F(x) + \lambda T(F(x) + \sum_{n=1}^{\infty} \lambda^n T^n F(x))$$

$$= F(x) + \sum_{n=1}^{\infty} \lambda^n T^n F(x) = u(x) \tag{7.14}$$

which proves this theorem.

The next theorem gives sufficient conditions for the convergence of this iterative process.

Theorem 3: If $F(x)$ is bounded on $[a,b]$ (i.e. $|F(x)| < m$ for some $m > 0$) and $|K(x,z)| < M$ for all $x,z \in [a,b]$ then the series (7.13) converges for

$$|\lambda| < \frac{1}{M(b-a)} \quad . \tag{7.15}$$

We shall not prove this theorem.

Example 1: Use the iterative method to solve

$$y(x) = 1 + \lambda \int_0^2 (1- \frac{3}{4}\, xz)y(z)dz \tag{7.16}$$

Solution: Starting with $y_0(x) = F(x) = 1$ we obtain

$$TF(x) = \int_0^2 (1 - \frac{3}{4}\, xz)dz = 2 - \frac{3}{2}\, x$$

$$T^2F(x) = \int_0^2 (1 - \frac{3}{4}\, xz)(2 - \frac{3}{2}\, z)dz = 1$$

and hence $T^3F = TF$, etc. It follows then from (7.13) that the required solutions of (7.16) is given by

$$(u(x) = 1 + \lambda(2 - \frac{3}{2}\, x) + \lambda^2 + \lambda^3(2 - \frac{3}{2}\, x) + \lambda^4$$

$$+ \ldots = [1+\lambda(2 - \frac{3}{2}\, x)][1+\lambda^2 + \ldots]$$

$$= \frac{1+\lambda(2-\frac{3}{2}x)}{1-\lambda^2} \quad , |\lambda| < 1 \tag{7.17}$$

B. Solution by means of the Laplace transform

When the kernel $K(x,z)$ is in the form

$$K(x,z) = K(x-z) \tag{7.18}$$

the integral equation (7.7) with $p = 1$ can be represented, using the convolution operation, as

$$y = F + \lambda K * y \quad . \tag{7.19}$$

By applying the Laplace transform $\mathcal{L}$ to this equation (and remembering that $\mathcal{L}(f*g) = (\mathcal{L}f)(\mathcal{L}g)$) we obtain

$$\mathcal{L}(y) = \mathcal{L}(F) + \lambda\mathcal{L}(K)\mathcal{L}(y) \tag{7.20}$$

i.e.

$$\mathcal{L}(y) = \frac{\mathcal{L}(F)}{1-\lambda\mathcal{L}(K)} \quad . \tag{7.21}$$

Thus, $y(x)$ can be computed by standard techniques for the evaluation of the inverse Laplace transform.

Example 2: Solve

$$y(x) = \sin 2x - \int_0^x (x-z)y(z)dz \tag{7.22}$$

Solution: In this equation $\lambda = -1$ and $K(x,z) = K(x-z) = x - z$, i.e. $K(x) = x$. Applying the Laplace transform we obtain

$$\mathcal{L}(y) = \mathcal{L}(\sin 2x) - \mathcal{L}(x)\mathcal{L}(y) \quad .$$

But

$$\mathcal{L}(\sin 2x) = \frac{2}{s^2+4}, \quad \mathcal{L}(x) = \frac{1}{s^2} \quad .$$

Hence

$$\mathcal{L}(y) = \frac{\mathcal{L}(\sin 2x)}{1+\mathcal{L}(x)} = \frac{2s^2}{(s^2+4)(s^2+1)} = \frac{2}{3}\left[\frac{4}{s^2+4} - \frac{1}{s^2+1}\right].$$

Evaluating the inverse this yields

$$y(x) = \frac{4}{3} \sin 2x - \frac{2}{3} \sin x. \tag{7.23}$$

C. Equations with separable Kernel.

Definition: The Kernel of an integral equation is said to be separable if

$$K(x,z) = \sum_{i=1}^{n} \varphi_i(x)\psi_i(z) \quad , \quad n < \infty \quad . \tag{7.24}$$

When the Kernel of a Fredholm equation (with $p = 1$) is separable its solution is equivalent to the solution of a system of linear equations. To see this we substitute (7.24) in (7.6) and obtain

$$y(x) = F(x) + \lambda \int_a^b \sum_{i=1}^{n} \varphi_i(x)\psi_i(z)y(z)dz$$

$$= F(x) + \lambda \sum_{i=1}^{n} c_i\varphi_i(x) \tag{7.25}$$

where

$$c_i = \int_a^b \psi_i(z)y(z)dz \quad . \tag{7.26}$$

It follows then from (7.25) that the solution of the original equation will be known if we can compute the

c_i's. To evaluate these constants multiply equation (7.25) by $\psi_m(x)$ (where m is fixed but arbitrary) and integrate over [a,b]. We obtain

$$c_m = b_m + \lambda \sum_{i=1}^{n} a_{mi} c_i \tag{7.27}$$

where

$$b_m = \int_a^b f(x)\psi_m(x)dx,$$

$$a_{mi} = \int_a^b \varphi_i(x)\psi_m(x)dx. \tag{7.28}$$

Applying this procedure with $m = 1,\ldots,n$ we obtain a system of linear equations which can be written in the form

$$(I-\lambda A)\mathbf{c} = \mathbf{b} \tag{7.29}$$

where $A = (a_{mi})$, $\mathbf{b} = (b_i)$ and $\mathbf{c} = (c_i)$. Since the matrix A and the vector **b** are known equation (7.29) represents a system of linear equations for **c** which can be solved by standard techniques.

Example 3: Solve the homogeneous equation

$$y(x) = \lambda\int_0^{2\pi} \sin(mx+nz)y(z)dz \tag{7.30}$$

where m,n are integers.

Solution: Since

$$\sin(mx+nz) = \sin mx \cos nz + \cos mx \sin nz$$

the kernel of eq. (7.30) is separable with

$$\varphi_1(x) = \sin mx \quad , \quad \varphi_2(x) = \cos mx$$

$$\psi_1(z) = \cos nz \quad , \quad \psi_2(z) = \sin nz \quad . \tag{7.31}$$

Hence from (7.25)

$$y(x) = \lambda(c_1 \text{sim}x + c_2 \cos mx) \quad . \tag{7.32}$$

Multiplying (7.32) by $\psi_1(x)$, $\psi_2(x)$ and integrating over $[0,2\pi]$ we obtain

$$c_1 = \lambda c_2 \pi \delta_{mn} \quad , \quad c_2 = \lambda c_1 \pi \delta_{mn}$$

where δ_{mn} is the Kronecker delta. Thus $y(x) = 0$ unless $m = n$. In the latter case we have (eq. (7.28)).

$$a_{11} = a_{21} = 0 \quad , \quad a_{12} = a_{21} = \pi .$$

Using (7.29) this can be rewritten in matrix form as

$$A = \begin{bmatrix} 0 & \pi \\ \pi & 0 \end{bmatrix} \quad , \quad b = 0 \quad \text{and} \quad c = \begin{bmatrix} c_1 \\ c_2 \end{bmatrix} . \tag{7.33}$$

It follows then that (7.29) in this particular case reduces to a homogeneous system. This system has a nontrivial solution for c only if $1/\lambda$ is an eigenvalue of A, i.e. $\lambda = \frac{1}{\pi}, -\frac{1}{\pi}$. Thus, we obtain the following two (independent) solutions;

1. $\lambda = \frac{1}{\pi}$. Hence $c_1 = c_2$ and

$$y(x) = c(\sin mx + \cos mx), c \text{ arbitrary constant.}$$

2. $\lambda = -\frac{1}{\pi}$. Hence $c_1 = -c_2$ and

$$y(x) = c(\sin mx - \cos mx) \quad . \tag{7.34}$$

Finally we observe that though in practice most Kernels are not separable we can approximate them by such a Kernel and obtain an approximate solution for the integral equation.

EXERCISE 7

1. Show that if $p(x) > 0$ on $[a,b]$ then equations (7.6), (7.7) can be transformed into an equation of the second kind, i.e. with $p = 1$.

Hint: Divide (7.6) (or (7.7)) by $\sqrt{p(x)}$ and define $\tilde{y} = \sqrt{p(x)y(x)}$

2. Use the Laplace transform to solve

a. $y(x) = \sin x + \int_0^x (x-z)^2 y(z)dz$

b. $y(x) = x^2 + \int_0^x (x-z)y(z)dz$.

3. Solve by iterations and by reduction to a system of linear equations the following integral equations

a. $y(x) = k + \lambda \int_0^1 xzy(z)dz$, k is a constant

b. $y(x) = 1 + \lambda \int_0^1 |x-z|y(z)dz \qquad 0 < \lambda \ll 1$

4. Solve

$$y(x) = ax + \lambda \int_0^{2\pi} \cos(x+z)y(z)dz$$

5. Obtain an approximate solution for

$$y(x) = x + \lambda \int_0^1 \cos(xz)y(z)dz \qquad 0 < \lambda \ll 1$$

Hint: Use a truncated Taylor expansion for $\cos(xz)$.

BIBLIOGRAPHY

1. I Stakgold - Green's functions and Boundary Value Problems, Wiley, New York, NY (1979).
2. F. B. Hildebrand - Methods of applied mathematics, 2nd edition Prentice Hall, Englewood Cliffs, N.J. (1965).
3. E. L. Ince - Ordinary differential equations, Dover, New York, NY (1956).
4. G. F. Roach - Green's functions, 2nd edition, Cambridge University Press, London 1982.
5. A Hochstadt - Integral equations, Wiley, New York, NY 1973.
6. A. J. Jerri - Introduction to integral equations with applications, M. Dekker, New York, NY (1985).

CHAPTER 7 PERTURBATION THEORY

Many mathematical models of real life systems are approximations. These approximations are generally made to simplify the model equations and make them mathematically tractable. However, after the initial successes of such approximations in predicting the general behavior of the system under consideration attention must turn to an examination of the effects of these approximations on the general behavior of the system and its equilibrium states. In most cases, however, the new additional terms in the model equations lead to nonlinearities which preclude any attempt to solve these equations in a closed form. As a result a practical technique had evolved to estimate the effect of these additional terms through some perturbation expansion. In this chapter we shall give a brief introduction to this branch of applied mathematics mostly through examples.

1. PRELIMINARIES.

In perturbation theory we shall find it necessary to solve inhomogeneous differential equations viz.

$$Lu = f(t) \quad , \quad L = \sum_{i=1}^{n} a_i(t) \frac{d^i}{dt^i} \quad . \tag{1.1}$$

The general solution of such equations is given by a sum

$$u = v + u_p$$

where v is the general solution of the homogeneous equation $Lv = 0$ and u_p is some particular solution of equation (1.1). Since we assume that the reader is well versed with the solution of homogeneous equation we shall review here briefly the necessary background to solve for u_p.

First if $v_1, \ldots, v_n$ is a fundamental set of solutions for the equation $Lv = 0$ then it is possible to compute u_p by variation of parameters, i.e. assume u_p in the form

$$u_p = c_1(t)v_1(t) + \ldots + c_n(t)v_n(t) \tag{1.2}$$

substitute in (1.1) and obtain a system of equations for the c_i's whose solution yields u_p.

Example 1: Let v_1, v_2 be a fundamental set of solutions for

$$\ddot{u} + p\dot{u} + qu = f(t), \ \dot{u} = \frac{du}{dt} \tag{1.3}$$

To find u_p we let

$$u_p = c_1(t)v_1(t) + c_2(t)v_2(t) \tag{1.4}$$

and substitute in (1.3). We obtain the system

$$\dot{c}_1 v_1 + \dot{c}_2 v_2 = 0$$

$$\dot{c}_1 \dot{v}_1 + \dot{c}_2 \dot{v}_2 = f(t) \quad . \tag{1.5}$$

whose solution provide us with the required answer. We point out, however, that in actual practice this method is rarely used for the inhomogeneous equations that appear in perturbation theory. This is due to the fact that in this context $f(x)$ is usually related to the solutions of $Lv = 0$ or an equation that is closely related to it. Hence it is possible in many cases to find u_p directly by making a "proper ansatz". We demonstrate this by several examples.

Example 2: To find u_p for

$$\ddot{u} + k^2 u = a \sin \omega t, \quad k \neq \omega \tag{1.6}$$

we let $u_p = A \sin \omega t$ where A is undetermined parameter. Substituting this in equation (1.6) we obtain

$$A(-\omega^2+k^2) = a \tag{1.7}$$

and hence

$$u_p = \frac{a}{k^2-\omega^2} \sin \omega t \quad . \tag{1.8}$$

If, however, $k = \omega$ we try $u_p = At \cos kt$. We obtain $A = -\frac{a}{2k}$, i.e.

$$u_p = -\frac{at}{2k} \cos kt.$$

Example 3: Find u_p for

$$\ddot{u} + \dot{u} = c(1-e^{-t}) \quad . \tag{1.9}$$

Solution: Since $u = a$ and $u = ae^{-t}$ are solutions of the homogeneous equation we try to find u_p in the form

$$u_p = At + Bte^{-t} \tag{1.10}$$

substituting this in (1.9) we obtain $A = c$, $B = c$ hence

$$u_p = ct(1+e^{-t}) \tag{1.11}$$

This procedure can be applied in many instances to partial differential equations.

Example 4: Find a particular solution for

$$\nabla^2 u = f(r) \tag{1.12}$$

on the unit disk in R^2.

Solution: In polar coordinates

$$\nabla^2 u = \frac{1}{r}\frac{\partial}{\partial r}\left(r\frac{\partial u}{\partial r}\right) + \frac{1}{r^2}\frac{\partial^2 u}{\partial\theta^2} \tag{1.13}$$

we, therefore, try $u_p = g(r)$. Substituting this in (1.12) we obtain

$$rg''(r) + g'(r) = rf(r) \tag{1.14}$$

which can be solved using the methods described above.

Example 5: Find a particular solution for

$$\nabla^2 u = -r\sin^2\theta = -\frac{1}{2}r\sin 2\theta \tag{1.15}$$

on the unit disk in R^2.

Solution: Since the right-hand side of (1.15) depends on θ and r we try

$$u_p = g(r)\sin 2\theta \tag{1.16}$$

substituting this in (1.15) we obtain

$$r^2g'' + rg' - 4g = -\frac{r^3}{2} \quad . \tag{1.17}$$

The homogeneous part of (1.17) is an Euler type equation and hence we try to solve it in the form $g_p = r^\alpha$ which yield $\alpha = \pm 2$. To find a particular solution we try $g_p = cr^3$ and obtain after substitution $c = -\frac{1}{10}$. Hence

$$u_p = [Ar^2 + Br^{-2} - \frac{1}{10}r^3]\sin 2\theta \quad . \tag{1.18}$$

However, if we assume that u is bounded on the unit disk we must set $B = 0$ since r^{-2} has a singularity at $r = 0$. Thus finally

$$u_p = (Ar^2 - \frac{1}{10}r^3)\sin 2\theta \tag{1.19}$$

Remark: The general solution of $\nabla^2 u = 0$ on the disk $r \leq a$ is given by

$$u(r,\theta) = a_0 + \sum_{n=1}^{\infty} r^n(a_n\cos n\theta + b_n\sin n\theta), \tag{1.20}$$

where a_0, a_1, b_i $i = 1,2,\ldots$ are arbitrary.

2. SOME BASIC IDEAS-REGULAR PERTURBATIONS:

When a system of differential equations represent a model for some real life system the parameters and variables in the equations have some dimensional dependence, e.g., length, time, mass, etc.. Consequently, the equations under consideration will change if the units of these quantities are changed.

Example 1: Suppose that the model equation for some physical system is

$$\frac{d^2x}{dt^2} = 10x + x^2 \qquad (2.1)$$

where the time t and the distance x are measured in seconds and meters respectively. If, however, one changes the unit length to kilometers, i.e. introduce $\bar{x} = x/1000$ then (2.1) will transform into

$$\frac{d^2\bar{x}}{dt^2} = 10\bar{x} + 1000\bar{x}^2 \quad . \qquad (2.2)$$

Thus, it might seem as though the relative importance of the linear and quadratic terms in (2.1) and (2.2) is different and this might lead to errors in judgment regarding the qualitative behavior of the system.

To avoid such possible pitfalls especially when several variables, parameters and units are involved it is advantageous to bring the model equation into a nondimensional form. In this form the parameters and variables in the equations are pure numbers and, therefore, are independent of any changes in scale of the physical units.

Although we shall not attempt to formalize this process its basic steps can be described as follows;

1. List all relevant variables and parameters in the model with their dimension.

2. For each variable x_i form a combination p_i of the parameters which has the same dimension as x_i.

3. Reformulate the model equations in term of the nondimensional variables

$$\tilde{x}_i = x_i/p_i \quad .$$

To illustrate this process we consider the following example;

Example 2: The basic model equation for a spring-mass system is

$$m \frac{d^2x}{dt^2} + b \frac{dx}{dt} + kx = 0 \tag{2.3}$$

In this equation x, t have the dimensions of length and time respectively while the parameters m, b, k have the dimensions of mass, mass/time and mass/(time)^2 (remember that all terms in an equation must have the same dimensionality). Hence we can nondimensionalize the time variable by introducing

$$\tau = \omega_0 t \quad , \qquad \omega_0 = \sqrt{\frac{k}{m}} \quad .$$

Equation (2.3) transforms then into

$$\frac{d^2x}{dt^2} + \beta \frac{dx}{d\tau} + x = 0 \tag{2.4}$$

where $\beta = b/(m\omega_0)$ is a pure number.

Note that we did not scale x nor this is necessary since any transformation of the form $\bar{x} = \alpha x$ will not affect equation (2.3)(or (2.4)). Thus, equation (2.4) is invariant with respect to any changes in scale in x.

Remark: Even if one chooses to work with the dimensional form of an equation it is expedient to choose units so that the magnitude of the quantities in question is neither too large nor too small. This is especially important when the equations are solved numerically.

Thus, if one considers Kapler's laws for the earth-sun system it is natural to define the unit mass as the earth's mass and the average distance between the earth and the sun as the unit length (rather than deal with grams, cms, etc.).

As a result of this nondimensionalization process some of the parameters in the model equations might turn out to be small ($\ll 1$) while others will be large (≥ 1). The essence of perturbation theory (or "asymptotic expansions") is to investigate how the addition of the terms with the small parameters affects the solution of the equations without these terms.

To introduce the basic ideas of perturbation theory in general and the method of regular perturbations in particular we consider the following simple example (which can be solved also exactly).

Example 3: Consider a system which was modeled initially by the equation

$$\dot{x} + 2x = 0 \quad , \quad x(0) = 1 \tag{2.5}$$

but this was refined later to

$$\dot{x} + 2x + \epsilon x^2 = 0 \quad , \quad x(0) = \cosh \epsilon \tag{2.6}$$

where ϵ, the perturbation parameter, satisfies $0 < \epsilon \ll 1$.

Since ϵ is a small parameter the basic idea of perturbation theory is to find the solution of equation (2.6) in terms of a power series in ϵ, i.e. in the form

$$x(t,\epsilon) = x_0(t) + \epsilon x_1(t) + \epsilon^2 x_2(t) + \ldots \quad . \tag{2.7}$$

Substituting (2.7) in equation (2.6) we obtain

$$(\dot{x}_0+2x_0) + \epsilon(\dot{x}_1+2x_1+x_0^2) + \epsilon^2(\dot{x}^2+2x_2+2x_0x_1)$$

$$+ \ldots = 0 \tag{2.8}$$

$$x_0(0) + \epsilon x_1(0) + \epsilon^2 x_2(0) + \ldots = \cosh\epsilon$$

$$= 1 + \frac{\epsilon^2}{2} + \frac{\epsilon^4}{24} + \ldots \tag{2.9}$$

Equations (2.8), (2.9) can be considered as polynomials in ϵ and hence each coefficient of ϵ^n must be zero.

Thus, we obtain the following coupled system of equations

$$\begin{array}{llll} \epsilon^0 & \dot{x}_0 + 2x_0 = 0 & , & x_0(0) = 1 \\ \epsilon^1 & \dot{x}_1 + 2x_1 = -x_0^2 & , & x_1(0) = 0 \\ \epsilon^2 & \dot{x}_2 + 2x_2 = -2x_0x_1 & , & x_2(0) = \frac{1}{2} \end{array} \tag{2.10}$$

etc. We observe that although (2.20) form a coupled system of differential equations the coupling is trivial and the equations can be solved in sequential order.

The solutions we obtain are the following:

$$x_0(t) = e^{-2t} \tag{2.11}$$

$$x_1(t) = \frac{1}{2}(e^{-4t}-e^{-2t}) \tag{2.12}$$

$$x_2(t) = \frac{1}{4}(3e^{-2t}+e^{-6t}-2e^{-4t}) \quad . \tag{2.13}$$

Hence,

$$x(t) = e^{-2t} + \frac{\epsilon}{2}(e^{-4t}-e^{-2t}) + \frac{\epsilon^2}{4}(3e^{-2t}+e^{-6t}-2e^{-4t}) + \ldots \quad . \tag{2.14}$$

From (2.14) we see that the coefficient of each ϵ^n in this expansion is bounded and hence since $0 < \epsilon << 1$ the contribution of the (n+1) term is small compared to the nth term. In view of this we may truncate the expansion after the second term and consider this as an approximate solution to equation (2.6).

Remarks: 1. If by chance we obtain $x_1(t) = 0$ in the perturbation expansion we then use $x_0(t) + \epsilon^2 x_2(t)$ as the approximate solution. In a more general way; if $x_1(t) = \ldots = x_{n-1}(t) = 0$ then the approximate solution is taken to be $x_0(t) + \epsilon^n x_n(t)$.

2. Note that the perturbation expansion replaced equation (2.6) which is nonlinear by a system of nonhomogeneous, but linear equations. This is one of

the major advantages of this approach since there is no general technique to solve nonlinear equations.

The technique outlined above for one equation can also be used to derive a perturbation solution to a system of equations as we demonstrate in the following example:

Example 4: Solve the system

$$\frac{dx}{dt} = -2x + y + \epsilon y^2$$
$$\frac{dy}{dt} = x - 2y + \epsilon x^2 \tag{2.15}$$
$$x(0) = y(0) = 1$$

to first order in ϵ where $0 < \epsilon << 1$.

Solution: Substituting the perturbation expansion

$$x = x_0 + \epsilon x_1 + \dots$$
$$y = y_0 + \epsilon y_1 + \dots$$

in the system (2.15) we obtain to order ϵ^0

$$\frac{dx_0}{dt} = - 2x_0 + y_0$$
$$\frac{dy_0}{dt} = x_0 - 2y_0 \tag{2.16}$$
$$x_0(0) = y_0(0) = 1$$

and to order ϵ^1

$$\frac{dx_1}{dt} = - 2x_1 + y_1 + y_0^2$$
$$\frac{dy_1}{dt} = x_1 - 2y_1 + x_0^2 \tag{2.17}$$
$$x_1(0) = y_1(0) = 0 \quad .$$

To solve these systems we can either use standard techniques for the solution of systems of differential equations or in this special case observe that the equations and the boundary conditions are symmetric in x_0, y_0 and x_1, y_1. Thus, we infer that

$$x_0(t) = y_0(t) \quad , \quad x_1(t) = y_1(t) \quad .$$

Hence,

$$x_0(t) = y_0(t) = e^{-t}$$

$$x_1(t) = y_1(t) = e^{-t} - e^{-2t} \quad . \tag{2.18}$$

Therefore, to first order in ϵ the solution of the system (2.15) is given by

$$x(t) = y(t) = e^{-t} + \epsilon(e^{-t} - e^{-2t}) \quad . \tag{2.19}$$

Example 5: Bending of light rays by a star

Until the advent of Einstein's theory of General Relativity it was thought that light rays move always in straight lines. One of the major predictions of general relativity was that light rays will bend by the gravitational field of a star (or, in fact, any material body) and the verification of this prediction constituted a major triumph for this theory.

To see how this prediction was made we note that according to general relativity the trajectory of a light ray coming from infinity near a star is given by

$$w'' + w = \epsilon w^2 \quad , \quad w(\tfrac{\pi}{2}) = 1/r_0 \tag{2.20}$$

where $w = 1/r(\theta)$ and r is the distance to the star center (see Fig. 1). Writing out the usual perturbation expansion for w

$$w(\theta, \epsilon) = w_0(\theta) + \epsilon w_1(\theta) + \ldots \tag{2.21}$$

and substituting in (2.20) we obtain to first order in ϵ

$$\epsilon^0; \qquad w_0'' + w_0 = 0 \quad , \quad w_0(\pi/2) = 1/r_0 \tag{2.22}$$

$$\epsilon^1; \qquad w_1'' + w_1 = w_0^2, \quad w_1(\pi/2) = 0 \quad . \tag{2.23}$$

Solving equations (2.22)-(2.23) we obtain

$$w = \frac{1}{r_0} \sin\theta + \frac{\epsilon}{2r_0^2}(1 + \frac{1}{3} \cos 2\theta) = \ldots \; . \tag{2.24}$$

The first term in (2.24) represents a straight line trajectory while the second the deviation from this path. For large distances (i.e. asymptotically) $w = 0$ and $0 < \theta << 1$ (or $\pi - \theta << 1$). Hence the deviation angle δ of the ray from $\theta = \frac{\pi}{2}$ to $\theta \sim 0$ must satisfy $(\sin\theta \sim \delta$ at $w = 0)$

$$\delta = \frac{-2\epsilon}{3r_0} .$$

It should be observed, however, that the total deflection of the light ray is given by the angle between the two asymptotes at $\pm\infty$. Since the system is symmetric with respect to the x-axis we infer that the total deviation of the ray is 2δ.

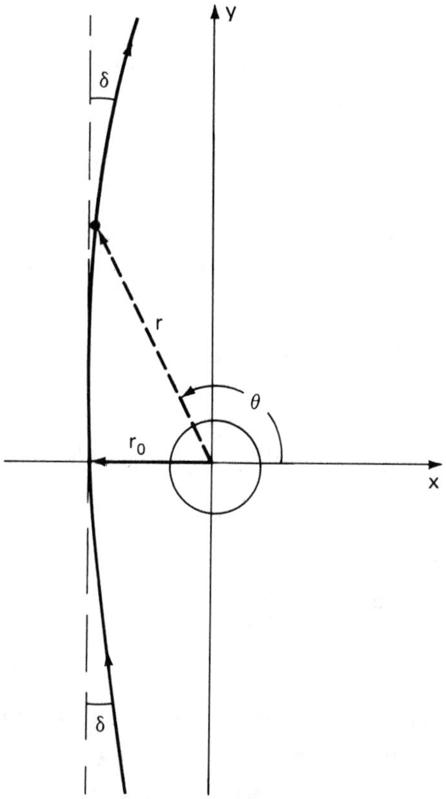

Figure 1: The deflection of light rays by a star according to general relativity.

EXERCISE 1

1. Use perturbation expansion to solve the following differential equations.

   a. $\dot{x} + \epsilon x + 2x^3 = 0$, $x(0) = \cos\epsilon$

   b. $\ddot{x} + 2x = e^{\epsilon t}$, $x(0) = x(1) = 0$

   c. $(1+\epsilon t^2)\ddot{x} + x = t^2$, $x(0) = \epsilon$, $x(1) = 1$

2. The equation of motion of a particle of mass m in a free fall to the surface of the earth is given by

$$m \frac{d^2x}{dt^2} = - \frac{GMm}{(R+x)^2} \qquad (2.11)$$

where G is the constant of gravity M is the Earth's mass and R its radius. If we now introduce

$$\tau = \frac{Rt}{(GM)^{1/2}}$$

equation (2.11) will reduce to

$$\frac{d^2x}{d\tau^2} = - \frac{1}{(1+\epsilon x)^2} \text{ , } \epsilon = 1/R \quad .$$

Use regular perturbations to obtain an approximate solution for $x(\tau)$. Note that although $x_i(t)$ contains powers of t it is still bounded since we are interested only in a finite time interval.

3. According to the theory of General Relativity the equation of motion of the planet Mercury around the sun is given by

$$\frac{d^2w}{d\theta^2} + w = \alpha(1+\epsilon w^2) \quad .$$

Use regular first order perturbations to find the approximate effect of the term $\epsilon\alpha w^2$. Note that this term represents the difference between the classical and general relativistic equations of motion.

4. Use regular perturbations to solve the following system to first order in ϵ

$$\frac{dx}{dt} = x + 2y + \epsilon xy$$
$$\frac{dy}{dt} = 4x + 3y - \epsilon xy \quad .$$

Consider the following initial conditions;

a. $x(0) = y(0) = 0$

b. $x(0) = 0, \quad y(0) = 1 \quad .$

5. Compare the exact solution of equation (2.6) with the approximate solution given by equation (2.14).
6. Find a second nondimensional form for equation (2.7). Under what circumstances should one use this form rather than the one given by the text.
7. Carry out the solution of equations (2.22)-(2.23).

3. SINGULAR PERTURBATIONS

Regular perturbations sometimes fail to yield a solution in which $y_n(x)$ are bounded for all values of the independent variables under consideration. As an example of such a situation we consider the equation for a spring-mass system with a small nonlinearity in the force, i.e.

$$\ddot{x} + k^2x + \epsilon x^3 = 0 \ , \ x(0) = 0, \ \dot{x}(0) = ka \qquad (3.1)$$

(Note that the force exerted by a spring is an odd function of the displacement and, therefore, a correction term cannot be given by ϵx^2). To obtain an approximate solution to this equation we substitute

$$x(t) = x_0(t) + \epsilon x_1(t) + \epsilon^2 x_2(t) + \dots \quad . \qquad (3.2)$$

Using the procedure given in the previous section this leads to the following (trivially) coupled system of equations;

$$\begin{aligned}
&\epsilon^0 \quad \ddot{x}_0 + k^2x_0 = 0 \ , \ x_0(0) = 0 \ , \ \dot{x}_0(0) = ka \\
&\epsilon^1 \quad \ddot{x}_1 + k^2x_1 = -x_0^3 \ , \ x_1(0) = 0 \ , \ \dot{x}_1(0) = 0 \qquad (3.3) \\
&\epsilon^2 \quad \ddot{x}_2 + k^2x_2 = -3x_0^2x_1 \ , \ x_2(0) = 0 \ , \ \dot{x}_2(0) = 0 \ ,
\end{aligned}$$

etc.

Solving for x_0 we obtain

$$x_0(t) = a\sin kt$$

and hence $x_1(t)$ must satisfy

$$\ddot{x}_1 + k^2x_1 = -a^3\sin^3kt$$

$$= \frac{a^3}{4}(\sin3kt - 3\sin kt) \tag{3.4}$$

whose solution is

$$x_1(t) = \frac{-a^3}{32k^2}(\sin3kt+9\sin kt) + \frac{3a^3}{8k}t\cos kt\ . \tag{3.5}$$

However, if we try to write the approximate solution of (3.1) as

$$x(t) = a\sin kt - \frac{\epsilon a^3}{32k^2}(\sin3kt+9\sin kt-12kt\cos kt) \tag{3.6}$$

we see immediately that the coefficient of ϵ (i.e., $x_1(t)$) is not bounded for all t. (The term $t\cos kt$ is called a "secular term"). Moreover, this secular term in $x_1(t)$ will propagate itself to higher order terms in the perturbation expansion so that none of the x_k's is bounded for all t. As a result no approximate solution can be obtained through truncation of the series, i.e., the perturbation expansion breaks down.

To get rid of the secular terms and obtain a regular perturbation expansion we use "straining methods". These consist (in their simplest form) of applying a perturbation expansion on both the independent and dependent variables, i.e.,

$$t = \tau + \epsilon f_1(\tau) + \epsilon^2 f_2(\tau) + \dots \tag{3.7}$$

$$x(\tau) = x_0(\tau) + \epsilon\, x_1(\tau) + \epsilon^2x_2(\tau) + \dots \tag{3.8}$$

with $f_i(0) = 0$ (so that $t = 0$ will correspond to $\tau = 0$). In practice, however, the expansion (3.7) is too general (in most cases) and is replaced by a simpler one viz.

$$t = \tau(1 + \epsilon b_1 + \epsilon^2 b_2 + \dots), \tag{3.9}$$

where the b_i's are constants to be determined so that the secular terms disappear from the perturbation expansion. In order to substitute (3.8), (3.9) in (3.1) we must first note that

$$\dot{x} = \frac{dx}{dt} = \frac{dx}{d\tau} \cdot \frac{d\tau}{dt} = \frac{dx}{d\tau}(1+\epsilon b_1 + \epsilon^2 b_2 + \dots)^{-1}$$

$$= \frac{dx}{d\tau}(1 - \epsilon b_1 + \dots) = x'(1 - \epsilon b_1 + \dots) \tag{3.10}$$

Remark: To obtain the equality before the last in equation (3.10) we use a Taylor expansion on $g(\epsilon) = (1 + \epsilon b_1 + \epsilon^2 b_2 + \dots)^{-1}$ around $\epsilon = 0$.

Similarly,

$$\ddot{x} = \frac{d}{dt}\left(\frac{dx}{d\tau}(1-\epsilon b_1+\dots)\right) = \frac{d^2x}{d\tau^2}(1-\epsilon b_1+\dots)^2 =$$

$$= \frac{d^2x}{d\tau^2}(1-2\epsilon b_1+\dots) = x''(1-2\epsilon b_1+\dots) . \tag{3.11}$$

Substituting (3.8), (3.11) in (3.1) we obtain

$$(x_0''+\epsilon x_1''+\dots)(1-2\epsilon b_1+\dots) + k^2(x_0+\epsilon x_1+\dots)$$

$$+ \epsilon(x_0+\epsilon x_1 + \dots)^3 = 0 \tag{3.12}$$

$$x_0(0) + \epsilon x_1(0) + \dots = 0 \tag{3.13}$$

(since $t = 0$ corresponds to $\tau = 0$). The second initial condition yields

$$ka = \dot{x}(t=0) = x'(0)(1-\epsilon b_1+\dots)$$

$$=(x_0'(0) + \epsilon x_1'(0)+ \dots)(1-\epsilon b_1+\dots) . \tag{3.14}$$

Hence we obtain the following equations to order ϵ (i.e., for x_0, x_1)

$$x_0''+k^2x_0 = 0, \; x_0(0) = 0, \; x_0'(0) = ka \tag{3.15}$$

$$x_1''+k^2x_1 = 2b_1x_0''-x_0^3, \; x_1(0) = 0, \; x_1'(0) = b_1ka \tag{3.16}$$

Thus from (3.15)

$$x_0(\tau) = a\sin k\tau$$

and, therefore, $x_1(\tau)$ must satisfy

$$x_1'' + k^2x_1 = -(2b_1k^2a\sin k\tau + a^3\sin^3k\tau$$

$$= -(2b_1k^2 + \frac{3a^2}{4})a\sin k\tau + \frac{a^3}{4}\sin 3k\tau. \quad (3.17)$$

Now the source of the secular term in the solution of this equation is the first term in the right-hand side of (3.17), to eliminate it we choose

$$b_1 = \frac{-3a^2}{8k^2}$$

and hence

$$x_1(\tau) = \frac{-a^3}{32k^2}(9\sin k\tau + \sin 3k\tau) \quad . \quad (3.18)$$

This expression of $x_1(\tau)$ is uniformly bounded and gives, therefore, a legitimate approximation for $x(t)$ which is given by;

$$x = a\sin k\tau - \frac{\epsilon a^3}{32k^2}(9\sin\tau + \sin 3k\tau) \quad (3.19)$$

$$t = (1 - \epsilon\frac{3a^2}{8k^2})\tau \quad (3.20)$$

As a second example of singular perturbations we now consider equation (3.1) with an additional "soft" resonance term, i.e. we consider a system which is modelled by the equation

$$\ddot{x} + k^2x + \epsilon x^3 = \epsilon F\cos\omega t, \ x(0) = A,$$

$$\dot{x}(0) = 0 \quad . \quad (3.21)$$

To obtain an approximate solution to this equation using a singular perturbation expansion we substitute (3.8), (3.9) and use the expansion

$$\cos\omega\tau(1+\epsilon b_1+\ldots) = \cos\omega\tau - (\epsilon b_1\omega\tau)\sin\omega\tau + \ldots \quad (3.22)$$

(where we assumed that $|\epsilon b_1\omega\tau| << 1$) to obtain

$$\epsilon^0; \quad x_1'' + k^2x_0 = 0 \ , \ x_0(0) = A, \ \dot{x}_0(0) = 0 \quad (3.23)$$

$$\epsilon^1; \qquad x_1'' + k^2x_1 = 2b_1x_0'' - x_0^3 + F\cos\omega\tau, \quad x_1(0) = \dot{x}_1(0) = 0 \quad . \tag{3.24}$$

Hence,

$$x_0 = A\cos\omega\tau \tag{3.25}$$

and

$$x_1'' + k^2x_1 = 2b_1k^2\cos k\tau - A^3\cos^3 k\tau + F\cos\omega\tau$$

$$= -(2b_1k^2 + \frac{3A^2}{4})A\cos k\tau - \frac{A^3}{4}\cos 3k\tau + F\cos\omega\tau . \tag{3.26}$$

Letting $b_1 = -\frac{3A^2}{8k^2}$ (to nullify the secular term) we obtain for x_1

$$x_1 = \frac{A^3}{32k^2}(\cos 3k\tau - \cos k\tau) + \frac{F}{k^2-\omega^2}(\cos\omega\tau - \cos k\tau) \tag{3.27}$$

we now see that when $\omega \approx k$ the second (resonance term will be large and the perturbation expansion will fail since the terms in x_1 will be of the same order as x_0. Under these circumstances it is better to approximate $\cos\omega\tau$ in (3.26) by $\cos k\tau$ and let

$$b_1 = -\frac{3A^2}{8k^2} - \frac{F}{Ak^2} \tag{3.28}$$

we then obtain the following approximation for x to order ϵ;

$$x = A\cos k\tau + \frac{\epsilon A^3}{32k^2}(\cos 3k\tau - \cos k\tau) \tag{3.29}$$

where

$$t = \tau(1 - \frac{3A^2}{8k^2} - \frac{F}{Ak^2}) \quad . \tag{3.30}$$

EXERCISE 2

1. Continue the examples given above and show that one can obtain a regular perturbation expansion up

to second order in ϵ (i.e., without secular terms) by a proper choice of b_2.

2. It is well known that the Van der Pol equation

$$\ddot{x} - \epsilon(1-x^2)\dot{x} + kx = 0 \quad x(0) = 0$$
$$\dot{x}(0) = a(\epsilon) = a_0 + \epsilon a_1 + \epsilon^2 a_2 + \ldots$$

admits a limit cycle. Obtain a regular perturbation expansion of this solution by the method of strained variables and a proper choice of the a_i's.

Hint: Since the solution we are looking for is periodic all $x_i(\tau)$ should be chosen as periodic.

3. Use the method of strained variables to obtain a perturbation expansion to first order in ϵ for the following equations:

   a. $\ddot{x} + \epsilon\dot{x} + kx = 0$, $x(0) = 0$, $\dot{x}(0) = 1$
   (spring-mass system with "small" damping)

   b. $\ddot{x} + \epsilon\dot{x}^3 + kx = 0$, $x(0) = 0$, $\dot{x}(0) = 1$.

4. Redo exercise 3 with initial conditions

$$x(0) = a, \quad \dot{x}(0) = 0$$

5. Solve equation (3.1) with the same initial condition as in exercise 4.

4. BOUNDARY LAYERS

In many physical problems there exist very thin regions in which some physical variables undergoes sharp changes. These regions are called boundary layers since they were first discovered and treated in the theory of fluid motions. To see how an approximate solution in such regions of nonuniformity can be obtained by singular perturbation theory we examine the following protype problem;

Example: Consider

$$\epsilon\ddot{x} + \dot{x} + kx = 0, \quad x(0) = 0, \quad x(a) = A, \quad A \neq 0$$
$$0 < \epsilon << 1 . \qquad (4.1)$$

To gain some insight into this problem we attempt, at first, to obtain an approximate solution via regular perturbations. Thus, if we write

$$x(t) = x_0(t) + \epsilon x_1(t) + \epsilon^2 x_2(t) + \dots \qquad (4.2)$$

and substitute in (4.1) we obtain the following system of coupled equations;

$$\epsilon^0 \quad \dot{x}_0 + kx_0 = 0, \quad x_0(0) = 0, \; x_0(a) = A \qquad (4.3)$$

$$\epsilon^1 \quad \dot{x}_1 + kx_1 = -\ddot{x}_0, \; x_1(0) = x_1(a) = 0 \qquad (4.4)$$

$$\epsilon^2 \quad \dot{x}_2 + kx_2 = -\ddot{x}_1, \; x_2(0) = x_2(a) = 0 \qquad (4.5)$$

etc.

We now see that though the original equation (4.1) is of second order (with two boundary conditions) the perturbation expansion yields a system of first order equations with two boundary conditions for which no solution exist in general. Thus, the general solution of (4.3) is

$$x_0 = Ce^{-kt}$$

but since $x_0(0) = 0$ it follows that $C = 0$ and, therefore, $x_0(a) = A$ cannot be satisfied.

To see the origin of this peculiar behavior and how we can remedy it we note that the exact solution of (4.1) is

$$x(t) = \frac{Ae^{(a-t)/2\epsilon}\sinh\omega t}{\sinh\omega a}, \quad \omega = \sqrt{1-4k\epsilon}/2\epsilon \qquad (4,6)$$

This expression shows that though $\epsilon\ddot{x}$ can be neglected compared to $\dot{x}$ and x for large t (e.g., $t > a + 2\epsilon$) there exists a small interval near the origin where $\ddot{x}$ is so large that $\epsilon\ddot{x}$ is of the same order as $\dot{x}$ and x and hence can not be neglected in the zero order approximation of the differential equation.

In view of this phenomena it is natural to seek a solution of our problem in two separate regions; (1) an inner region $0 < t < t_0$ where t_0 is small and (2) an exterior solution i.e., a solution in the domain $t_0 < t$ (Note that the exact value of t_0 remains unspecified).

Denoting the outer solution by w we note that in this region w, $\dot{w}$ and $\ddot{w}$ are of the same order and hence $\epsilon\ddot{w}$ is small and negligible as compared to the other terms in the differential equation. Therefore, in the outer region equation (4.1) can be correctly approximated by regular perturbation expansion viz.

$$\dot{w}_0 + kw_0 = 0 \quad , \quad w_0(a) = A \tag{4.7}$$

$$\dot{w}_1 + kw_1 + \ddot{w}_0 = 0 \quad , \quad w_1(a) = 0 \tag{4.8}$$

etc.

(Note that $t = 0$ is outside the region under consideration and, therefore, we do not have to take into account the boundary condition at this point).

From (4.7), (4.8) we obtain

$$w_0 = Ae^{k(a-t)}$$

$$w_1 = Ak^2(a-t)e^{k(a-t)}$$

i.e. to first order in ϵ the outer solution is given by

$$w = Ae^{k(a-t)} + \epsilon Ak^2(a-t)e^{k(a-t)} \quad . \tag{4.9}$$

To obtain a similar perturbation expansion in the inner region we must make first some transformation on the inner solution u so that u, $\dot{u}$ and $\ddot{u}$, be of the same order of magnitude. (So that the term multiplied by ϵ be small compared to the other terms in the differential equation.). This objective can be accomplished by a magnification of the time scale. Thus, we introduce

$$\tau = \epsilon^p t \tag{4.10}$$

where p is a parameter to be determined later. Substituting (4.10) in (4.1) we obtain

$$\epsilon^{1+2p} u'' + \epsilon^{p} u' + ku = 0, \quad u(0) = 0 \tag{4.11}$$

where primes denote differentiation with respect to τ. From (4.11) we see that u'' and u' will have the same coefficient if $1 + 2p = p$ i.e., $p = -1$. (Thus, $t = \epsilon$ corresponds to $\tau = 1$). Equation (4.1) then takes the form

$$u'' + u' + k\epsilon u = 0 \quad , \quad u(0) = 0 \tag{4.12}$$

Regular perturbation on (4.12) yield

$$u_0'' + u_0' = 0 \quad , \quad u_0(0) = 0$$

$$u_1'' + u_1' = -ku_0, \quad u_1(0) = 0$$

etc.

The solution of this equation to first order in ϵ is given by;

$$u_0 = c_1(1-e^{-\tau}) \tag{4.13}$$

$$u_1 = k[c_2-c_1\tau-(c_2+c_1\tau)e^{-\tau}] \quad . \tag{4.14}$$

Our last task in solving our problem is now to determine the constants c_1, c_2 which appear in the expression for the inner solution. This can be done by requiring that the outer and inner solutions blend "smoothly" at the edge of the boundary layer and several "principles" were formulated in the past to this end. The oldest of such principles is due to Prandtl and it requires that one should set the inner limit of the outer solution to be equal to the outer limit of the inner solution, i.e.,

$$\lim_{t\to 0} w(t,\epsilon) = \lim_{\tau\to\infty} u(\tau,\epsilon) \quad . \tag{4.15}$$

Applying this principle to the zeroth order perturbation solution of our problem this principle requires that

$$\lim_{t\to 0} Ae^{k(a-t)} = \lim_{\tau\to\infty} c_1(1-e^{-\tau}) \tag{4.16}$$

and hence $c_1 = Ae^{ak}$. However, it is easy to verify that this principle can not be used to determine c_1, c_2 in the first order perturbation expansion. To handle such higher order perturbation expansions M. Van Dyke formulated a generalized matching principle.

To apply this principle to the perturbation expansion in the inner and outer regions to order ϵ^n viz. match

$$w(t) = w_0(t) + \epsilon w_1(t) + \dots + \epsilon^n w_n(t)$$

with

$$u(\tau) = u_0(\tau) + \epsilon u_1(\tau) + \dots + \epsilon^n u_n(\tau)$$

we must express $w(t)$ in terms of τ and $u(\tau)$ in terms of t then expand the resulting expressions in powers of ϵ up to ϵ^n and finally match the coefficients of ϵ^k, $k = 0,\dots,n$ in the two expressions to determine the redundant constants.

As an example we now use this principle to determine c_1, c_2 in equations (4.13)-(4.14) by matching the inner and outer solutions to order ϵ.

Rewriting the outer solution (4.9) in terms of τ and expanding in ϵ we obtain

$$\begin{aligned} w(\tau) &= Ae^{ak}\left[e^{-\epsilon k\tau} + \epsilon k^2(a-\epsilon\tau)e^{-\epsilon k\tau}\right] \\ &= Ae^{ak}[1-\epsilon k\tau+\epsilon k^2 a+O(\epsilon^2)] \quad . \end{aligned} \tag{4.17}$$

Similarly for the inner solution we have

$$\begin{aligned} u(t) &= c_1(1-e^{-t/\epsilon}) + \epsilon k[c_2 - c_1 t/\epsilon \\ &\quad - (c_2 + c_1 t/\epsilon)e^{-t/\epsilon}] = c_1(1-kt) + \epsilon k c_2 + O(\epsilon^2) \end{aligned} \tag{4.18}$$

(Note that in the last expression $e^{-t/\epsilon}$ for any fixed $t > 0$ converge to 0 faster than any powers of ϵ and, therefore, is equivalent to 0).

Hence,

$$c_1(1-kt) + \epsilon k c_2 = Ae^{ak}[(1-kt) + \epsilon k^2 a] \quad . \tag{4.19}$$

This yields

$$c_1 = Ae^{ak} \quad , \quad c_2 = Aake^{ak} \quad . \tag{4.20}$$

EXERCISE 4

For the following two differential equations find inner, outer and matched solutions. In each case use first principles rather than substitution in formulas derived above!

1. $\epsilon\ddot{x} + (1+bt)\dot{x} + kx = 0 \quad x(0) = 0 \quad x(a) = 1$
2. $\epsilon\ddot{x} + b\dot{x} + x^2 = 0 \quad x(0) = 1/2 \quad x(a) = 1$

Remark: In each case assume the existence of a boundary layer.

5. OTHER PERTURBATION METHODS

In this section we discuss two methods for the derivation of approximate solutions to systems of ODEs. These methods are especially powerful when one attempts to determine the existence and properties of limit cycles.

A. Harmonic approximations.

In this technique we assume that the equation under consideration possess an approximate periodic solution with a unique frequency ω. To determine the amplitude and frequency of this solution we substitute it in the original equations and neglect all the "Harmonics" (i.e. terms with frequency different from ω) that appear in this process. We illustrate this procedure through examples.

Example 1: To find whether Rayleigh's equation

$$\ddot{u} + \epsilon(1-k\dot{u}^2)\dot{u} + u = 0, \ \dot{u}(0) = 0 \tag{5.1}$$

has an approximate periodic solution of the form

$$u = A\cos\omega t \tag{5.2}$$

(where A, ω are to be determined) we substitute (5.2) in (5.1) to obtain;

$$-A\omega^2\cos\omega t - \epsilon A\omega\sin\omega t(1-kA^2\omega^2\sin^2\omega t) + A\cos\omega t$$
$$= A(1-\omega^2)\cos\omega t - \epsilon A\omega\sin n\omega t(1 - \frac{3kA^2\omega^2}{4})$$
$$- \frac{\epsilon kA^3\omega^3}{4}\sin 3\omega t = 0 \qquad (5.3)$$

(we used the identity $\sin^3\omega t = \frac{3\sin\omega t - \sin 3\omega t}{4}$).

Neglecting the higher harmonic term in (5.3) we see that (5.2) is a solution of (5.1) if and only if

$$\omega = 1, \quad A = \frac{2}{\sqrt{3k}} \quad . \qquad (5.4)$$

Hence we conclude that equation (5.1) admits an approximate limit cycle with these amplitude and frequency. Note, however, that no inference regarding the stability of this limit cycle can be made from this technique.

Example 2: Find if Van der Pol's equation

$$\ddot{u} - \epsilon(1-u^2)\dot{u} + k^2u = 0, \quad \dot{u}(0) = 0 \qquad (5.5)$$

admits an approximate limit cycle of the form $u = A\cos\omega t$.

Solution: First we note that the proposed solution satisfy the initial condition. To see if we can adjust A, ω so that $A\cos\omega t$ is a solution of (5.5) we substitute it in this equation to obtain

$$A(k^2-\omega^2)\cos\omega t + \epsilon(1-A^2\cos^2\omega t)A\omega\sin n\omega t\omega t$$
$$= A(k^2-\omega^2)\cos\omega t + \epsilon A\omega(1 - \frac{A^2}{4})\sin n\omega t$$
$$- \frac{\epsilon A^3\omega}{4}\sin 3\omega t = 0 \quad . \qquad (5.6)$$

Neglecting the higher order harmonic we deduce

$$A = 2, \quad \omega = k \qquad (5.7)$$

i.e. $u = 2\cos 2t$ is an approximate limit cycle of this equation.

Example 3: Beats phenomena.

Consider two weakly coupled oscillators of the same natural frequency whose equations of motion are

$$\ddot{x} + \omega_0^2 x = \epsilon y$$
$$\ddot{y} + \omega_0^2 y = \epsilon x \qquad 0 < \epsilon \ll 1 . \qquad (5.8)$$

Substituting

$$x = Ae^{i\omega t} \quad , \quad y = Be^{i\omega t}$$

we obtain

$$A(\omega_0^2-\omega^2) = \epsilon B$$
$$B(\omega_0^2-\omega^2) = \epsilon A \quad .$$

These two equations have nonzero solutions for A,B if and only if

$$(\omega_0^2-\omega^2)^2 = \epsilon^2$$

i.e.

$$\omega_{1,2}^2 = \omega_0^2 \pm \epsilon \quad .$$

Since $\epsilon \ll 1$ we can approximate this relation by

$$\omega_{1,2} = \omega_0 \pm \epsilon/2\omega_0 + o(\epsilon^2).$$

For $\omega_1 : A = B$ while for $\omega_2 : A = -B$.

Hence to order ϵ the general solution of the system (5.8) is

$$x = A_1 e^{i\omega_1 t} + A_2 e^{i\omega_2 t}$$
$$y = A_1 e^{i\omega_1 t} - A_2 e^{i\omega_2 t} \qquad (5.9)$$

and the amplitudes of the two oscillators are given by

$$|x| = |A_1| \cdot |1 + \frac{A_2}{A_1} e^{-i\epsilon t/\omega_0}|$$
$$(5.10)$$
$$|y| = |A_1| \cdot |1 - \frac{A_2}{A_1} e^{-i\epsilon t/\omega_0}| \quad .$$

We infer then that the system will display the phenomena of beats, i.e., the amplitude of the first oscillator is maximum when that of the second is minimum and vise versa.

B. Method of Averages.

To motivate and describe the basic ideas of this method we consider a class of equations in the form

$$\ddot{u} + k^2 u + \epsilon f(u,\dot{u}) = 0, \quad |\epsilon| << 1 \quad . \tag{5.11}$$

To begin with we observe that when $\epsilon = 0$ the solution of (5.8) is given by

$$u = A\cos(kt+\varphi), \quad \dot{u} = -Ak\sin(kt+\varphi) \tag{5.12}$$

where A, φ are constants. Motivated by this observation we now attempt to find an approximate solution to (5.8) with $\epsilon \neq 0$ in the form;

$$u(t) = A(t)\cos(kt+\varphi(t)) \tag{5.13}$$

$$\dot{u}(t) = -A(t)k\sin(kt+\varphi(t)) \tag{5.14}$$

(note that $\dot{u}(t)$ is not the true derivative of $u(t)$) where we assume that $A(t)$, $\varphi(t)$ are slowly varying functions of t.

Computing $\dot{u}(t)$ using (5.13) and comparing with (5.14) we obtain the compatibility condition

$$\dot{A}\cos(kt+\varphi) - A\dot{\varphi}\sin(kt+\varphi) = 0 \quad . \tag{5.15}$$

Furthermore, if we substitute (5.13), (5.14) in (5.11) (using $\ddot{u} = \frac{d}{dt}(\dot{u})$) it follows that

$$\dot{A}\sin(kt+\varphi) + A\dot{\varphi}\cos(kt+\varphi) = \frac{\epsilon f(u,\dot{u})}{k} \tag{5.16}$$

Solving (5.12), (5.13) for $\dot{A}, \dot{\varphi}$ then yields

$$\dot{A} = \frac{\epsilon f(u,\dot{u})}{k}\sin(kt+\varphi) \tag{5.17}$$

$$\dot{\varphi} = \frac{\epsilon f(u,\dot{u})}{kA}\cos(kt+\varphi) \quad . \tag{5.18}$$

Since both $\dot{A}, \dot{\varphi}$ are proportional to ϵ where $|\epsilon| << 1$ these equations confirm our hypothesis that $A(t)$, $\varphi(t)$ are slowly varying functions of time.

We now come to the central idea of the method of averages. Since equations (5.17), (5.18) are rather complicated in general we approximate them by their time average over one period of the system, i.e., over

$[0, 2\pi/k]$. Note, however, that A, φ are to be considered as constants while taking these averages. Thus,

$$\dot{A} = \frac{\epsilon}{2\pi k}\int_0^{2\pi} \sin\psi f(A\cos\psi, -Ak\sin\psi)d\psi \quad (5.19)$$

$$\dot{\varphi} = \frac{\epsilon}{2\pi kA}\int_0^{2\pi} \cos\psi f(A\cos\psi, -Ak\sin\psi)d\psi \quad (5.20)$$

where $\psi = kt+\varphi$.

The solution of these averaged equations provides then an approximate solution of the form (5.13) to equation (5.11).

Example 4: For Rayleigh's equation

$$\ddot{u} + \epsilon(1-\dot{u}^2)\dot{u} + u = 0 \quad (5.21)$$

we have $f(u,\dot{u}) = (1-\dot{u}^2)\dot{u}$, $k = 1$ and, therefore,

$$\dot{A} = \frac{\epsilon}{2\pi}\int_0^{2\pi} \sin\psi(1-A^2\sin^2\psi)(-A\sin\psi) = \frac{\epsilon}{2}A(\frac{3}{4}A^2-1) \quad (5.22)$$

$$\dot{\varphi} = \frac{\epsilon}{2\pi A}\int_0^{2\pi} \cos\psi(1-A^2\sin^2\psi)(-A\sin\psi)d\psi = 0 \ . \quad (5.23)$$

Thus, $\varphi = \text{const}$ and we can compute A by elementary integration methods.

C. Stability of limit cycles.

The solution (5.13)-(5.14) represents an approximate limit cycle of equation (5.11) if $\dot{A} = 0$. From equation (5.19) we infer then that this happens when A is a root of

$$H(A) = \frac{\epsilon}{2\pi k}\int_0^{2\pi} \sin\psi f(A\cos\psi, -Ak\sin\psi)d\psi \ . \quad (5.24)$$

To determine whether such a limit cycle with $A = a$ is stable we apply a small perturbation to a, i.e. consider $A = a + x(t)$. The limit cycle under consideration is asymptotically stable if it can be shown that $x(t) \to 0$ as $t \to \infty$.

But

$$\dot{A} = \dot{x}(t) = H(a+x(t)) \approx H'(a) \cdot x(t) \qquad (5.25)$$

(remember we assume $H(a) = 0$).

Hence,

$$x(t) = Ce^{H'(a)t} \qquad . \qquad (5.26)$$

We conclude, therefore, that the limit cycle with $A = a$ is

1. asymptotically stable if $H'(a) < 0$
2. unstable if $H'(a) > 0$

Example 5: For Rayliegh's equation (see(5.18)) we have from (5.19)

$$H(A) = \frac{\epsilon}{2} A(\frac{3}{4} A^2-1) \qquad (5.27)$$

Hence the equation has a limit cycle at $A = 0$ (critical point) and $A = \frac{2\sqrt{3}}{3}$.

But

$$H'(A) = \frac{\epsilon}{2} (\frac{9}{4} A^2-1) \qquad , \qquad (5.28)$$

and, therefore, if $\epsilon > 0$ then $H'(0) < 0$ and $H'\left[\frac{2\sqrt{3}}{3}\right] > 0$, i.e., the critical point $a = 0$ is asymptotically stable and the limit cycle with $a = \frac{2\sqrt{3}}{3}$ is unstable.

EXERCISE 5

1. Apply the method of harmonic approximation to the following equations
   a. $\ddot{u}+ku+\epsilon u^3 = 0$, $\dot{u}(0) = 0$ (Duffing's equation)
   b. $\ddot{u}+ku+\epsilon u^5 = 0$, $\dot{u}(0) = 0$
   c. $\ddot{u}+\epsilon\dot{u}^3+ku = 0$, $\dot{u}(0) = 0$
   d. $\ddot{u}+ku+\epsilon(au^5+bu^3) = 0$, $\dot{u}(0) = 0$

   where a,b are arbitrary constants.
2. Apply the method of averages to the equations of exercise 1.

3. Apply the method of averages to the Van der Pol's equation

$$\ddot{u} - \epsilon(1-u^2)\dot{u} + k^2 u = 0$$

4. Discuss the application of the method of averages to equations of the form

$$\ddot{u} + b\dot{u} + \epsilon f(u,\dot{u}) = 0.$$

Consider $b > 0$ and $b < 0$.

5. Generalize the results of the previous exercise to equations of the form

$$\ddot{u} + b\dot{u} + ku + \epsilon f(u,\dot{u}) = 0$$

6. Use the method of averages to discuss the stability of the limit cycle solution to the Van der Pol's and Duffing's equations.

7. Compare the averaged and exact solutions of the following equations.

a. $\dot{u} = \epsilon u \sin kt$

b. $\dot{u} = \epsilon(u-u^2)\cos^2 kt$

Hint: Use the same procedure which was applied in the text to convert equations (5.16)-(5.17) to equation (5.19)-(5.20).

*6. PERTURBATIONS AND PARTIAL DIFFERENTIAL EQUATIONS

The basic techniques of regular and singular perturbations which were studied in previous sections can be applied without change to partial differential equations. However, for these equations a perturbation in the boundary conditions might be as important as a perturbation in the equations themselves. In this section we demonstrate the application of perturbation methods to partial differential equations through several examples.

Example 1 (regular perturbations): Find to first order in ϵ the solution of

$$\nabla^2 u + \epsilon r\left(\frac{\partial u}{\partial r}\right)^2 = 0 \tag{6.1}$$

on the unit disk with the boundary condition

$$u(1,\theta) = \sin\theta \ . \tag{6.2}$$

Solution: Applying the usual perturbation expansion

$$u(r,\theta) = u_0(r,\theta) + \epsilon u_1(r,\theta) + \epsilon^2 u_2(r,\theta) + \ldots \tag{6.3}$$

to equation (6.1) we obtain the following system of coupled equations

$$\epsilon^0: \qquad \nabla^2 u_o = 0 \quad , \qquad u_o(1,\theta) = \sin\theta$$

$$\epsilon^1: \qquad \nabla^2 u_1 = -r\left[\frac{\partial u_0}{\partial r}\right]^2, \qquad u_1(1,\theta) = 0 \tag{6.4}$$

etc.

The solution of the first two equations in this system is given by

$$u_0 = r\sin\theta$$

$$u_1 = \frac{1-r^3}{18} + \frac{1}{10}(r^3-r^2)\cos 2\theta \ .$$

Hence, to first order in ϵ

$$u = r\sin\theta + \epsilon\left[\frac{1-r^3}{18} + \frac{1}{10}(r^3-r^2)\cos 2\theta\right] + O(\epsilon^2) \tag{6.5}$$

Remarks: 1. The Laplace operators in polar coordinates is given by

$$\nabla^2 u = \frac{\partial^2 u}{\partial r^2} + \frac{1}{r}\frac{\partial u}{\partial r} + \frac{1}{r^2}\frac{\partial^2 u}{\partial \theta^2}$$

2. Note that $u_1(r,\theta)$ is bounded since the region in which the solution is applicable is the unit disk in R^2.

Example 2: (Boundary layers)

Consider the first order PDE

$$\epsilon(u_t+u_x) + u = \sin kt, \quad t > 0, \ -\infty < x < \infty \tag{6.6}$$

with the initial condition

$$u(x,0) = \varphi(x) \ . \tag{6.7}$$

Since ϵ appears as a coefficient of the highest derivatives in equation (6.6) we do expect the solution to exhibit a boundary layer and, therefore, consider a perturbation solution for the outer and inner regions in t.

For the outer region (where the initial condition (6.7) is not effective we apply regular perturbation, i.e.,

$$u(x,t) = u_0(x,t) + \epsilon u_1(x,t) + \dots$$

and obtain the following system of equations

$$u_0 = \sin kt$$

$$u_m = -\left[\frac{\partial u_{m-1}}{\partial t} + \frac{\partial u_{m-1}}{\partial x}\right] \quad , \quad m \geq 1 \tag{6.8}$$

Hence,

$$u_{2m} = k^{2m}(-1)^m \sin kt, \quad u_{2m+1} = k^{2m+1}(-1)^{m+1}\cos kt \ . \tag{6.9}$$

Turning our attention to the inner region we apply the transformation

$$\tau = \epsilon^P t$$

to equation (6.6) and obtain

$$\epsilon^{1+p}u_\tau + \epsilon u_x + u = \sin(k\epsilon^{-p}\tau) \quad . \tag{6.10}$$

We can now equalize the coefficients of u_τ and u by choosing $p = -1$ and this leads to

$$u_\tau + u + \epsilon u_x = \sin(k\epsilon)\tau \quad . \tag{6.11}$$

Applying a perturbation expansion

$$u(x,\tau) = u_0(x,\tau) + \epsilon u_1(x,\tau) + \dots$$

to (6.11) leads to

$$\frac{\partial u_0}{\partial \tau} + u_0 = 0 \quad , \quad u_0(x,0) = \varphi(x) \tag{6.12}$$

$$\frac{\partial u_1}{\partial \tau} + u_1 = k\tau - \frac{\partial u_0}{\partial x} \ , \quad u_1(x,0) = 0 \tag{6.13}$$

etc. (Note that $\sin(k\epsilon\tau) = k\epsilon\tau - \frac{(\epsilon k\tau)^3}{6} + \dots$).

Hence,

$$u_0(x,\tau) = \varphi(x)e^{-\tau}$$

$$u_1(x,\tau) = k(\tau-1) + [(k-\tau\varphi'((x))]e^{-\tau} .$$

Thus, to first order in ϵ the inner solution is given by

$$u(x,\tau) = \varphi(x)e^{-\tau} + \epsilon\{k(\tau-1) + (k-\tau\varphi'(x))e^{-\tau}\}.$$

Note that this solution contains no arbitrary constants.

Example 3: (Boundary Perturbations)

In many practical situations it is important to obtain a solution of a BVP on a region whose boundary is not smooth. As an example we now consider the "almost" unit disk in R^2 whose boundary is given by

$$r = 1 + \epsilon\cos k\theta \quad 0 < \epsilon << 1, \ k \ \text{an integer}.$$

In the interior of this region we now want to solve the following BVP

$$\nabla^2 u = 0 \tag{6.14}$$

$$u(1+\epsilon\cos k\theta,\theta) = \varphi(\theta) \tag{6.15}$$

where $\varphi(\theta)$ is a given function. To obtain a perturbation solution to this problem we write

$$u(r,\theta) = u_0(r,\theta) + \epsilon u_1(r,\theta) + \ldots \tag{6.16}$$

and apply a Taylor expansion in ϵ to (6.15)

$$u(1+\epsilon\cos k\theta,\theta) = u(1,\theta) + (\epsilon\cos k\theta)\,\frac{\partial u}{\partial r}\,(1,\theta)$$

$$+ (\epsilon\cos k\theta)^2\,\frac{\partial^2 u}{\partial r^2}\,(1,\theta) + \ldots = u_0(1,\theta)$$

$$+ \epsilon(u_1(1,\theta) + \cos k\theta\,\frac{\partial u_0}{\partial r}\,(1,\theta)) + \ldots . \tag{6.17}$$

Inserting (6.16), (6.17) in (6.14), (6.15) respectively and equating powers of ϵ we obtain the following system of equations

$$\nabla^2 u_0 = 0 \quad , \ u_0(1,\theta) = \varphi(\theta) \tag{6.18}$$

$$\nabla^2 u_1 = 0 \quad , \ u_1(1,\theta) = -\cos k\theta\,\frac{\partial u_0}{\partial r}\,(1,\theta) \tag{6.19}$$

etc. If e.g., $\varphi(\theta) = \cos\theta$ then

$$u_0(r,\theta) = r\cos\theta \tag{6.20}$$

and

$$u_1(r,\theta) = \frac{-1}{2}\,[r^{k+1}\cos(k+1)\theta + r^{k-1}\cos(k-1)\theta] \tag{6.21}$$

i.e. to first order in ϵ

$$u(r,\theta) = r\cos\theta - \frac{\epsilon}{2}\,[r^{k+1}\cos(k+1)\theta + r^{k-1}\cos(k-1)\theta]. \tag{6.22}$$

EXERCISE 6

In exercises 1, 2, and 3 derive a perturbation expansion and solve to first order in ϵ the following:

1. $u_t = u_{xx} + (\epsilon\cos x)u_x \qquad 0 \le x \le \pi$

   $u(x,0) = \sin x, \quad u(0,t) = 0, \quad u(\pi,t) = 0$

2. $u_{tt} = u_{xx} + (\epsilon\cos x)u_x \qquad 0 \le x \le \pi$

   $u(0,t) = u(\pi,t) = 0,$

   $u(x,0) = \sin kx, \quad u_t(x,0) = 0.$

3. $\nabla^2 u + \epsilon u^k = 0$

   on the unit disk with $u(1,\theta) = 1$

Remark: Note that for $k = 1$ this equation has a closed form solution viz.

$$u = \frac{J_0(\mu r)}{J_0(\mu)} \qquad , \quad \mu = \epsilon^{1/2}$$

where J_0 is the zeroeth order Bessel function.

4. Show that the first order inner and outer solutions derived for equation (6.6) satisfy Prandtl matching principles to order ϵ (assume $\epsilon\tau$ is small!).

5. Show that Equation (6.6) has an exact solution which is given by

$$u(x,t) = \frac{1}{1+\epsilon^2}\,(\sin kt - \epsilon\cos kt) + \left[\frac{\epsilon}{1+\epsilon^2} + \varphi(x-t)\right]e^{-t/\epsilon}$$

6. Write a computer program to compare the exact solution of equation (6.6) with the approximate one derived above (to first order in ϵ) when $\epsilon = 10^{-2}$, 10^{-5} and $\varphi(x) = \sin(mx)$.

7. Find a first order perturbation solution to

$$\nabla^2 u = 0$$

in the region; $0 \leq x \leq \pi$

$$y = \epsilon x \quad \text{and} \quad y = \pi$$

if $u(0,y) = 0$, $u(\pi,y) = 0$, $u(x,\pi) = 0$ and $u(x,\epsilon x) = \cos kx$.

8. Solve to first order in ϵ the following boundary value problem

$$\nabla^2 u + \epsilon \frac{\partial u}{\partial r}\frac{\partial u}{\partial \theta} = 0$$

on the unit disk in R^2 with the boundary conditions

a. $u(1,\theta) = \sin k\theta$

b. $u(1,\theta) = \cos k\theta$ $\qquad k = 1,2,\ldots$.

9. Repeat the previous exercise when the boundary is not smooth, i.e.

$$u(1+\epsilon\cos m\theta) = \sin k\theta$$

*7. PERTURBATION OF EIGENVALUE PROBLEMS

In many practical situations especially Quantum mechanics one encounters the following situation: Given a differential operator H_0 whose eigenvalues and eigenfunctions can be evaluated exactly we are asked to find (or estimate) the eigenvalues and eigenfunctions of the operator $H_0 + \epsilon V$ where $0 < \epsilon << 1$. In other words we are given $\psi_n^{(0)}$, $E_n^{(0)}$ so that

$$H_0\psi_n^{(0)} = E_n^{(0)}\psi_n^{(0)} \tag{7.1}$$

and we want to find approximation to ψ_n, E_n so that

$$(H_0+\epsilon V)\psi_n = E_n\psi_n. \tag{7.2}$$

Remarks: 1. We assume implicitly that ψ_n are "well behaved" and satisfy certain boundary conditions.
2. The set of eigenvalues might be continuous rather than discrete. In this section we consider only the discrete case.

In the following we solve the problem posed above subject to the condition that $\left\{\psi_n^{(0)}\right\}$ form an orthonormal basis for the Hilbert space under consideration, i.e.

(1) $$\int_V \psi_n^*(\mathbf{x})\psi_m(\mathbf{x})d\mathbf{x} = \delta_{nm} \tag{7.3}$$

where * denotes complex conjugation and the integration is over the whole coordinate space.
(2) for any function ψ in the space of functions there exist c_n's so that

$$\psi(\mathbf{x}) = \Sigma\, c_n\psi_n(\mathbf{x}) \tag{7.4}$$

for almost all $\mathbf{x} \in V$.
We now divide our discussion of the problem into two cases:
Case 1: The nondegenerate case, i.e., all $E_n^{(0)}$ are simple (no two eigenfunctions have the same eigenvalue).

To solve our problem in this case we perform a perturbation expansion on both ψ_n and E_n, i.e.,

$$\psi_n = \psi_n^{(0)} + \epsilon\psi_n^{(1)} + \epsilon^2\psi_n^{(2)} + \ldots \tag{7.5}$$

$$E_n = E_n^{(0)} + \epsilon E_n^{(1)} + \epsilon^2 E_n^{(2)} + \ldots \tag{7.6}$$

substituting these expressions in (7.2) we obtain

$$\begin{aligned}(H_0+\epsilon V)[\psi_n^{(0)}+\epsilon\psi_n^{(1)}+\epsilon^2\psi_n^{(2)} + \ldots] \\ = [E_n^{(0)}+\epsilon E_n^{(1)}+\epsilon^2 E_n^{(2)} + \ldots] \\ [\psi_n^{(0)}+\epsilon\psi_n^{(1)}+\epsilon^2\psi_n^{(2)} + \ldots] \quad .\end{aligned} \tag{7.7}$$

Equating powers of ϵ we now obtain the following system of equations

$$H_0\psi_n^{(0)} = E_n^{(0)}\psi_n^{(0)} \tag{7.8}$$

$$H_0\psi_n^{(1)} + V\psi_n^{(0)} = E_n^{(0)}\psi_n^{(1)} + E_n^{(1)}\psi_n^{(0)} \tag{7.9}$$

$$H_0\psi_n^{(2)} + V\psi_n^{(1)} = E_n^{(0)}\psi_n^{(2)} + E_n^{(1)}\psi_n^{(1)} + E_n^{(2)}\psi_n^{(0)} \tag{7.10}$$

etc. The first of these equations holds by assumption hence for first order perturbation we must solve only equation (7.9). To do so we now use the fact that $\{\psi_n^{(0)}\}$ form a basis for the Hilbert space under consideration to write

$$\psi_n^{(1)} = \sum_m c_{nm}\psi_n^{(0)} \tag{7.11}$$

substituting this expression in (7.9) gives;

$$H_0 \sum_m c_{nm}\psi_m^{(0)} + V\psi_n^{(0)} = E_n^{(0)} \sum_m c_{nm}\psi_m^{(0)} + E_n^{(1)}\psi_n^{(0)} \tag{7.12}$$

but since $\psi_m^{(0)}$ are eigenfunctions of H_0 this can be rewritten as

$$\sum_m c_{nm}E_m^{(0)}\psi_m^{(0)} + V\psi_n^{(0)} = E_n^{(0)} \sum_m c_{nm}\psi_m^{(0)} + E_n^{(1)}\psi_n^{(0)}. \tag{7.13}$$

Multiplying this expression by $\psi_k^{(0)*}$ and integrating over the whole space (using the orthonormality of $\psi_n^{(0)}$) we obtain

$$\sum_m c_{nm}E_m^{(0)}\delta_{mk} + V_{nk} = E_n^{(0)} \sum_m c_{nm}\delta_{mk} + E_n^{(1)}\delta_{nk} \tag{7.14}$$

where

$$V_{nk} = \int \psi_k^{(0)*}(x)V(x)\psi_n^{(0)}(x)dx \quad .$$

Using the properties of δ_{mk} (7.14) reduces to

$$c_{nk}E_k^{(0)} + V_{nk} = E_n^{(0)}c_{nk} + E_n^{(1)}\delta_{nk} \tag{7.15}$$

for $k = n$ we obtain from (7.15)

$$E_n^{(1)} = V_{nn} \tag{7.16}$$

while for $k \neq n$ we obtain

$$c_{nk} = \frac{V_{nk}}{E_n^{(0)} - E_k^{(0)}} \quad . \tag{7.17}$$

Notice that c_{nn} remains undetermined but since ψ_n must have a unit norm, i.e.,

$$\int_V \psi_n^* \psi_n dx = 1 \quad . \tag{7.18}$$

we can use this additional equation to determine this coefficient. To summarize: First order perturbations yield the following corrections for the eigenvalues and eigenfunctions of $H = H_0 + \epsilon V$

$$E_n = E_n^{(0)} + \epsilon V_{nn} \tag{7.19}$$

$$\psi_n = \psi_n^{(0)} + \epsilon \sum_m c_{nm} \psi_m^{(0)} \tag{7.20}$$

Case 2: The degenerate case:

In this case there exist several eigenfunctions which belong to the same eigenvalue and, therefore, part of the procedure which we followed to derive the first order corrections in the nondegenerate case breaks down. To see this we note that equation (7.17) will require in the degenerate case a division by zero since for some k's

$$E_n^{(0)} = E_k^{(0)}$$

The source of this trouble comes from equation (7.15) which can be rewritten as

$$c_{nk}[E_k^{(0)} - E_n^{(0)}] + V_{nk} = E_n^{(1)} \delta_{nk} \quad . \tag{7.21}$$

This equation can be satisfied only if $V_{nk} = 0$ when $E_k^0 = E_n^{(0)}$.

To see how we can overcome this difficulty we consider a situation in which

$$E_n^{(0)} = E_{n+1}^{(0)} = \ldots = E_{n+m}^{(0)} \tag{7.22}$$

((m+1) - fold degeneracy of the eigenvalue $E_n^{(0)}$).

The corresponding eigenfunctions of $H_0, \{\psi_n^{(0)}, \ldots, \psi_{n+m}^{(0)}$, span a vector space W of dimension $(m+1)$. To overcome the difficulty posed by equation (7.21) we now assume (and this assumption is justified in most applications) that there exist in W an orthonormal basis of functions $\varphi_0, \ldots, \varphi_m$ which are eigenfunctions of V, i.e.,

$$V\varphi_k = \lambda_k \varphi_k \qquad k = 0, \ldots, m \quad . \tag{7.23}$$

These functions are said to diagonalize V on the space W since

$$V_{ij} = \int_V \varphi_j^*(x) V \varphi_i(x) dx = \lambda_i \delta_{ij} \tag{7.24}$$

(Note that φ_k, $k = 0, \ldots, m$ are still eigenfunctions of the H_0 since they are linear combinations of ψ_{n+i} $i = 0, \ldots, m$). If we now replace the functions $\psi_n^{(0)}, \ldots, \psi_{n+m}^{(0)}$ in the original basis of the Hilbert space by $\varphi_0, \ldots, \varphi_m$ and follow the perturbation procedure for the nondegenerate case we shall encounter no difficulties with equation (7.21) in view of equation (7.24). Thus, we obtain to first order in ϵ the following approximations;

$$E_{n+k} = E_n^{(0)} + \epsilon\lambda_k \quad , \quad k = 0, \ldots, m \tag{7.25}$$

$$\psi_{n+k} = \varphi_k + \epsilon \sum_{i \neq n} \frac{V_{ni}}{E_n^{(0)} - E_i^{(0)}} \psi_i^{(0)} \tag{7.26}$$

where the summation in (7.26) extends only over those ψ's whose eigenvalue $E_i^{(0)}$ is different from $E_n^{(0)}$.

Remarks:

1. Note that the perturbation usually removes the degeneracy in the eigenvalues at least partially.

2. In general one can not find exact eigenfunctions of V in W but one can find $\varphi_k \epsilon W$

which satisfy on the whole eigenfunction space $V\varphi_k = \lambda_k\varphi_k$ + terms in the orthogonal complement of W. In most cases, however, these additional "off subspace" terms are small and can be neglected.

3. Many cases of physical interest require perturbations of a degenerate spectrum as we illustrate in the next section.

EXERCISE 7

1. Show that c_{nn} is an arbitrary pure imaginary number. (In Quantum mechanics this is equivalent to a choice of a phase in the Wave function which has no physical relevance).
2. Use equation (7.10) to evaluate the second order perturbation term for E_n.
3. Calculate first order perturbation corrections to the eigenvalues and eigenfunctions of the following equation

$$y'' - k^2x^2y + \epsilon x^3y = Ey$$

Remark: The equation $y'' - k^2x^2y = Ey$ represents, in Quantum mechanics, a Harmonic oscillator in one dimension.

*8. THE ZEEMAN AND STARK EFFECTS.

In this section we consider two examples to illustrate the perturbation technique which was presented in the previous section. Both examples are related to the Quantum mechanical treatment of the Hydrogen atom subject to external forces which are treated as perturbations on its free states (i.e., its states with no external forces). We begin our discussion by considering these (free) states.

A. The (free) Hydrogen Atom.

In Quantum mechanics the (time independent) Schrödinger equation whose eigenfunctions describe the states of the Hydrogen atom is (we ignore spin)

$$\left[-\frac{\hbar^2}{2m_e}\nabla^2 - \frac{e^2}{r}\right]\psi(\mathbf{x}) + E\psi(\mathbf{x}) = 0 \tag{8.1}$$

where $h = 2\pi\hbar$ is Planck constant, m_e is the electron mass, e the electron charge and E (the eigenvalue) is the energy of the state.

This equation can be solved in spherical coordinates by separation of variables.

If we consider only the discrete spectrum of equation (8.1), which describe the bounded states of the electron in the atom, we obtain the following eigenfunctions and eigenvalues;

$$\psi_{n,\ell,m} = R_{n\ell}(r)P_\ell^m(\cos\theta)e^{im\varphi} \tag{8.2}$$

$$0 \le \ell \le n-1, \quad -\ell \le m \le \ell$$

$$E_n = -\left[\frac{e^2}{\hbar}\right]^2 \frac{m_e}{2n^2} \quad . \tag{8.3}$$

In these equations n,ℓ,m are integers, $n \ge 1$, $P_\ell^m(\cos\theta)$ are the associated Legendre polynomials

$$P_\ell^m(u) = \frac{(1-u^2)^{\frac{m}{2}}}{2^\ell\cdot\ell!}\frac{d^{\ell+m}}{du^{m+\ell}}(u^2-1)^\ell \tag{8.4}$$

and $R_{n\ell}$ are the solutions of

$$\left[\frac{d^2}{dr^2} + \frac{2m_e}{\hbar^2}\frac{e^2}{r} - \frac{\ell(\ell+1)}{r^2} + \frac{2m_eE_n}{\hbar^2}\right] R_{n\ell} = 0 \tag{8.5}$$

(these can be expressed in terms of the confluent hypergeometric series). The functions $\psi_{n,\ell,m}$ are normalized so that

$$\int_{R^3} |\psi_{n,\ell,m}|^2 d\mathbf{x} = 1 \quad . \tag{8.6}$$

From equations (8.2),(8.3) we infer that the eigenvalues of equation (8.1) are degenerate (except n = 1) since they depend only on n and, therefore, it follows that to each eigenvalue there correspond

$$\sum_{\ell=0}^{n-1} (2\ell+1) = n^2 \quad \text{states.}$$

In the following we investigate the effect of an external magnetic and electric fields on these eigenvalues and their degeneracy.

B. Zeeman effect.

When one applies a weak and uniform magnetic field in the z-direction on the Hydrogen atom the Schrödiner equation which describes the system (equation (8.1)) is modified by the addition of a perturbation

$$V = -\frac{eH}{2m_e}\frac{\hbar}{i}\frac{\partial}{\partial\varphi} + \frac{e^2H^2}{8m_e}(x^2+y^2) \tag{8.7}$$

where H is the strength of the magnetic field. (Note that the first term in (8.7) is an operator.) However, since we assume that the magnetic field is weak the second term in equation (8.7) can (and will) be neglected. Under these assumptions the matrix of the perturbation V is already diagonal in the basis $\{\psi_{n,\ell,m}\}$ since

$$V\psi_{n,\ell,m} = -\frac{eH}{2m_e}m\hbar\,\psi_{n,\ell,m} \; . \tag{8.8}$$

Moreover, since $\psi_{n,\ell,m}$ are orthogonal to each other we obtain that

$$V_{n,\ell,m,n',\ell',m'} = -\frac{eH}{2m_e}m\hbar\delta_{nn'}\delta_{\ell\ell'}\delta_{mm'} \; . \tag{8.9}$$

Hence each eigenvalue E_n will now split into

$$E_{n,m} = E_n - \frac{eH}{2m_e}m\hbar, \quad -(n-1) \le m \le (n-1). \tag{8.10}$$

Thus, the perturbation due to the magnetic field has reduced the degeneracy of the spectrum but did not eliminate it completely.

Example: Discuss explicitly the Zeeman effect on $n = 2$.

Solution: The eigenvalue E_2 in the hydrogen atom is four fold degenerate since $\psi_{2,1,1}$, $\psi_{2,1,0}$, $\psi_{2,1,-1}$ and $\psi_{2,0,0}$ correspond to it. Under the effect of a magnetic field this level will split into three

1. $E = E_2 - \frac{eH\hbar}{2m_e}$ with $\psi_{2,1,1}$,
2. $E = E_2 + \frac{eH\hbar}{2m_e}$ with $\psi_{2,1,-1}$,
3. $E = E_2$ with $\psi_{2,1,0}$ and $\psi_{2,0,0}$ (two fold degenerate).

C. Stark effect.

When a uniform electric field in the z-direction is applied to the Hydrogen atom equation (8.1) is modified by the addition of the perturbation

$$V = e\mathcal{E}z = -e\mathcal{E}r\cos\theta \tag{8.11}$$

where ϵ is the strength of the electric field.

It can be easily seen now that the functions $\psi_{n,\ell,m}$ are not eigenfunctions of V and hence in order to compute the effect of this perturbation one must find for each n a basis for the corresponding subspace in which V is diagonal.

We carry this out explicitly for $n = 2$.

The state $n = 2$ is four fold degenerate and hence in principle we have to find a basis which diagonalize the 4x4 matrix of V on this subspace. However, nonzero off diagonal elements in this matrix appear only between $\psi_{2,1,0}$ and $\psi_{2,0,0}$ and hence the problem reduces effectively to one in two dimensions. A direct calculations yields;

$$V_2 = \left[\begin{array}{c|c} \bigcirc & \begin{array}{c}\bigcirc\end{array} \\ \hline \bigcirc & \begin{array}{cc} 0 & 3e\mathcal{E}a \\ 3e\mathcal{E}a & 0 \end{array} \end{array}\right] \tag{8.12}$$

where $a = \hbar^2/(e^2 m_e)$ is called the Bohr radius.

The eigenvalues of V_2 are

$$\lambda_1 = \lambda_2 = 0, \quad \lambda_3 = 3e\mathcal{E}a, \quad \lambda_4 = -3e\mathcal{E}a \tag{8.13}$$

and the corresponding eigenvectors are

$$\begin{aligned} \varphi_1 &= \psi_{2,1,1} \quad , \quad \varphi_2 = \psi_{2,1,-1} \\ \varphi_3 &= \frac{1}{\sqrt{2}} (\psi_{2,1,0} + \psi_{2,0,0}) \\ \varphi_4 &= \frac{1}{\sqrt{2}} (\psi_{2,1,0} - \psi_{2,0,0}) \quad . \end{aligned} \tag{8.14}$$

Schematically we can describe the splitting of the $n = 2$ level as follows:

$$\begin{array}{lcl} E_2 + 3e\mathcal{E}a & \text{———————} & \varphi_3 \\ E_2 & \text{———————} & \varphi_1, \ \varphi_2 \\ E_2 - 3e\mathcal{E}a & \text{———————} & \varphi_4 \end{array}$$

Stark effect on $n = 2$

(Thus one of the eigenvalues is still two fold degenerate).

EXERCISE 8

1. Show that equations (8.1),(8.2) give the eigenfunctions and eigenvalues of (8.1).

Remark: The hydrogen atom is treated in many books on Quantum mechanics, the reader is encouraged to consult these books regarding this problem.

2. Carry out the computation of V_2 in eq. (8.12)

Hint: Note that P_ℓ^m is even if ℓ is even and odd if ℓ is odd. Also note that $\cos\theta \approx P_1$. For the computation of the off diagonal elements the explicit form of $\psi_{2,1,0}$ and $\psi_{2,0,0}$ is needed.

$$\psi_{2,1,0} = \frac{1}{4(2\pi a_0^3)^{1/2}} \cdot \frac{r}{a_0} \exp(-r/2a_0)\cos\theta \quad (8.15)$$

$$\psi_{2,0,0} = \frac{1}{4(2\pi a_0^3)^{1/2}} \left(2 - \frac{r}{a_0}\right)\exp(-r/2a_0) \quad (8.16)$$

3. Discuss the perturbation of

$$(\nabla^2 + k^2 r^2)\psi = E\psi \quad (8.17)$$

by a term of the form $V = az$.

Remark: Equation (8.17) represent a Harmonic oscillator in 3-dimensions

4. Discuss the Zeeman effect for $n = 3$.

BIBLIOGRAPHY

1. V. I. Arnold - Geometrical methods in the theory of ordinary differential equations, Springer-Verlag, 1983.
2. K. W. Chung and F. A. Howes - Nonlinear singular perturbation phenomena, Springer-Verlag, 1984.
3. N. N. Bogolinbov and V. A. Mitropolsky - Asymptotic methods in the theory of nonlinear oscillations, Hindustan Pub. Co., 1961.
4. A. H. Nayfeh - Perturbation Methods, J. Wiley, 1973.
5. T. Kato - A short introduction to perturbation theory, Springer-Verlag, 1982.
6. J. Kevorkian and J. D. Cole - Perturbation methods in applied mathematics, Springer-Verlag, 1980.
7. A. T. Fromhold, Jr. - Quantum Mechanics for applied physics and engineering, Academic Press, 1981.
8. A. H. Nayfeh - Problems in Perturbations, J. Wiley, 1985.
9. L. I. Schiff - Quantum Mechanics, 3rd edition, McGraw Hill, 1968.
10. R. Adler, et al - Introduction to General Relativity, 2nd edition, McGraw Hill, 1975.

CHAPTER 8. PHASE DIAGRAMS AND STABILITY

1. GENERAL INTRODUCTION

In most realistic models of scientific and engineering systems one is confronted with the necessity to analyze and solve systems of nonlinear differential equations. However, since the search for exact analytic solutions of such systems is, in most instances hopeless, it is natural to inquire in retrospect what is the most crucial information that has to be extracted from these equations. One discovers then that many such systems have transient states which are time dependent and equilibrium states which are time independent states. The equilibrium states are usually the most significant from a practical point of view and their stability against small perturbations and/or small changes in the system parameters is a central problem in the design and analysis of these systems.

We point out that the consideration of such small perturbations is natural and necessary since all mathematical models of real life systems use some approximations and idealizations. Consequently, the parameters that appear in the model equations as well as the equations themselves are approximate and, therefore, their predictions about the behavior of the real system are always subject to errors.

From a historical point of view the first attempts to analyze these equilibrium states without solving the model equations were due to H. Poincare and A. M. Liapounov. Their investigations showed that while it is very difficult (or impossible) to obtain an analytic solution for the transient states of the system there exist a simple, natural "qualitative" approach to the analysis of the equilibrium states. Today this type of analysis is an important and fruitful field of mathematical research. In this chapter, however, we give an elementary exposition of the basic ideas and results of this approach.

To motivate our study and introduce some of the basic ideas we consider the following model;

Example 1: For an ecological system with one species and abundance of food (e.g. fish in an artificial pool) the rate of change in the population $N(t)$, which we treat as a continuous variable, is due to the difference between the rate of reproduction and the rate of death. Since both of these processes are proportional to the size of the population we obtain

$$\frac{dN}{dt} = \text{rate of reproduction} - \text{rate of death}$$

$$= \alpha N - \beta N = (\alpha-\beta)N = aN. \tag{1.1}$$

However, if the amount of food available is limited there will be a competition for it with a corresponding decrease in the rate of change of the population which will be proportional to $N(N-1)/2 \sim N^2$ which represents the number of pairs in this population. It follows then that a model equation which describes such an ecological system is

$$\frac{dN}{dt} = aN - bN^2 = N(a-bN) \quad , \quad a,b > 0 \quad . \tag{1.2}$$

At equilibrium $\frac{dN}{dt} = 0$ and hence equation (1.2) has two equilibrium states $N = 0$ and $N = \frac{a}{b}$.

To study the stability of these equilibrium states we note that equation (1.2) though nonlinear can be integrated by elementary methods and we obtain

$$N(t) = \frac{aN_0e^{at}}{a-bN_0+bN_0e^{at}}$$

where $N_0 = N(0)$ is the initial population.

From equation (1.3) we infer that whenever $N_0 \neq 0$ the population $N(t)$ will approach a/b as $t \to \infty$ even if N_0 is very small. Thus, a small perturbation of the ecological system from $N = 0$ will cause its state to "run away" from this equilibrium while similar deviations from $N = a/b$ will decay as $t \to \infty$ and the system will return to its original state. We conclude, therefore, that the equilibrium state $N = 0$ is unstable while $N = a/b$ is (asymptotically) stable.

A graphical representation of this result is depicted in Figure 1 where we plotted some of the solution curves for the model as given by equation (1.3).

The objective of phase space techniques and stability theory is to derive these same results about the equilibrium points *without solving the differential equations under consideration*. This last feature is crucial in many instances since most realistic mathematical models lead to systems of nonlinear equations which cannot be solved analytically.

The basic idea of these techniques is to view equation (1.2) as an algebraic relation between N and dN/dt and remember that $\frac{dN}{dt} > 0$ (<0) implies that N is increasing (decreasing) function with time. Thus in the phase plane $N - \frac{dN}{dt}$ equation (1.2) is described by Figure 2 (remember that $N < 0$ is meaningless in this model).

We see from this figure that if $N < a/b$ then $\frac{dN}{dt} > 0$ and hence N is an increasing function. On the other hand if $N > a/b$ then $\frac{dN}{dt} < 0$ and N will decrease (these facts are indicated by arrows in the diagram). This is characteristic behavior of a stable equilibrium at a/b and unstable one at $N = 0$. We point out that these results about the stability of the equilibrium points were reached without any reference to the actual solutions of equation (1.2).

As another example of this technique we consider equation (1.2) with $a, b < 0$. The phase diagram for equation (1.2) with these parameters is given in Figure 3. We see that if $N < a/b$ then $\frac{dN}{dt} < 0$ and N will decrease to 0. On the other hand if $N > a/b$ then $\frac{dN}{dt} > 0$ and N will increase without bound. We conclude, therefore, that for these parameters $N = 0$ is a stable equilibrium while $N = a/b$ is unstable.

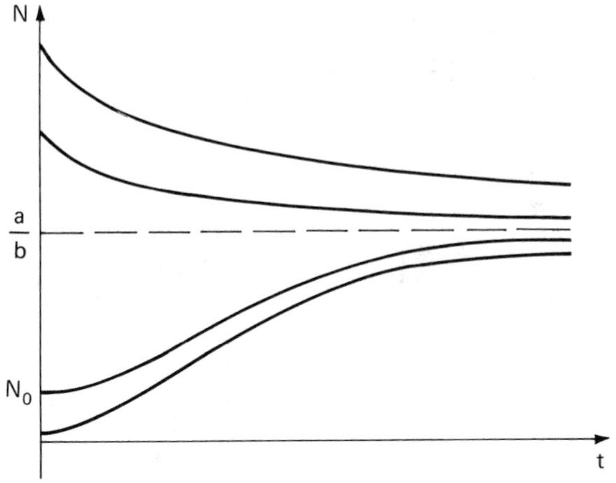

Figure 1: Solution Curves for Equation (1.2)

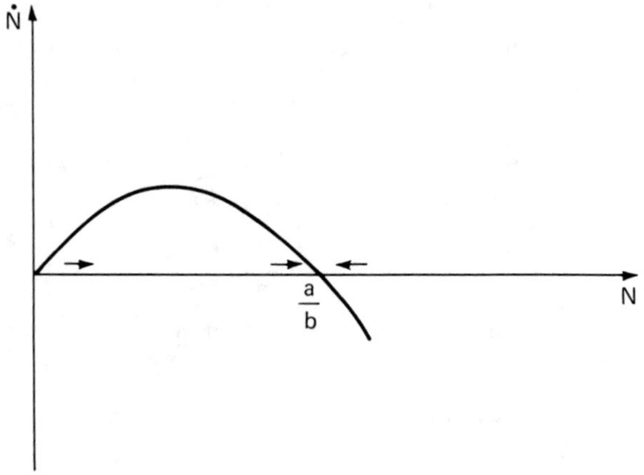

Figure 2; a,b > 0

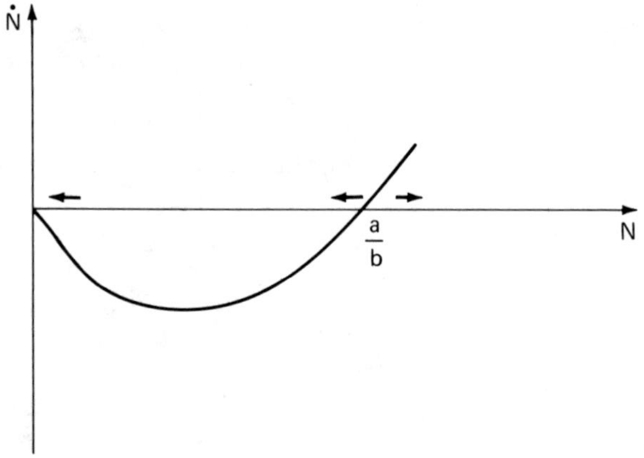

Figure 3: a,b < 0

EXERCISE 1

1. Solve the following equations in order to determine whether each of the equilibrium points is stable or unstable

   a. $\frac{dN}{dt} = bN^2 - aN$, a,b > 0 , N > 0

b. $\frac{dN}{dt} = aN + bN^2$, $a,b > 0$, $-\infty < N < \infty$

c. $\frac{dN}{dt} = aN + bN + c$, $a,b,c > 0$, $-\infty < N < \infty$.

2. Use phase space techniques to discuss the stability of the equilibrium points for

a. $\frac{dN}{dt} = aN^k(b-N)^m$

where $a,b > 0$ and $k,m > 0$. Assume $N \geq 0$.

b. $\frac{dN}{dt} = aN + bN^2 + cN^3$, $b^2 - 4ac > 0$

c. $\frac{dN}{dt} = N^2(aN^2+bN+c)$, $b^2 - 4ac > 0$

d. $\frac{dN}{dt} = (N-a)(N-b)(N-c)$

Pay special attention to the case where $a = b$ or $a = b = c$.

3. A model of two societies A,B where A exploits B was suggested by May and Noy-Meir. According to this model if N,x are the population sizes of A,B respectively then

$$\frac{dx}{dt} = ax(1-x/M) - cN \frac{x^2}{1+x^2}$$

In this equation the first term represents the natural growth of society B while the last represents the loss due to encounter between the two societies. Discuss the equilibrium states and their stability when N is small, moderate and large.

Hint: Plot the two terms in the equation above separately. The equilibrium states are represented by the intersection of the two curves.

2. SYSTEMS OF TWO EQUATIONS

Mathematical models which consist of two first order equations appear in many applications. In particular many mechanical and electrical systems are modelled by one ordinary differential equation of second order which is equivalent to such a system. In this section we demonstrate through several examples how to apply phase space techniques to determine the stability of the equilibrium states of such systems.

Example 1: Two competing species.

Model: In this model we consider an ecological system which consists of two species of fish in a lake which consume the same organic food as well as members of the other species. If only one of these species existed in the lake then its population will be governed by equation (1.2). However, due to the adverse effect of the second species there will be a decrease in the rate of change of the population of each of these species which we assume to be proportional to the product of the population's size in each instance. If $F_1(t)$, $F_2(t)$ represent the fish population of the two species at time t the ecological system under consideration will be modelled by

$$\frac{dF_1}{dt} = F_1(a_1-b_1F_1-c_1F_2) \tag{2.1}$$

$$\frac{dF_2}{dt} = F_2(a_2-b_2F_2-c_2F_1) \tag{2.2}$$

where a_i, b_i, c_i, $i = 1,2$ are nonnegative constants.

Equilibrium States

The equilibrium states of the system (2.1)-(2.2) are given by the simultaneous solutions of

$$F_1(a_1-b_1F_1-c_1F_2) = 0 \tag{2.3}$$

$$F_2(a_2-b_2F_2-c_2F_1) = 0 \tag{2.4}$$

(remember: at equilibrium $\frac{dF_1}{dt} = \frac{dF_2}{dt} = 0$) which are;

(1) $F_1 = F_2 = 0$

(2) $F_1 = 0$, $F_2 = \frac{a_2}{b_2}$

(3) $F_1 = \frac{a_1}{b_1}, \quad F_2 = 0$

(4) the intersection of the lines

$$a_1 - b_1F_1 - c_1F_2 = 0 \tag{2.5}$$

$$a_2 - b_2F_2 - c_2F_1 = 0 \tag{2.6}$$

if such an intersection exists in the first quadrant of the $F_1 - F_2$ plane (obviously negative populations are meaningless).

Stability Analysis

To analyze the equilibrium points with respect to their stability, using phase space techniques, we must first find the regions in the $F_1 - F_2$ plane for which $\frac{dF_1}{dt} \gtrless 0$, $\frac{dF_2}{dt} \gtrless 0$. To accomplish this we note that $\frac{dF_1}{dt} > 0$ when

$$F_1(a_1 - b_1F_1 - c_1F_2) > 0 \quad .$$

But $F_1 \geq 0$ and, therefore, $\frac{dF_1}{dt}$ is positive if and only if

$$a_1 - b_1F_1 - c_1F_2 > 0 \quad , \quad F_1 > 0 \quad . \tag{2.7}$$

Similarly $\frac{dF_1}{dt}$ is negative if and only if

$$a_1 - b_1F_1 - c_1F_2 < 0 \quad , \quad F_1 > 0 \quad . \tag{2.8}$$

It follows then that in the $F_1 - F_2$ plane the dividing line between the regions in which $\frac{dF_1}{dt}$ is positive and negative is given by

$$a_1 - b_1F_1 - c_1F_2 = 0 \tag{2.9}$$

(see Figure 4). Similar analysis for $\frac{dF_2}{dt}$ shows that

1. $\frac{dF_2}{dt} > 0$ if and only if

$$a_2 - b_2F_2 - c_2F_1 > 0 \quad , \quad F_2 > 0 \tag{2.10}$$

2. $\frac{dF_2}{dt} < 0$ if and only if

$$a_2 - b_2F_2 - c_2F_1 < 0 \quad , \quad F_2 > 0 \qquad (2.11)$$

3. The two regions in which $\frac{dF_2}{dt}$ is positive or negative are separated by the line

$$a_2 - b_2F_2 - c_2F_1 = 0 \qquad (2.12)$$

(see Figure 5).

The relative position of the two lines (2.9), (2.12) can be any of the four shown in Figure 6 (depending on the values of a_i, b_i, c_i, $i = 1,2$) and for each of these possibilities the stability of the corresponding equilibrium points can be inferred as follows; starting from an initial point near an equilibrium point E we trace the approximate path of the system in the $F_1 - F_2$ plane using the arrows which indicate at each point whether F_1, F_2 are increasing or decreasing there. If all such trajectories move towards the equilibrium point then E is asymptotically stable. If on the other hand *even one* such trajectory tends to move away from E then it is unstable.

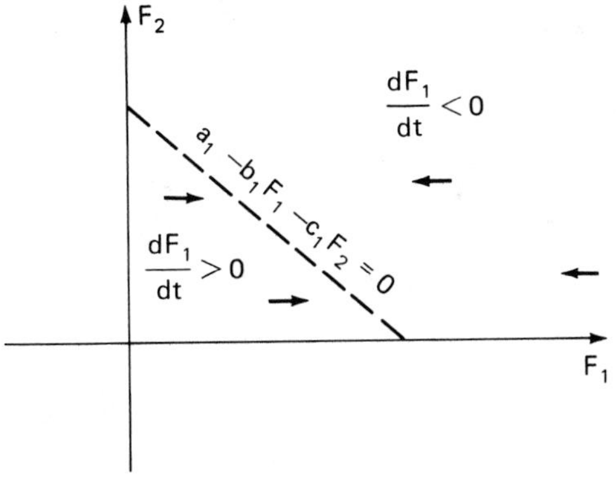

Figure 4

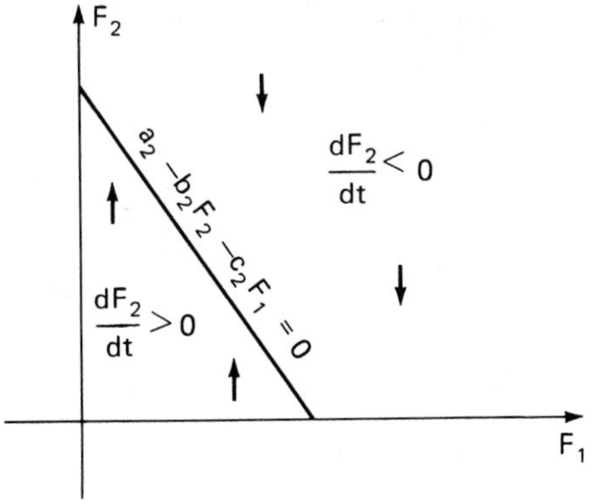

Figure 5

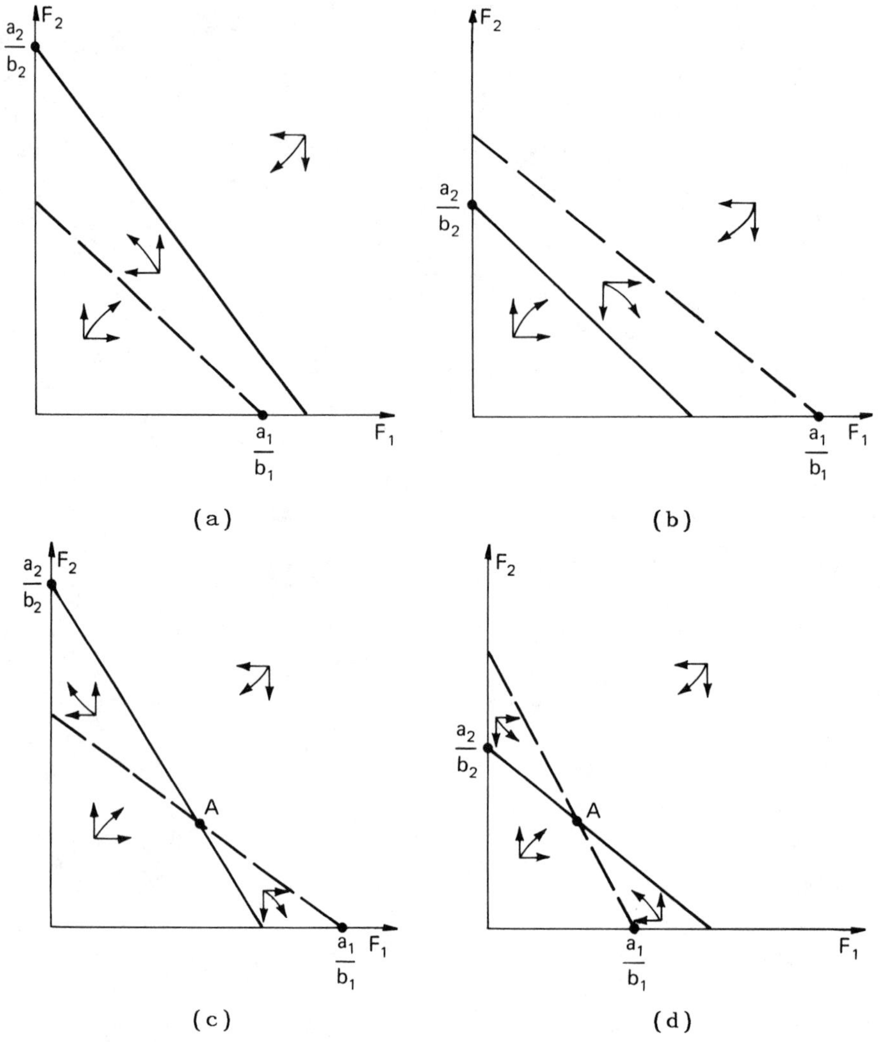

Figure 6

In the border case where the trajectories do not approach nor run away from E we say that it is stable. (More precise definitions will follow in Section 3.) Using this approach to analyze the situation depicted in Figure 6 (a) we find that

1. $(0,0)$ is unstable

2. $\left[\frac{a_1}{b_1}, 0\right]$ is unstable in the y-direction and hence unstable,

3. $\left[0, \frac{a_2}{b_2}\right]$ is a stable equilibrium point.

Similarly for the situation described by Figure 6 (d) we find that $(0,0)$ $\left[\frac{a_1}{b_1}, 0\right]$ and $\left[0, \frac{a_2}{b_2}\right]$ are unstable while the equilibrium point A given by the intersection of the two lines is stable. Ecologically this means that the two species can coexist only with populations given by the coordinates of the stable equilibrium point A.

Example 2: Arms level of two nations (Richardson's model).

Model: In this example we consider two nations A,B and their arms level x,y respectively. To derive the equations which determine x and y we assume that three processes contribute to the rate of change in these variables viz. reaction, constraint and grievance.

The first of these processes constitutes the reaction of each nation to the arms level of the other. In the following we assume that its contribution to the rate of change in the state vector $\begin{bmatrix} x \\ y \end{bmatrix}$ is given by $\begin{bmatrix} a_1 y \\ a_2 x \end{bmatrix}$, $a_1, a_2 > 0$. Thus this process leads to an increase in the arms level of each nation by linear reaction to the arms level of the other.

The second process "constraints" which is due to economic and political considerations induces a decrease in the arms level of the two nations which we assume to be linearly related to the arms level of each nation, i.e. its contribution to the rate of change in $\begin{bmatrix} x \\ y \end{bmatrix}$ is given by $-\begin{bmatrix} b_1x \\ b_2y \end{bmatrix}$, $b_1, b_2 > 0$.

The third process "grievance" is related to the historic and current political competition between the two nations. Its contribution to the rate of change in $\begin{bmatrix} x \\ y \end{bmatrix}$ is modelled by $\begin{bmatrix} h_1 \\ h_2 \end{bmatrix}$ where $h_1, h_2 > 0$ if there is historic grievance and $h_1, h_2 < 0$ if the nations stand in friendly relation.

Finally we assume that the three processes are additive to each other and hence the model equations for the arms level of the two nations are given by

$$\frac{dx}{dt} = a_1y - b_1x + h_1 \tag{2.13}$$

$$\frac{dy}{dt} = a_2x - b_2y + h_2 \tag{2.14}$$

Equilibrium states:

To find the equilibrium states of the model given by equations (2.14)-(2.14) we first observe that negative x or y are meaningless within the context of this model and hence we must consider only the first quadrant in the $x - y$ plane. Furthermore, at equilibrium $\frac{dx}{dt} = \frac{dy}{dt} = 0$ and, therefore, the equilibrium states of this model must satisfy simultaneously the equations

$$a_1y - b_1x + h_1 = 0 \tag{2.15}$$

$$a_2x - b_2y + h_2 = 0 \quad . \tag{2.16}$$

In phase space these equations represent the lines

$$(E_1) \qquad y = \frac{b_1}{a_1}x - \frac{h_1}{a_1} \tag{2.17}$$

$$(E_2) \qquad y = \frac{a_2}{b_2} x + \frac{h_2}{b_2} \qquad (2.18)$$

and we observe that both lines have positive slope.

Stability Analysis

The nature and stability of the equilibrium states of this model depend on the mutual position of the lines (2.17)-(2.18) which is depicted in Figure 7.

To analyze this figure we observe that in Figure 7 (a) the y-intercept of (2.17) is negative while the y-intercept of (2.18) is positive. Furthermore, the slope of the first line is larger than that of the second and hence the two lines intersect in the first quadrant. It is clear from the direction of the arrows indicated on this diagram that the equilibrium state represented by the intersection of these two lines is stable.

While the situations shown in Figures 7(b) and 7(c) contain no equilibrium state still the two figures represent two completely different real life situations viz. arms race in Figure 7(b) and disarmament in Figure 7(c). As to the equilibrium point in Figure 7(d) it is easy to deduce that it represents an unstable equilibrium. Furthermore, we point out that there is a line called "separatrix" (shown as dashed line in this figure) so that different outcomes to the arms levels of the two nations arise from trajectories which commence on either side of this line. Thus, for trajectories below the separatrix there will be a disarmament process while for those above it the nations will engage in an arms race. We conclude then

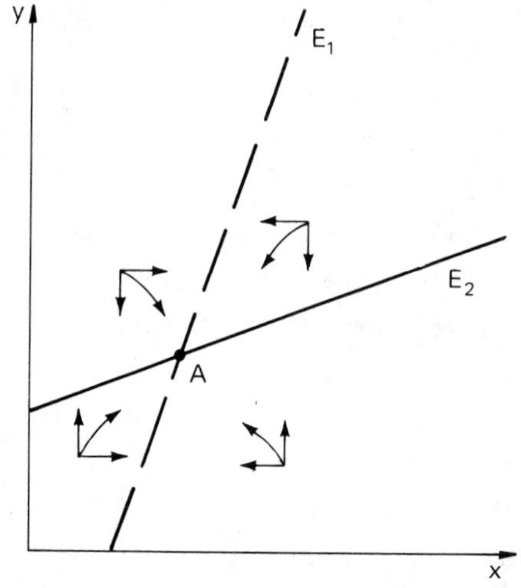

(a) A is a stable equilibrium
$h_1, h_2 > 0$, $b_1 b_2 > a_1 a_2$

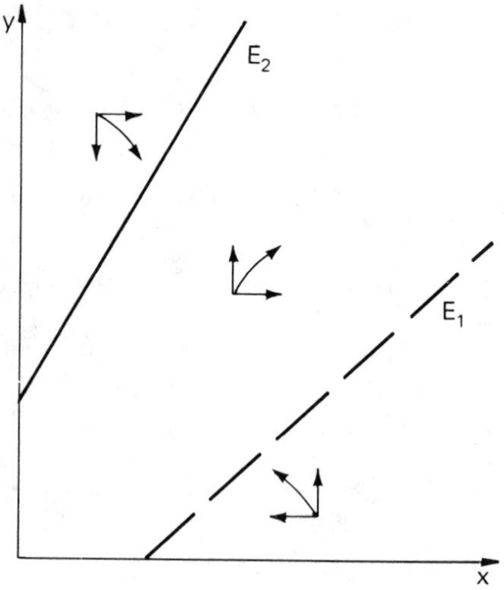

(b) Arms race
$h_1, h_2 > 0$, $b_1 b_2 < a_1 a_2$

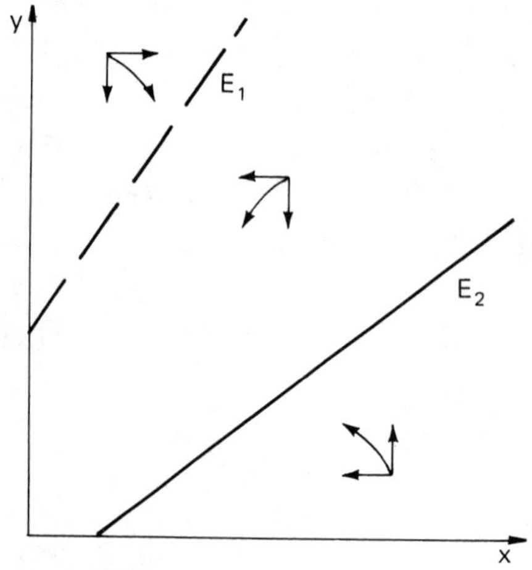

(c) Disarmament
$h_1, h_2 < 0$, $b_1 b_2 > a_1 a_2$

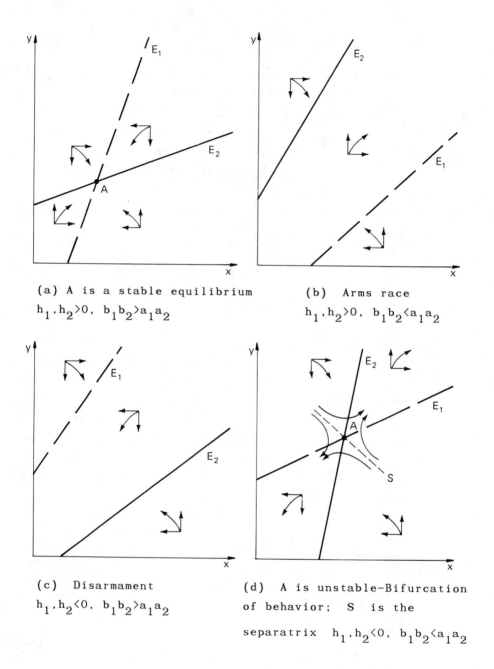

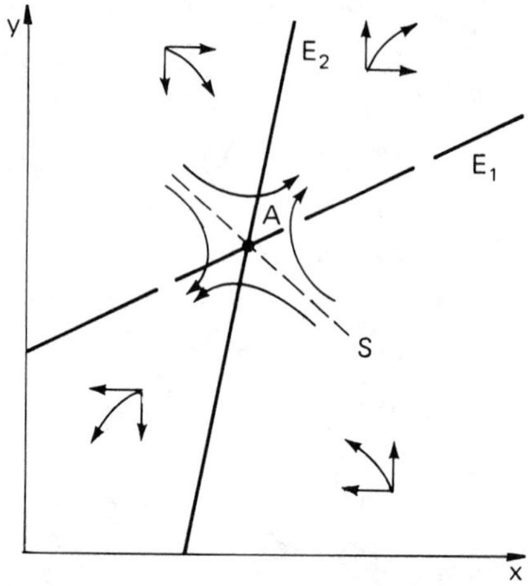

(d) A is unstable-Bifurcation of behavior; S is the separatrix $h_1, h_2 < 0$, $b_1 b_2 < a_1 a_2$

Figure 7 Richardson Process

that this line represents a "bifurcation" in the arms level and the political relationship between the two nations.

Example 3: An Exchange Model

Model: In this model we consider two producers A, B of two commodities (whose unit value is the same) which by treaty have to exchange q percent of their production. Thus, if $x, y > 0$ are the productions of A, B respectively then A receives from B qy units and retains $px = (1-q)x$ of his own production. Similarly B receives from A qx units of A's production and retains py of his own product. Furthermore, we assume that the utility functions (=gain) for A, B respectively are of the form

$$u_A = \ln(1+px+qy) - bx \tag{2.19}$$

$$u_B = \ln(1+qx+py) - by \quad . \tag{2.20}$$

In these formulas the first expressions with the ln function represent a utility which increases with x and y but with decreasing margins i.e. the gain per additional unit decreases with increasing values of x and y. As to the second terms $-bx$, $-by$ they represent the desire on the part of A and B to do less work.

Finally, to obtain differential equations for the rate of change in x and y we assume that each producer changes his *production* in the direction of maximum utility i.e.

$$\frac{dx}{dt} = \frac{\partial u_A}{\partial x}, \quad \frac{dy}{dt} = \frac{\partial u_B}{\partial y} \quad . \tag{2.21}$$

Thus we infer that

$$\frac{dx}{dt} = \frac{p}{1+qx+py} - b \tag{2.22}$$

$$\frac{dy}{dt} = \frac{p}{1+px+qy} - b \quad . \tag{2.23}$$

Equilibrium States:

At equilibrium $\frac{dx}{dt} = \frac{dy}{dt} = 0$ and hence x,y must satisfy simultaneously

$$px + qy = \frac{p}{b} - 1 \tag{2.24}$$

$$qx + py = \frac{p}{b} - 1 \quad . \tag{2.25}$$

We infer then that at equilibrium

$$x(p-q) = y(p-q) \tag{2.26}$$

i.e. if $p \neq 1/2$ then there exists a unique equilibrium which is given by

$$x = y = \frac{p}{b} - 1$$

Stability Analysis:

We first observe that when $p < b$ the lines (2.24)-(2.25) do not enter the first quadrant of the x-y plane and, therefore, the equilibrium point is devoid of any real life meaning. The rates of change in x and y in the first quadrant under these circumstances are indicated in Figure 8(a) and we conclude that under these conditions both A and B will cease their production.

When $p \geq b$ the lines (2.24)-(2.25) intersect in the first quadrant of the x-y plane and the stability of the equilibrium point depends on whether $p>q$ or $p<q$ as indicated in Figures 8(b), 8(c). We again point out that for the unstable equilibrium depicted in Figure 8(c) a separatrix line exists so that trajectories which start above it will cause y to exploit x while those which start lower will cause x to exploit y.

Applications:

The model described above can be applied to many real life situations which come close to its set up. We consider two such applications.

1. Communist commune: In this case the individual producer gives to the commune (or government) his total production i.e. $q \cong 100\%$ and receives from the government its necessities regardless of his production. Obviously in this case $p = 1 - q < b$ (whatever b is) and our model leads us to predict that the individual producer will cease its production. This explains why these communes are such a total economic failure.
2. Tax system in the U.S.: In this system each taxpayer pays to the government part of his/her income in return for some services. We see from our model that if $q > 1/2$ the system will lead to exploitation, i.e. some citizens will exploit the system while others will be exploited by it. Thus such a system will lead to social unrest and be unstable. On the other hand if $q < 1/2$ the system will be socially stable. This helps to explain why the maximum federal tax in the U.S. (at the present time) is less than 50% and why the government is willing to let taxpayers deduct state and local taxes from their income.

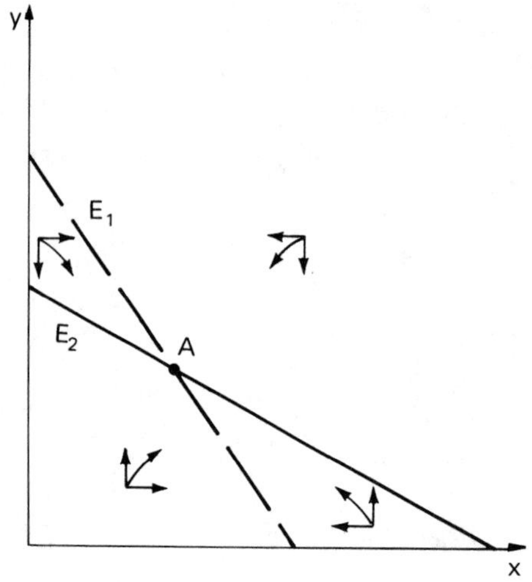

(b) p>b, p>q
stable equilibrium

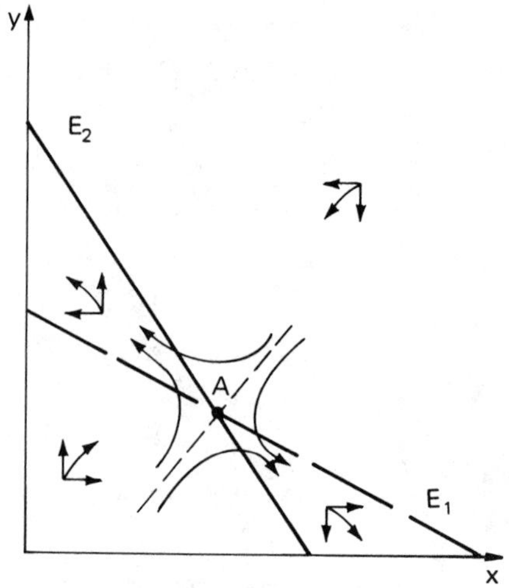

(c) p>b, p<q A is
unstable expoita-
tion outcome

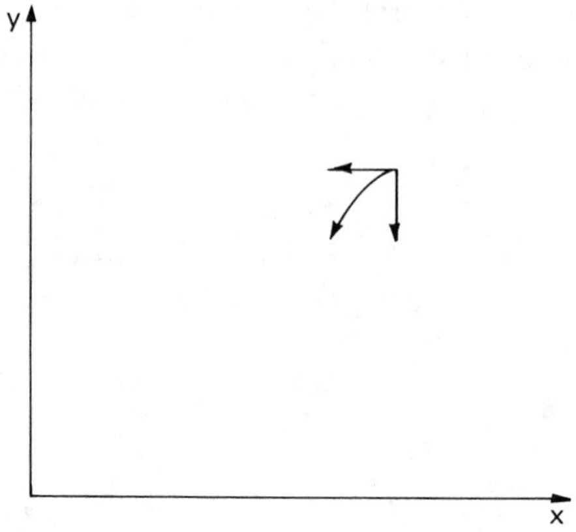

(a) p<b No production outcome

Figure 8: Exchange Model

EXERCISE 2

1. Show that in Fig. 6(b) the equilibrium states $(0,0)$ and $\left[0, \frac{a_2}{b_2}\right]$ are unstable while $\left[\frac{a_1}{b_1}, 0\right]$ is stable.

2. Show that in Fig. 6(c) the equilibrium states $(0,0)$ and A are unstable while $\left[0, \frac{a_2}{b_2}\right]$ and $\left[\frac{a_1}{b_1}, 0\right]$ are stable. This result means that under these circumstances only one species will survive in this ecosystem. This is called the "principle of competitive exclusion".

3. Show that the signs of $\frac{dx}{dt}, \frac{dy}{dt}$ as shown by the arrows in Fig. 7 are correct.

4. Repeat ex. 3 for Fig. 8.

5. Discuss the stability of the equilibrium states for the following systems of ODE's.

   a. $\frac{dx}{dt} = x - 2xy, \quad \frac{dy}{dt} = y - xy$

   b. $\frac{d^2\theta}{dt^2} + \frac{g}{L} \sin \theta = 0$ (nonlinear pendulum)

   where L is the length of the pendulum and g is the acceleration of gravity.

   c. $m \frac{d^2x}{dt} + b \frac{dx}{dt} + kx = 0, \quad m, b, k > 0$

   (spring mass system with damping). What happens if $b < 0$?

   d. $\frac{d^2x}{dt^2} - \mu(1-x^2) \frac{dx}{dt} + Kx = 0, \quad \mu > 0$

   (Van der Pol equation)

   e. $\frac{d^2\theta}{dt^2} + b \frac{d\theta}{dt} + \frac{g}{L} \sin \theta = 0, \quad b > 0$

   (damped pendulum).

3. SOME GENERAL THEORY.

In the previous sections of this chapter we introduced the basic concepts of stability theory in an intuitive way. In this section we reintroduce these concepts in a more general and formal context.

Definition 1: A system of ordinary differential equations

$$\frac{d\mathbf{x}}{dt} = \mathbf{F}(\mathbf{x},t) = \begin{bmatrix} f_1(\mathbf{x},t) \\ \vdots \\ f_n(\mathbf{x},t) \end{bmatrix}, \quad \mathbf{x} \in R^n. \tag{3.1}$$

is called *autonomous* if $\mathbf{F}(\mathbf{x},t) = \mathbf{F}(\mathbf{x})$ i.e. the indepedent variable t does not appear explicitely in $\mathbf{F}$.

We observe (although this will have no bearing on our present discussion) that a nonautonomous system is equivalent to an autonomous system with an additional equation. In fact if we define

$$\mathbf{y} = \begin{bmatrix} \mathbf{x} \\ t \end{bmatrix}, \quad \mathbf{G}(\mathbf{y}) = \begin{bmatrix} \mathbf{F}(\mathbf{x},t) \\ 1 \end{bmatrix} \tag{3.2}$$

and introduce the extraneous variable s then

$$\frac{d\mathbf{y}}{ds} = \mathbf{G}(\mathbf{y}) \tag{3.3}$$

is equivalent to (3.1).

In the following we consider only autonomous systems. Such systems are referred to as *dynamical systems*.

Definition 2: The phase space of the system

$$\frac{d\mathbf{x}}{dt} = \mathbf{F}(\mathbf{x}) \tag{3.4}$$

is the space R^n with coordinates $(x_1,\ldots,x_n)$.

We point out that in classical mechanics where the equations of motion for a system of particles are of second order the phase space is defined as $\left[\mathbf{x}, \frac{d\mathbf{x}}{dt}\right]$. This is not in conflict with out definition since when a second order differential equation is converted to a system of first order equations one has to introduce $\frac{dx}{dt}$ as a variable.

Example 1: The equations of motion for a point particle of mass m under the influence of an external force $F = F(x)$ is

$$m \frac{d^2x}{dt^2} = F(x) \tag{3.5}$$

which is equivalent to

$$\frac{dx}{dt} = v, \quad m \frac{dv}{dt} = F(x) \quad . \tag{3.6}$$

Hence the phase space of equation (3.5) is $(x, v) \equiv \left[x, \frac{dx}{dt}\right]$.

Definition 3: A critical point ($\equiv$ equilibrium state) of the system (3.4) is a point x_0 such that $F(x_0) = 0$.

We note that a critical point can be isolated, i.e. there is a neighborhood of x_0 in which no other critical point exists or it might be part of a continuous set of critical points. In general it is rather difficult to analyze the behavior of a system with a continuous set of critical points. Therefore, we consider in the following only systems with isolated critical points.

Example 2: The system

$$\frac{dx}{dt} = ax + by, \quad \frac{dy}{dt} = cx + dy \tag{3.7}$$

has a continuous set of critical points when $ad - bc = 0$ since $\frac{dx}{dt} = \frac{dy}{dt} = 0$ imply

$$ax + by = 0, \quad cx + dy = 0 \quad . \tag{3.8}$$

This system of equations have infinitely many solutions when the determinant of the coefficients matrix is zero. On the other hand this analysis shows that (0,0) is an isolated critical point of the system (3.7) when $ad - bc \neq 0$.

The following is a generalization of this example.

Theorem 1: Let $\mathbf{F(x)}$ be analytic and $\mathbf{x}_0$ a critical point of the system (3.4). A sufficient condition for $\mathbf{x}_0$ to be isolated is that the Jacobian matrix of $\mathbf{F(x)}$ at $\mathbf{x}_0$

$$J(\mathbf{x}_0) = \left[\frac{\partial f_i}{\partial x_j}\right] \tag{3.9}$$

be nonsingular.

Proof: Since $\mathbf{F(x)}$ is analytic we can use a Taylor series expansion to approximate it around the critical point $\mathbf{x}_0$.

$$\begin{aligned} \mathbf{F(x)} &= \mathbf{F}(\mathbf{x}_0) + J(\mathbf{x}_0)(\mathbf{x} - \mathbf{x}_0) + 0(|\mathbf{x} - \mathbf{x}_0|^2) \\ &= J(\mathbf{x}_0)(\mathbf{x} - \mathbf{x}_0) + 0(|\mathbf{x} - \mathbf{x}_0|^2) \quad . \end{aligned} \tag{3.10}$$

Therefore, a second critical point in the vicinity of $\mathbf{x}_o$ must satisfy

$$J(\mathbf{x}_o)(\mathbf{x} - \mathbf{x}_0) = 0 \quad . \tag{3.11}$$

But $J(\mathbf{x}_0)$ is nonsingular which implies that the only solution of this system is $\mathbf{x} = \mathbf{x}_0$. Hence $\mathbf{x}_0$ is isolated. Note, however, that this theorem set only a sufficient but not necessary condition for $\mathbf{x}_0$ to be isolated.

Example 3: The only critical point of the system

$$\frac{dx}{dt} = x^2, \quad \frac{dy}{dt} = y^2 \tag{3.12}$$

is $(0,0)$ although $J(0,0)$ is singular.

We also observe that for several practical reasons it is convenient to translate the critical point under consideration to the origin by the transformation $\mathbf{u} = \mathbf{x} - \mathbf{x}_0$. The equations of the system will then take the form

$$\frac{d\mathbf{u}}{dt} = \mathbf{G}(\mathbf{u}) \tag{3.13}$$

where $\mathbf{G}(\mathbf{u}) = \mathbf{F}(\mathbf{u} + \mathbf{x}_0)$

Example 4: Consider the system

$$\frac{dx}{dt} = 1 - xy, \quad \frac{dy}{dt} = x - y^3 \tag{3.14}$$

whose critical points are (1,1) and (-1,-1). To translate (1,1) to the origin we perform the transformation

$$u = x - 1, \quad v = y - 1 \tag{3.15}$$

and the resulting form of the system (3.14) is

$$\frac{du}{dt} = -u - v - uv$$
$$\frac{dv}{dt} = u - 3v - 3v^2 - v^3 \quad . \tag{3.16}$$

Similarly if we wish to consider the stability of the critical point (-1,-1) we perform the transformation

$$u = x + 1, \quad v = y + 1 \quad . \tag{3.17}$$

The resulting form of the system (3.14) is

$$\frac{du}{dt} = u + v - uv$$
$$\frac{dv}{dt} = u - 3v + 3v^2 - v^3 \tag{3.18}$$

Definition 4: A trajectory (≡ orbit, path) of the system (3.4) is a solution of this system $\mathbf{x} = \mathbf{x}(t)$.

Sometimes such a trajectory is written as $\mathbf{x}(t,\mathbf{c})$ where $\mathbf{x}(0) = \mathbf{c}$.

Definition 5: Let $\mathbf{x}_0$ be a critical point of the system (3.4). We say that

1. $\mathbf{x}_0$ is stable if for any given $\epsilon > 0$ there exists a $\delta > 0$ so that if $|\mathbf{x}(0) - \mathbf{x}_0| < \delta$ then $|\mathbf{x}(t) - \mathbf{x}_0| < \epsilon$ for all $t > 0$.

2. $\mathbf{x}_0$ is asymptotically stable if it is stable and furthermore,

$$\lim_{t\to\infty} |\mathbf{x}(t) - \mathbf{x}_0| = 0 \quad . \tag{3.19}$$

Thus, $\mathbf{x}_0$ is stable if whenever the initial state of the system is close to the equilibrium point then its trajectory will remain close to it for all times. On the other hand if these trajectories approach $\mathbf{x}_0$ as $t \to \infty$ then $\mathbf{x}_0$ is asymptotically stable.

Definition 6: A critical point that is neither stable nor asymptotically stable is called unstable.

Definition 7: The integral curves of the system (3.4) are the solutions of the system

$$\frac{dx_1}{f_1} = \dots = \frac{dx_n}{f_n} \quad . \tag{3.20}$$

Thus the integral curves of (3.4) are the trajectories of this system parameterized in terms of the x_i's rather than in terms of the "extraneous" variable t.

Example 5: Find the trajectories and integral curves of

$$\frac{dx}{dt} = ax, \quad \frac{dy}{dt} = by \tag{3.21}$$

Solution: The trajectories of the system are

$$x = c_1 e^{at}, \quad y = c_2 e^{bt} \tag{3.22}$$

where c_1, c_2 are arbitrary constants. The corresponding integral curves are $y^a = cx^b$.

Example 6: Find the integral curves for the equation of the linearized Pendulum (no damping).

Solution: The equation of motion for the pendulum is

$$\frac{d^2\theta}{dt^2} = \frac{-g}{L} \sin\theta = \omega^2 \sin\theta \quad . \tag{3.23}$$

For small θ we can appoximate $\sin\theta$ as θ so that equation (3.23) reduces to the linear equation

$$\frac{d^2\theta}{dt^2} + \omega^2\theta = 0 \quad . \tag{3.24}$$

This is equivalent to the system

$$\frac{d\theta}{dt} = \varphi$$
$$\frac{d\varphi}{dt} = -\omega^2\theta \quad . \tag{3.25}$$

The integral curves of this system satisfy

$$\frac{d\theta}{d\varphi} = \frac{d\varphi}{-\omega^2\theta} \tag{3.26}$$

and hence

$$\omega^2\theta^2 + \varphi^2 = c^2$$

where c is a constant. We conclude, therefore, that the integral curves of the linearized pendulum are ellipses (see Figure 9). We also infer from this result that the critical point $(0,0)$ of the system (3.25) is *stable* but not asymptotically stable. This result is obvious from a physical point of view since the system contains no damping.

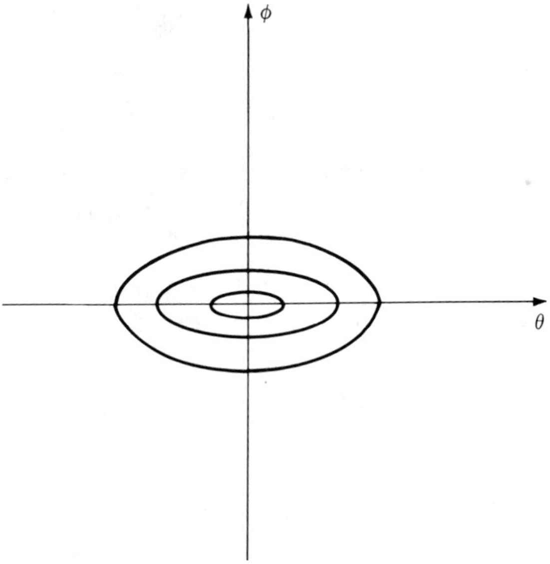

Figure 9
The integral curves of (3.25) are ellipses

EXERCISE 3

1. Find the form of the following systems when each of their critical points is translated to the origin.

a. $\frac{dx}{dt} = (x-1)(x-2)(y-2)$

$\frac{dy}{dt} = (x-3)(y-1)^2$

b. $\frac{dx}{dt} = \sin\pi x \cos\pi y$

$\frac{dy}{dt} = (\sin\pi y)^2 - 1$

2. Find the trajectories, integral curves and the stability of the critical point at the origin for the following systems

a. $\frac{dx}{dt} = -x, \quad \frac{dy}{dt} = 3x + 2y$

b. $\frac{dx}{dt} = x + y, \quad \frac{dy}{dt} = x - y.$

c. $\frac{dx}{dt} = mz - ny$

$\frac{dy}{dt} = nx - kz$

$\frac{dz}{dt} = ky - mx$

3. Find the integral curves and the stability of the critical points for the following systems.

a. $\ddot{x} = -x^3$

b. $m\ddot{x} + b\dot{x} + kx = 0, \quad m,b,k > 0$

what happens if $b = 0$ or $b < 0$?

c. $\ddot{\theta} + \omega^2\sin\theta = 0$

(nonlinear pendulum).

4. ALMOST LINEAR SYSTEMS

When it is impossible to find the trajectories or integral curves of a nonlinear dynamical system it is also impossible to deduce the stability of its critical points directly. To overcome this difficulty it is desirable to approximate the given system near a critical point by a linear set of equations and attempt to deduce the stability of this critical point from the solutions of this approximation. In this section we treat dynamical systems for which such approximation schemes is valid, i.e. yield correct results regarding the stability of the critical point (with few exceptions).

Definition 1: Let $\mathbf{x}_0$ be an isolated critical point of the system

$$\frac{d\mathbf{x}}{dt} = \mathbf{F}(\mathbf{x}) = \begin{bmatrix} f_1(\mathbf{x}) \\ \vdots \\ f_n(\mathbf{x}) \end{bmatrix} \quad . \tag{4.1}$$

We say that the system is almost linear at $\mathbf{x}_0$ if the Jacobian of $\mathbf{F}$ at $\mathbf{x}_0$ is nonsingular, i.e.

$$\det J(\mathbf{x}_0) = \left| \frac{\partial f_i}{\partial x_j}(\mathbf{x}_0) \right| \neq 0 \quad . \tag{4.2}$$

Example 1: The system

$$\frac{dx}{dt} = (x-1)(x-2)^2$$

$$\frac{dy}{dt} = y \tag{4.3}$$

is almost linear at the critical point $(1,0)$ but is not so at the critical point $(2,0)$.

To treat systems which are almost linear at the critical point $\mathbf{x}_0$ we first translate this point to the origin and then take the Taylor expansion of each $f_i(\mathbf{x})$ around this point

$$f_i(x) = \sum_{j=1}^{n} a_{ij} x_j + O(|\mathbf{x}|^2) \quad . \tag{4.4}$$

It follows then that the linear approximation to the original system at $\mathbf{0}$ is given by

$$\frac{d\mathbf{x}}{dt} = A\mathbf{x} \tag{4.5}$$

where $A = J(\mathbf{0})$ is a constant coefficient matrix.

We observe that in view of (4.2) not all the coefficients $a_{i1}, \dots, a_{in}$ in the expansion (4.4) are zero and, therefore, there exists a neighborhood of **0** where the linear terms are dominant in the corresponding equation. This can be considered as the root cause for the close relationship between the stability of the linear system (4.5) at the origin and the original almost linear system.

Example 2: The system

$$\frac{dx}{dt} = x + y^2 - 2, \quad \frac{dy}{dt} = y - x^2$$

has a critical point at $x = 1$, $y = 1$. The determinant of the Jacobian at this point is

$$\det J(1,1) = \begin{vmatrix} \frac{\partial f_1}{\partial x} & \frac{\partial f_1}{\partial y} \\ \frac{\partial f_2}{\partial x} & \frac{\partial f_2}{\partial y} \end{vmatrix} (1,1) = \begin{vmatrix} 1 & 2 \\ -2 & 1 \end{vmatrix} \neq 0. \quad (4.6)$$

Hence this system is almost linear at this critical point. To compute the linear approximation we first translate the critical point to the origin by the transformation.

$$\bar{x} = x - 1, \quad \bar{y} = y - 1 \quad . \qquad (4.7)$$

The system now takes the form (after dropping the bars)

$$\frac{dx}{dt} = x + 2y + y^2, \quad \frac{dy}{dt} = y - 2x - x^2 \qquad (4.8)$$

and the resulting linear approximation at the origin is

$$\frac{dx}{dt} = x + 2y \ , \quad \frac{dy}{dt} = -2x + y \quad . \qquad (4.9)$$

In matrix notation this takes the following form:

$$\frac{d}{dt}\begin{bmatrix} x \\ y \end{bmatrix} = \begin{bmatrix} 1 & 2 \\ -2 & 1 \end{bmatrix}\begin{bmatrix} x \\ y \end{bmatrix} \qquad (4.10)$$

We now characterize the stability of the critical point at the origin for the system (4.5).

Theorem 1: The critical point at the origin of the system (4.5) is

(1) Asymptotically stable if all the eigenvalues of A have negative real parts.

(2) Stable if all the eigenvalues of A have nonpositive real parts and every eigenvalue of A which has a zero real part is a simple zero of the characteristic polynomial of A.

(3) Unstable if (1) and (2) are false.

Example 3: The critical point at the origin of the system

$$\frac{d\mathbf{x}}{dt} = \begin{bmatrix} 3 & 2 & -1 \\ 0 & -1 & 0 \\ 0 & 2 & -2 \end{bmatrix} \mathbf{x} \tag{4.11}$$

is unstable since the eigenvalues of A are 3, -1, -1.

As to the relationship between the stability of the systems (4.1) and (4.5) at the critical point **0** we have the following theorem:

Theorem 2: If the critical point at the origin of the system (4.5) is asymptotically stable or unstable then the same is true for the critical point of the original system. However, if the origin is stable critical point for the system (4.5) then the stability of the original system at this point is indeterminate, i.e., cannot be deduced from that of the linearized system.

Example 4: The system

$$\frac{d}{dt}\begin{bmatrix} x \\ y \end{bmatrix} = \begin{bmatrix} 0 & 1 \\ -1 & 0 \end{bmatrix}\begin{bmatrix} x \\ y \end{bmatrix} - b\begin{bmatrix} 0 \\ x^2 y \end{bmatrix} \tag{4.12}$$

is almost linear at the critical point **0**. Its linearization at this point is given by

$$\frac{d}{dt}\begin{bmatrix} x \\ y \end{bmatrix} = \begin{bmatrix} 0 & 1 \\ -1 & 0 \end{bmatrix}\begin{bmatrix} x \\ y \end{bmatrix} = A\mathbf{x} \quad . \tag{4.13}$$

Since the eigenvalues of A are $\pm i$ it follows from theorem 1 that **0** is a stable critical point of the system (4.13). However, it can be shown that the original system (4.12) is asymptotically stable if $b > 0$ and unstable if $b < 0$. (See exercise 6 in Section 7).

EXERCISE 4

1. Show that the solution $x = 0$ of

$$a_n x^{(n)} = a_{n-1} x^{(n-1)} + \ldots + a_1 x' + a_0 x = 0 \qquad (4.14)$$

is stable if and only if all the roots of the polynomial

$$p(\lambda) = a_n \lambda^n + \ldots + a_0$$

have nonpositive real parts and all roots with zero real parts are simple. (What happens if such a root is not simple).

Hint: Solve equation (4.14).

For the following exercises find the critical points at which the system is almost linear and discuss their stability.

2. $$\frac{dx}{dt} = (x - 1)(x - 2)^3(y - 3)$$
$$\frac{dy}{dt} = (x - 1)^2(y - 2)$$

3. $$\frac{dx}{dt} = y(\sin x + 1)$$
$$\frac{dy}{dt} = x(\cos y + 1)$$

4. $\ddot{x} + b(x^2 - 1)\dot{x} + kx = 0, \quad b > 0$.
(Van der Pol's equation)

5. $$\frac{dx}{dt} = a(y - x)$$
$$\frac{dy}{dt} = bx - y - xz$$
$$\frac{dz}{dt} = -cz + xy, \quad a,b,c > 0$$
(Lorentz's equations)

6. $\ddot{x} + b\dot{x} - x + x^3 = 0$
(Duffin's equation)

5. ALMOST LINEAR SYSTEMS IN R^2

The theorems stated in the previous section specify the general relationship between the stability of an almost linear system in R^n and its linearization at the critical point. However, for almost linear systems in R^2 it is possible to perform additional analysis of the systems behavior at the critical point in terms of its phase diagram or "phase portrait". the purpose of this section is to carry out this classification and analysis.

Let the system

$$\frac{dx}{dt} = F(x,y), \quad \frac{dy}{dt} = G(x,y) \tag{5.1}$$

have an isolated critical point at the origin. If this system is almost linear at **0** we can take the Taylor expansion of F,G around this point and rewrite equation (5.1) as

$$\frac{dx}{dt} = ax + by + f(x,y)$$

$$\frac{dy}{dt} = cx + dy + g(x,y) \tag{5.2}$$

where, since the Jacobian of (5.1) is nonsingular at **0**, $ad - bc \neq 0$. Moreover, f,g are of order $|\mathbf{x}|^2$ and, therefore,

$$\lim_{r\to 0} \frac{f(x,y)}{r} = \lim_{r\to 0} \frac{g(x,y)}{r} = 0 \tag{5.3}$$

where $r = \sqrt{x^2+y^2}$. Thus the linear approximation to the system (5.1) near the origin is given by

$$\frac{dx}{dt} = ax + by, \quad \frac{dy}{dt} = cx + dy, \quad ad - bc \neq 0. \tag{5.4}$$

To solve this system we assume a trial solution of the form

$$x = A\exp(st) \qquad y = B\exp(st) \tag{5.5}$$

(same s for both x and y). Substituting (5.5) in (5.4) yields then

$$(a - s)A + bB = 0 \tag{5.6}$$

$$cA + (d - s)B = 0 \quad . \tag{5.7}$$

Equations (5.6)-(5.7) form a system of linear homogeneous equations for the coefficients A, B. A nontrivial solution for these coefficients exists if and only if the determinant of the coefficients of this system vanishes, i.e.

$$\begin{vmatrix} a - s & b \\ c & d - s \end{vmatrix} = s^2 - (a+d)s + (ad-bc) = 0. \quad (5.8)$$

We infer, therefore, that we obtain a nontrivial solution only if s is an eigenvalue of the coefficient matrix of the system (5.4) and $\begin{bmatrix} A \\ B \end{bmatrix}$ is the eigenvector related to this eigenvalue. The stability or instability as well as the phase portrait of the critical point at the origin depends, therefore, on the eigenvalues and eigenvectors of this coefficient matrix.

We now consider and classify all these possibilities.

Case 1: Equation (5.8) has real unequal roots of the same sign.

The trajectories of the system are given by

$$x = A_1 e^{s_1 t} + A_2 e^{s_2 t}$$

$$y = B_1 e^{s_1 t} + B_2 e^{s_2 t} \quad (5.9)$$

If $s_1, s_2 > 0$ the critical point at the origin is unstable. If on the other hand $s_1, s_2 < 0$ the critical point is asymptotically stable. A typical illustration of the integral curves of the system in this case is shown in Figure 10(a). A critical point with the phase portrait shown in this figure is called improper node (stable or unstable).

Case 2: Equation (5.8) has two real roots with opposite signs.

The solution of the system (5.4) is still given by equation (5.9). However, in this case the critical point is always unstable. Its phase portrait is given in Figure 10(b). A critical point with such a phase portrait is called a "saddle point".

Case 3: Equation (5.8) has two equal roots.

It is easy to infer from equation (5.8) that this root must be real and differenti from zero (since ad-bc $\neq$ 0). Hence the solution of the system (5.4) is given by

$$x = (A_1+A_2 t)e^{st}, \quad y = (B_1+B_2 t)e^{st} \quad . \tag{5.10}$$

It follows then that the critical point is asymptotically stable if $s < 0$ and unstable if $s > 0$. the phase portrait of such a critical point is improper or proper node (Figs. 10(a) or 10(c) respectively).

Case 4: Equation (5.8) has complex roots

$$s_{\pm} = \lambda \pm i\mu, \quad \lambda \neq 0.$$

The solution of the system (5.4) is

$$x = e^{\lambda t}(A_1 \cos\mu t + A_2 \sin\mu t)$$

$$y = e^{\lambda t}(B_1 \cos\mu t + B_2 \sin\mu t) \quad . \tag{5.11}$$

The critical point is unstable if $\lambda > 0$ and asymptotically stable if $\lambda < 0$. The phase portrait is a spiral, regardless of the stability of the critical point, as shown in Figure 10(d).

Case 5: Equation (5.8) has pure imaginary roots

$$s = \pm\, i\mu$$

The solution of the system (5.4) is still represented by equation (5.11) with $\lambda = 0$. The critical point is stable and its phase portrait as shown in Figure 10(e) is called a "center".

As to the relationship between the phase portraits, at the origin of the system (5.1) and its linearization (5.4) we have the following;

Theorem 1: The phase portraits at **0** of the systems

(5.1) and (5.4) is the same except in cases 3 and 5 where a spiral is possible as an additional phase portrait.

This result might be explained by noting that the additional terms in (5.2) might destroy the exact equality of the roots or add to them a small real part when they are purely imaginary as in case 5.

It should be also noted that when the roots of equation (5.8) are purely imaginary the stability of the critical point at 0 of the system (5.1) cannot be deduced from that of the system (5.4) in conformity with theorem 2 in the previous section.

EXERCISE 5

For the following systems classify and draw the phase portrait for the critical point at the origin.

1. $m\ddot{x} + b\dot{x} + cx = 0 \qquad m,b,c > 0$

   what happens if $b < 0$?

2. $\ddot{x} + b\dot{x} + kx^3 = 0, \quad k > 0$

   consider separately the case $b > 0$ and $b < 0$.

3. $\frac{dx}{dt} = \cos(x-2y)$

   $\frac{dy}{dt} = e^{x-y} - \sin x$

4. $\frac{dx}{dt} = y + 2y^2 - x^2$

   $\frac{dy}{dt} = y - 2x + y^3$

5. $\frac{dx}{dt} = 2y - y^3$

   $\frac{dy}{dt} = 2x$

(compare with the exact solution)

6. $\frac{dx}{dt} = y - 3x$

   $\frac{dy}{dt} = e^x - e^y \sin x$

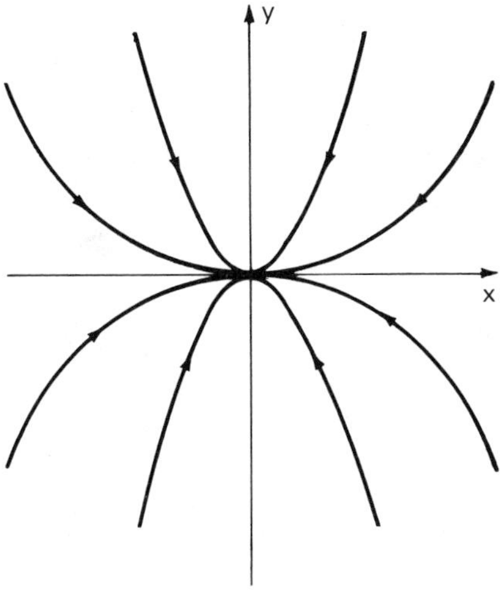

(a)
Improper node (stable)

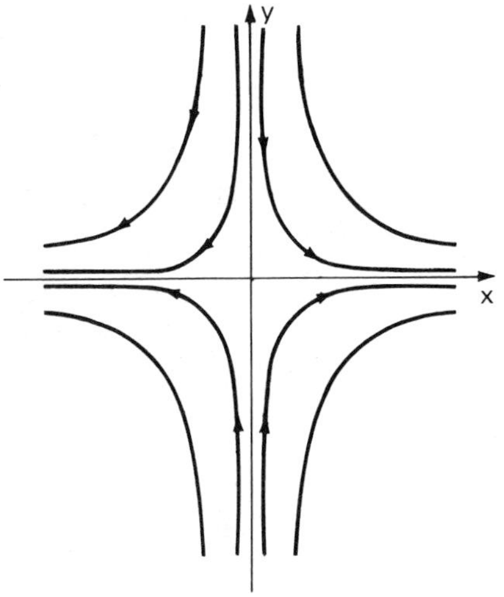

(b)
Saddle Point

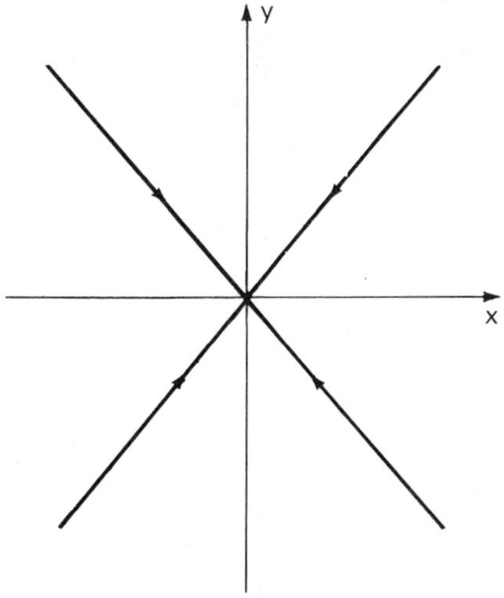

(c)
Proper Node (stable)

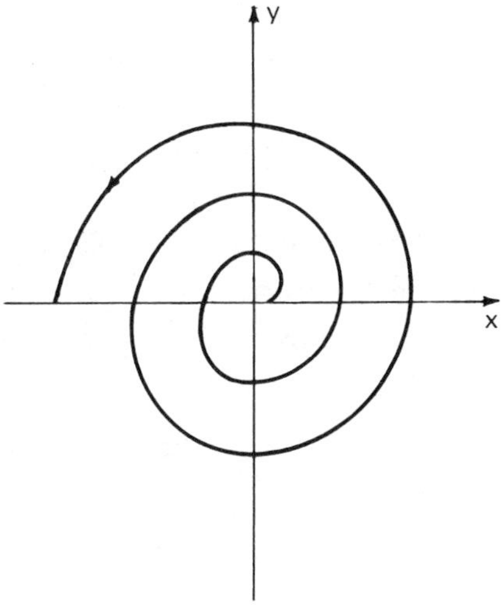

(d)
Spiral (unstable)

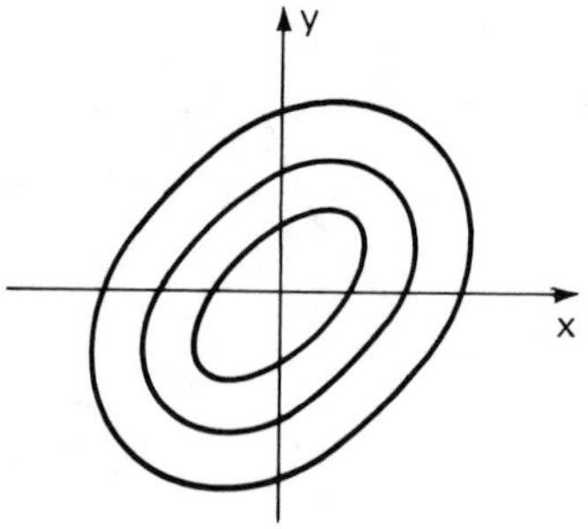

(e)
Center

Figure 10.

6. LIAPOUNOV DIRECT METHOD.

In previous sections of this chapter we discussed dynamical systems which are almost linear in the vicinity of their isolated critical points and described techniques to determine the stability of these equilibriums states. However, when a system is not almost linear in the neighborhood of the critical point or if it is a center in this approximation these techniques are not applicable. Furthermore, the linear approximation does not determine the size of the basin of stability, i.e. if $\mathbf{x}_0$ is asymptotically stable state of the system what is the "maximum perturbation" under which the solution returns to $\mathbf{x}_0$. To handle these problems Liapounov direct method (it is called so because no information about the solutions of the system is required) proved to be a powerful tool which has been applied successfully in many practical applications. In the following we describe the essence of this technique and discuss its application to gradient systems.

Definition 1: Let $H:R^n \to R$ be defined on a domain D containing the origin and $H(\mathbf{0}) = 0$.

1. H is said to be positive definite on D if $H(\mathbf{x}) > 0$ for all $\mathbf{x} \neq \mathbf{0}$ in D.
2. H is positive semidefinite in D if $H(\mathbf{x}) \geq 0$ for all $\mathbf{x} \in D$.

Similarly we can define negative definite and negative semidefinite functions.

Example 1: The function

$$H(x,y,z) = \sin(x^2+y^2+z^2) \tag{6.1}$$

is positive definite in the sphere $0 \leq x^2 + y^2 + z^2 < \pi$.

Example 2: The function

$$H(x,y) = (x-2y)^2 \tag{6.2}$$

is positive semidefinite in R^2 since $H(x,y) = 0$ on the line $x = 2y$.

Example 3: Show that the function

$$H(x,y) = ax^2 + bxy + cy^2 \tag{6.3}$$

is positive definite if $a > 0$ and $4ac - b^2 > 0$.

Solution: By adding and subtracting $\frac{b^2}{4a} y^2$ from the expression of $H(x,y)$ we obtain

$$H(x,y) - \left[\sqrt{ax} + \frac{b}{2\sqrt{a}} y\right]^2 + \frac{4ac-b^2}{4a} y^2 \tag{6.4}$$

which is obviously positive definite if $a > 0$ and $4ac - b^2 > 0$. Consider now the autonomous system

$$\frac{d\mathbf{x}}{dt} = \mathbf{F}(\mathbf{x}) = \begin{bmatrix} f_1(\mathbf{x}) \\ \vdots \\ f_n(\mathbf{x}) \end{bmatrix} \tag{6.5}$$

which has an isolated critical point at $\mathbf{x} = \mathbf{0}$.

Theorem 1: (Liapounov) Let $H(\mathbf{x})$ be positive definite function with continuous derivatives on a domain D containing the origin and let

$$\dot{H}(\mathbf{x}) = D_{\mathbf{F}}H(\mathbf{x}) = \text{grad}H\cdot\mathbf{F} = \sum_{i=1}^{n} \frac{\partial H}{\partial x_i} \cdot f_i \quad . \tag{6.6}$$

If on some domain containing the origin

1. $\dot{H}(\mathbf{x})$ is negative definite then $\mathbf{0}$ is asymptotically stable
2. $\dot{H}(\mathbf{x})$ is negative semidefinite then $\mathbf{0}$ is stable
3. $\dot{H}(\mathbf{x})$ is positive definite then $\mathbf{0}$ is unstable

Example 4: Discuss the stability of the equilibrium state $\mathbf{x} = \mathbf{0}$ for the system

$$\dot{x} = y + ax(x^2+y^2) \tag{6.7}$$

$$\dot{y} = -x + ay(x^2+y^2) \tag{6.8}$$

where a is a parameter.

Solution: Let $H = x^2 + y^2$ then

$$\dot{H}(x) = \text{grad}H\cdot\mathbf{F} = 2a(x^2+y^2)^2 \tag{6.9}$$

Therefore, from Liapounov theorem we infer that if $a < 0$ then the critical point at the origin is asymptotically stable. If $a = 0$ the point is stable while if $a > 0$ the point is unsable.

We infer from this example that the main step required for the application of Liapounov theorem is a proper choice of H (called Liapounov function) for which no constructive algorithmn is given in the theorem. Although several techniques were suggested in the past to overcome this difficulty most of the important applications of this method remain limited to those systems for which Liapounov function can be deduced on physical grounds viz. gradient and conservative systems.

Definition 2: A dynamical system

$$\frac{d\mathbf{x}}{dt} = \mathbf{F}(\mathbf{x}) \tag{6.10}$$

is called a gradient system if there exist $V(\mathbf{x})$ so that

$$\mathbf{F}(\mathbf{x}) = -\text{grad}V(\mathbf{x}). \tag{6.11}$$

$V(\mathbf{x})$ is called the potential function of the system.

Theorem 2: Let $\mathbf{x}_0$ be an isolated local minimum of $\mathbf{V}$. Then $\mathbf{x}_0$ is an asymptotically stable critical point of the system (6.10).

Proof: Since V is continuous and $\mathbf{x}_0$ is a local minima of V there exists a neighborhood of $\mathbf{x}_0$ in which

$$H(\mathbf{x}) = V(\mathbf{x}) - V(\mathbf{x}_0) \tag{6.12}$$

is positive definite and gradH = grad V $\neq$ 0. But

$$\dot{H}(\mathbf{x}) = \text{gradH}\cdot(-\text{gradH}) = -|\text{gradH}|^2 \quad . \tag{6.13}$$

Hence $\dot{H}(\mathbf{x})$ is negative definite on some neighborhood of $\mathbf{x}_0$ and by Liapounov theorem we infer that $\mathbf{x}_0$ is asymptotically stable.

Definition 3: (1) A force field **F** is said to be conservative if **F** = -gradV.

(2) A (mechanical) system is said to be conservative if the force field acting on the system is conservative.

Example 5: The equation of motion for a particle of mass m under the action of a force **F** is given by Newton's second law

$$m\,\frac{d^2\mathbf{x}}{dt^2} = \mathbf{F} \quad . \tag{6.14}$$

If **F** is conservative then

$$m\,\frac{d^2\mathbf{x}}{dt^2} = -\text{gradV} \quad . \tag{6.15}$$

Rewriting equation (6.15) as a system of first order equation we obtain

$$\frac{d\mathbf{x}}{dt} = \mathbf{v} \tag{6.16}$$

$$m\,\frac{d\mathbf{v}}{dt} = -\text{gradV} \quad . \tag{6.17}$$

We deduce from (6.16)-(6.17) that at the equilibrium states $(\mathbf{x}_0, \mathbf{v}_0)$ of such a particle $\mathbf{v}_0 = 0$ and $\text{gradV}(\mathbf{x}_0) = 0$, i.e. $\mathbf{x}_0$ corresponds to an extremum of V.

Theorem 3: The stable equilibrium states $(\mathbf{x}_0, 0)$ of a particle in a conservative force field correspond to the (isolated) local minima of V.

Proof: The total energy of the system under consideration is

$$E(\mathbf{x}, \mathbf{v}) = \frac{1}{2} m\mathbf{v}^2 + V(\mathbf{x}) \quad . \tag{6.18}$$

If $\mathbf{x}_0$ is a local minima of V then

$$H(\mathbf{x}, \mathbf{v}) = E(\mathbf{x}, \mathbf{v}) - E(\mathbf{x}_0, 0)$$

is positive definite on some domain around $\mathbf{x}_0$.

Moreover,

$$\begin{aligned} \dot{H}(x, v) &= \text{grad}H \cdot (\mathbf{v}, \frac{-1}{m} \text{grad}V) \\ &= (\text{grad}V, m\mathbf{v}) \cdot (\mathbf{v}, -\frac{1}{m}\text{grad}V) = 0 \quad . \end{aligned} \tag{6.19}$$

Hence by Liapounov theorem $(x_0, 0)$ is a stable equilibrium.

We remark that theorem 3 is a special case of Lagrange theorem which states that

Theorem: An equilibrium $(\mathbf{x}_0, 0)$ of a system in a conservative force field is stable if $\mathbf{x}_0$ is an isolated local minima of $V(\mathbf{x})$.

EXERICISE 6

1. Show that

$$H(x, y) = ax^2 + bxy + cy^2$$

is negative definite if $a < 0$ and $4ac - b^2 > 0$.

2. Let $f(0) = 0$, $f(x) > 0$ for $0 < x < a$ and $f(x) < 0$ for $-a < x < 0$. show that

$$H(x, y) = \frac{1}{2} y^2 + \int_0^x f(t)dt$$

is positive definite on

$$D = \{(x, y);\ -a < x < a,\ -\infty < y < \infty\}.$$

3. For the system

$$\ddot{x} + f(x) = 0$$

where $f(x)$ satisfies the assumptions of the previous exercise show that $x = 0$, $\dot{x} = 0$ is a stable critical point.

Hint: Let $H(x,\dot{x}) = \frac{1}{2}\dot{x}^2 + \int_0^x f(t)dt$.

4. Apply the results of ex. 3 to the nonlinear pendulum

$$\ddot{x} + \frac{g}{L}\sin x = 0$$

5. Use Liapounov function of the form $H = \frac{x^2}{2} y^4$ to show that the origin is an asymptotically stable state for the system

$$\frac{dx}{dt} = -2y^3$$

$$\frac{dy}{dt} = \frac{x}{2} - y^3 \quad .$$

Does this conclusion remain true if we linearize the system? Explain.

6. Show that the following two systems are asymptotically stable at **0**

a. $\frac{dx}{dt} = -x - 4xy^2, \quad \frac{dy}{dt} = -y - yx^2$

b. $\frac{dx}{dt} = -x^3 + 4xy^2, \quad \frac{dy}{dt} = -2x^2y - 4y^3$

Hint: Use Liapounov function of the form $H(x,y) = ax^2 + by^2$.

7. For the system

$$\frac{dx}{dt} = -y - xg(x,y)$$

$$\frac{dy}{dt} = x - yg(x,y)$$

Show that **0** is an asymptotically stable state when $g(x,y) > 0$ in some neighborhood of the critical point and unstable if $g(x,y) < 0$ in some neighborhood of **0** .

Hint: Note that this is ageneralization of Example 4.

7. PERIODIC SOLUTIONS (LIMIT CYCLES)

Although we considered in the previous sections only the stability of an isolated equilibrium point we encounter in practice many systems with periodic motions whose stability is important from a practical point of view. Thus, e.g. an important mathematical problem regarding the solar system is the stability of the periodic motion of the planets in this system. Since this motion is always subject to small perturbations (due to the gravitational fields of the planets, variations in the gravitational field of the sun, etc.) the answer to this question has important and obvious implications regarding the very existence of life on earth.

It turns out, however, that periodic motion is important not only from a celestial point of view but also in many engineering and scientific contexts (e.g telecommunication). In this section we discuss, therefore, some of the elementary techniques which determine the existence and stability of these limit cycles. First, however, we motivate our discussion with a few examples.

Example 1: Consider the following almost linear system with a critical point at $(0,0)$;

$$\frac{dx}{dt} = k^2x - y - x(x^2+y^2) \tag{7.1}$$

$$\frac{dy}{dt} = x + k^2y - y(x^2+y^2) \tag{7.2}$$

where k is a constant. Since the linear approximation of this system at the critical point O is given by

$$\frac{dx}{dt} = k^2x - y, \quad \frac{dy}{dt} = x + k^2y \tag{7.3}$$

it follows that O is an unstable spiral point.

However, we now show that this spiral does not go to infinity and furthermore the trajectories of the system (7.1)-(7.2) far away from the origin are directed inward.

To prove this we introduce polar coordinates

$$x = r\cos\theta \qquad y = r\sin\theta \tag{7.4}$$

and note that

$$x\frac{dx}{dt} + y\frac{dy}{dt} = r\frac{dr}{dt} \tag{7.5}$$

$$y\frac{dx}{dt} - x\frac{dy}{dt} = r^2\frac{d\theta}{dt} \quad . \tag{7.6}$$

The system (7.1)-(7.2) now takes the form

$$\frac{dr}{dt} = r(k^2-r^2) \quad , \quad \frac{d\theta}{dt} = -1 \quad . \tag{7.7}$$

Therefore, if $r > k$ then $\frac{dr}{dt} < 0$ and the trajectories of the system move inward if on the other hand $r < k$ then $\frac{dr}{dt} > 0$ and the trajectories move outward. Moreover for $r = k$ we have $\frac{dr}{dt} = 0$ which implies that

$$r = k = \text{const.} \quad , \quad \theta = -t + t_0 \tag{7.8}$$

is a periodic solution of this system. Furthermore trajectories of this system with initial condition $r_0 > k$ will spiral toward the periodic solution and the same is true for those with $r_0 < k$. Thus we infer that this periodic solution is a stable limit cycle.

Example 2: Systems with negative damping.

We first observe that for the damped spring-mass system

$$m\ddot{x} + b\dot{x} + kx = 0 \quad m,b,k > 0 \tag{7.9}$$

the term responsible for the damping effect is $b\dot{x}$ whenever $b > 0$. However, when $b < 0$ (negative damping) the effect of this term will be to increase the applitude of the oscillations. Now consider Reyliegh equation

$$\ddot{x} - b(1-\dot{x}^2)\dot{x} + kx = 0 \quad , \quad b,k > 0 \quad . \tag{7.10}$$

When $(\dot{x})^2 < 1$ the damping in this equation is negative but when $(\dot{x})^2 > 1$ the damping is positive. We infer, therefore, that the amplitude of the motion described by this equation will increase for small velocities and decrease for large ones. One can deduce (at least intuitively), therefore, that in between these two types of motions there is an oscillation of constant amplitude i.e. a periodic motion.

We now proceed to some more formal definitions and theorems regarding limit cycles.

Lemma: A trajectory of a dynamical system is closed if and only if it corresponds to a periodic solution of the system.

Proof: If the solution $\mathbf{x}(t)$ is periodic there exist T so that $\mathbf{x}(t + T) = \mathbf{x}(t)$. This obviously implies that the trajectory of the system in phase space is closed. The reverse is also obvious.

Definition 1: A closed trajectory C of a dynamical system which has nearby open trajectories spiraling towards it from both the inside and outside as $t \to \infty$ is called a *stable limit cycle*. If they spiral toward it from one side and spiral away from the other side we say that it is *semistable*. If nearly open trajectories spiral away from C on both sides then C is called *unstable limit cycle*. Finally if nearby trajectories neither approach nor recede from C we say that it is *neutrally stable*.

Example 3: The trajectories of the linear pendulum without damping are neutrally stable.

The question of existence and stability of limit cycles for a given dynamical system is a rather difficult mathematical problem in general. We discuss here only two classical results due to poincare and Bendixson for systems in two-dimensions (=planar systems).

Definition 2: A point $\mathbf{y}$ is a *limit point* of a trajectory $\mathbf{x}(t)$ of a dynamical system

$$\frac{d\mathbf{x}}{dt} = \mathbf{F}(\mathbf{x}) \tag{7.11}$$

if there exists a sequence $\{t_n\}$, $\lim_{n\to\infty} t_n = \pm\infty$ so that $\lim \mathbf{x}(t_n) = \mathbf{y}$.

Definition 3: A set C is called a limit set of (7.11) if each of its points is a limit point of some trajectory of this system.

Example 4: An asymptotically stable critical point of a dynamical system is obviously a limit point of all nearby trajectories.

Example 5: The trajectory $r = 1$ in Example 1 is a limit set for the system (7.1)-(7.2) since each of its points is a limit point for the trajectories that spiral towards $r = 1$.

Theorem 1: (Poincare-Bendixson) Let S be a nonempty compact limit set of the planar system

$$\frac{dx}{dt} = F(x,y) \quad , \quad \frac{dy}{dt} = G(x,y) \tag{7.12}$$

where $F,G \in C^1$. If S contains no critical points then it is a limit cycle.

The following theorem establishes the nonexistence of a limit cycle.

Theorem 2: If for the system (7.12) there exist a simply connected domain D where $\frac{\partial F}{\partial x} + \frac{\partial G}{\partial y}$ have the same sign then there is no limit cycle of (7.12) which lies entirely in D.

Example: If

$$F(x,y) = x^2 + y^2, \quad G(x,y) = x^2 + 3xy$$

we infer from Theorem 2 that no limit cycles of the system (7.12) lie entirely in $D = \{(x,y),\ x > 0\}$ since

$$\frac{\partial F}{\partial x} + \frac{\partial G}{\partial y} = 5x > 0 \quad \text{in} \quad D.$$

EXERCISE 7

1. Use the same analysis as in example 2 to show (intuitively) that the Van der Pol equation
$$\frac{d^2x}{dt^2} + \mu(1-x^2)\frac{dx}{dt} + kx = 0 \tag{7.13}$$
admits a limit cycle.

2. Show that the system
$$\frac{dx}{dt} = -y + xg(r) \quad , \quad \frac{dy}{dt} = x + yg(r) \tag{7.14}$$
has limit cycles which correspond to the roots of $g(r)$. Hint: Use a polar representation of the system (7.14).

3. Determine the periodic solutions and their stability for the system (7.14) if
   a. $g(r) = r(r-1)^2(r-2)(r-4)$
   b. $g(r) = r^2 - 2$
   c. $g(r) = \cos kr$, $k = 1,2...$

4. Under what conditions on $f(r)$ the following system admits a limit cycle?
$$\frac{dx}{dt} = x + y - xf(r)$$
$$\frac{dy}{dt} = -x + y - yf(r) \tag{7.15}$$
Hint: Use polar representation of the system

5. Determine the periodic solution and their stability for the system (7.15) if
   a. $f(r) = r^2 - k^2$
   b. $f(r) = \sin \pi r$.

6. Perform a qualitative analyis similar to example 2 to show that the solution $x = 0$ of
$$\ddot{x} + bx^2\dot{x} + kx = 0$$
is a symptotically stable if $b > 0$ and unstable if $b < 0$. Apply these results to the system (4.12) in section 4.

BIBLIOGRAPHY

1. W. E. Boyce and R. Diprima - Elementary Differential Equations and Boundary Value Problems, 3rd edition, J. Wiley and Sons, 1977.
2. N. Rouche, P. Habets and M. Laloy - Stability Theory by Liapunov's Direct Method, Springer-Verlag, 1977.
3. M. W. Hirsch and S. Smale - Differential Equations, Dynamical Systems and Linear Algebra, Academic Press, 1974.
4. M. C. Irwin - Smooth Dynamical Systems, Academic Press, 1980.
5. W. Szlenk - Introducing to the Theory of Dynamical Systems, J. Wiley, 1984.
6. J. Palis, Jr. and W. de Melo - Geometric Theory of Dynamical Systems, Springer-Verlag, 1982.
7. R. K. Miller and A. N. Michel - Ordinary Differential Equations, Academic Press, 1982.

CHAPTER 9. CATASTROPHES AND BIFURCATIONS

1. CATASTROPHES AND STRUCTURAL STABILITY

In the previous chapter we discussed various methods to analyze the stability of the equilibrium states of a dynamical system when the values of the system parameters are known and fixed. The objective of catastrophe and bifurcation theory is to investigate what happens to the type, number, and stability of the equilibrium states as a result of a continuous change in the system parameters. In other words catastrophe theory is concerned with the "dynamical analysis" of the equilibrium states as a function of the system parameters as compared to the "static analysis" of these states which were performed in the last chapter.

The motivation for such analysis stems from the fact that in many real life situations the values of the model parameters are not known accurately or might be actually (very) slowly varying functions of time which are approximated by constants to simplify the model equations.

To illustrate this type of dynamical analysis for the equilibrium states we reconsider Richardson's model which was treated in section 2 of the previous chapter.

Example 1: Richardson's model (simplified)

Richardson's model for the arms level of two nations is given by equations (2.13)-(2.14) of the previous chapter. It is apparent from these equations that the model (and hence the equilibrium states) depends on six parameters $a_i, b_i, h_i > 0$, $i = 1,2$ and, therefore, the *parameter space* P of this model is a subset of R^6. However, in order to simplify the dynamical analysis of these equations we symmetrize the model equations by assuming that $a_1 = a_2 = a$, $b_1 = b_2 = b$ and $h_1 = h_2 = h$. Using the analysis that was performed previously regarding the equilibrium states of this model we can now summarize the type and stability of the equilibrium states as a function of the model parameters in Figures 1,2.

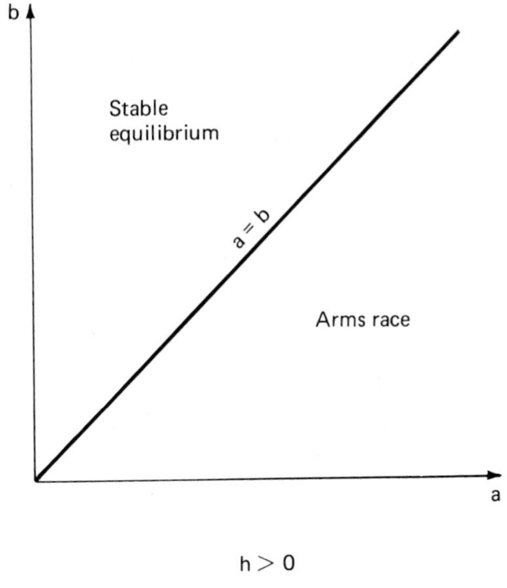

Figure 1
Outcome of Richardsonn's Model when $h > 0$

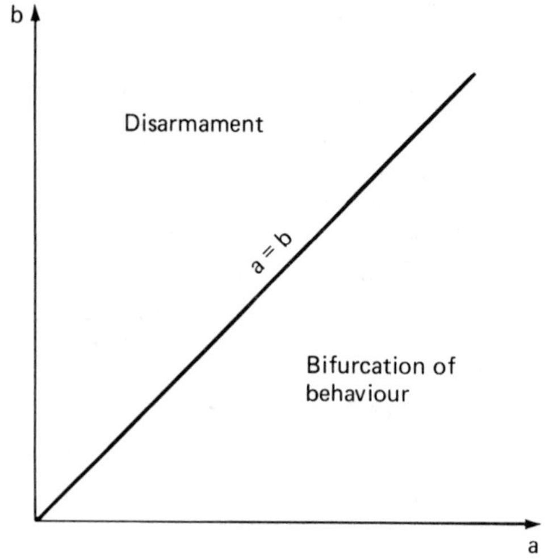

Figure 2
Outcome of Richardson's Model $h < 0$

We infer from these figures that if $|a-b| \gg 0$ the equilibrium state of this system will not change its *type* as a result of small changes in the values of the model parameters and thus be "*structurally stable*". On the other hand if $a \approx b$ then such a small change might cause a sudden transition in the type and stability of the equilibrium state of the system and this event will be referred to as a "Catastrophe".

To elaborate further on this concept suppose that initially $h > 0$, $\epsilon = b-a > 0$ and $|\epsilon| \ll 1$. From Figure 1 we deduce that with these parameters the arm levels of the two nations are in stable equilibrium. If, however, the system parameters undergo a small change as a result of which a becomes larger than b then the nature of the system behavior will change radically and the two nations will engage in an arms race. This radical change in the nature of the system behavior is termed as "*a Catastrophe event*". (Note carefully that the reverse change in behavior from an arms race to a stable equilibrium will also be considered by us as a Catastrophe event).

We now reintroduce formally the concepts which were discussed above.

Definition 1: A point p in the parameter set P of a system S is called structurally stable if there exists a neighborhood of p in which the equilibrium states of S which correspond to these parameters are of the same type. Otherwise p is called a Catastrophe point.

We now characterize the set K of all Catastrophe points in a system with m parameters $(a_1, \ldots, a_m) = \mathbf{a}$.

Theorem 1: The Catastrophe set K of

$$\frac{d\mathbf{x}}{dt} = F(\mathbf{x},\mathbf{a}) = \begin{bmatrix} f_1(\mathbf{x},\mathbf{a}) \\ \vdots \\ f_n(\mathbf{x},\mathbf{a}) \end{bmatrix} \tag{1.1}$$

consists of all the points $\mathbf{p}$ in the parameter space for which there exists a *critical* point $\mathbf{x}_0$ (which depends on $\mathbf{p}$) so that

$$\det J(\mathbf{x}_0,\mathbf{p}) = \begin{vmatrix} \frac{\partial f_1}{\partial x_1}, \ldots, \frac{\partial f_1}{\partial x_n} \\ \ldots \\ \frac{\partial f_n}{\partial x_1} \quad \ldots \quad \frac{\partial f_n}{\partial x_n} \end{vmatrix} (\mathbf{x}_0,\mathbf{p}) = 0$$

Example 1: (Richardson's model - continued)

Using the model equations

$$\frac{dx}{dt} = a_1 y - b_1 x + h_1 \tag{1.3}$$

$$\frac{dy}{dt} = a_2 x - b_2 y + h_2 \tag{1.4}$$

we infer that

$$\det J(\mathbf{x},\mathbf{p}) = \begin{vmatrix} -b_1 & a_1 \\ a_2 & -b_2 \end{vmatrix} = b_1 b_2 - a_1 a_2 \quad . \tag{1.5}$$

Hence the catastrophe set K is

$$K = \left\{ \left[a_1, a_2, b_1, b_2, h_1, h_2 \right],\ a_i, b_i > 0\ ,\ i = 1,2, a_1 a_2 - b_1 b_2 = 0 \right\} \quad . \tag{1.6}$$

Especially for the symmetrized model equations where $a_1 = a_2 = a$, $b_1 = b_2 = b$ and $h_1 = h_2 = h$ we obtain that

$$K = \{(a,a,h),\ a > 0\} \tag{1.7}$$

which confirms the intuitive results we obtained previously.

Example 2: (Exploitation model - see section 2 in previous chapter).

The differential equations which describe this model are;

$$\frac{dx}{dt} = \frac{p}{1+px+qy} - b \tag{1.8}$$

$$\frac{dy}{dt} = \frac{p}{1+qx+py} - b \tag{1.9}$$

where $q = 1-p$. Thus the model contains two parameters (p,b) and

$$\det J(x,p,b) = \frac{1}{(1+px+qy)^4} \begin{vmatrix} -p^2 & -pq \\ -pq & -p^2 \end{vmatrix} =$$

$$= \frac{p^2(p^2-q^2)}{(1+px+qy)^4} \quad . \tag{1.10}$$

Since the system (1.8)-(1.9) has a critical point only for $p > b$ we deduce that

$$K = \{(p,b),\ p > b,\ p = 1/2\} \quad . \tag{1.11}$$

(See Figure 3).

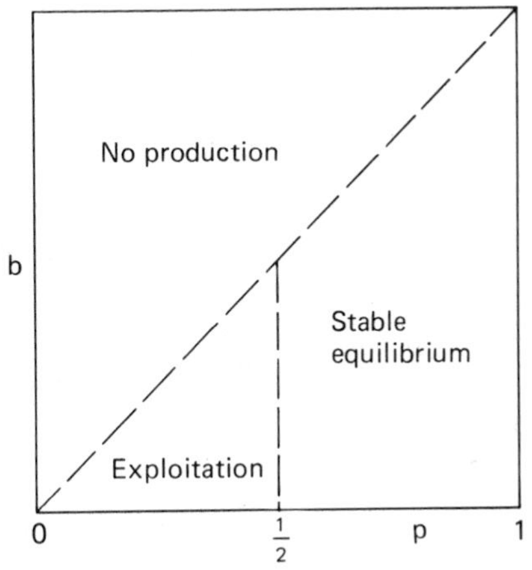

Figure 3

Results of the Exchange model in parameter space

The following is a specialization of theorem 1 to gradient systems;

Corollary 1: For gradient systems where $\mathbf{F}(\mathbf{x},\mathbf{a}) = -\mathrm{grad}V(\mathbf{x},\mathbf{a})$ the catastrophe set of the system (1.1) consists of all points $\mathbf{p}$ in parameter space for which;

1. There exists $\mathbf{x}_0$ so that $\mathrm{grad}V(\mathbf{x}_0,\mathbf{p}) = 0$
2. The determinant of the Hessian of V at $(\mathbf{x}_0,\mathbf{p})$ is zero, i.e.

$$\det \left[\frac{\partial^2 V}{\partial x_i \partial x_j}\right](x_0, p) = 0 \qquad (1.12)$$

Example 3: Consider a one dimensional gradient system where

$$V(x, a_1, a_2) = 2x^3 + a_2 x^2 + a_1 x \quad . \qquad (1.13)$$

A point x_0 is a critical point of V at (a_1, a_1) if

$$V'(x_0, a_1, a_2) = 6x_0^2 + 2a_2 x_0 + a_1 = 0 \quad . \qquad (1.14)$$

Furthermore the Hessian of V is zero at x_0 if

$$V''(x_0, a_1 a_2) = 12x_0 + 2a_2 = 0 \quad . \qquad (1.15)$$

Eliminating x_0 between (1.14) and (1.15) we find that the catastrophe set of the system is

$$K = \left\{(a_1, a_2);\ a_1 = \frac{a_2^2}{6}\right\}$$

which represents a parabola in the (a_1, a_2) plane.

To understand the "catastrophic meaning" of K in terms of the system behavior we observe that when $a_1 < \frac{a_2^2}{6}$, V has two extremum points but when $a_1 > \frac{a_2^2}{6}$, V has none. Thus on K the two extremum points coalesces and disappear. A system which might be initially at a stable equilibrium at the minimum of V when $a_1 < \frac{a_2^2}{6}$ will then move suddenly to $x = -\infty$ (the new "minimum" of V) when the parameters of the system move across the set K.

EXERCISE 1

1. Plot the function V given by equation (1.13) for different values of a_1, a_2. Pay special attention

to points near the catastrophe set $a_1 = \frac{a_2^2}{6}$.

2. Find and plot the catastophe sets for the following potentials:

   a. $V(x,a) = x^3 + a_1 x$

   b. $V(x,a_1,a_2) = x^4 + a_2 x^2 + a_1 x$

3. Find and discuss the meaning of the catastrophe set for the following ecological model

$$\frac{dF_1}{dt} = a_1 F_1 - cF_1F_2$$

$$\frac{dF_2}{dt} = a_2 F_2 - cF_1F_2, \quad a_1, a_2, \ c > 0.$$

2. CLASSIFICATION OF CATASTROPHE SETS.

As the reader might recall from elementary calculus the general formula for a quadratic surface is

$$a_1x^2 + a_2y^2 + a_3z^2 + a_4xy + a_5xz + a_6yz + a_7x + a_8y + a_9z + a_{10} = 0 \quad . \qquad (2.1)$$

At first glance this formula might lead us to believe that such a large family of surfaces will exhibit an infinitely many (or at least a very large number) "types" of surfaces. Actually, however, it turns out that all quadratic surfaces can be classified, up to rotations and translations, in terms of eleven canonical surfaces.

A similar classification attempt to obtain a "canonical representation" for the catastrophe sets of gradient systems with up to four parameters was carried out in 1975 by R. Thom with the following surprising result;

Theorem 1: For gradient systems with at most four parameters (but any number of variables) there are essentially only seven possible types of catastrophe sets. The canonical form of the potential for each of these catastrophe sets is summarized in Table 1.

Thoms's List of Seven Elementary Catastrophes

Name	Potential Function
Fold	$x^3 + ax$
Cusp	$x^4+a_2x^2+a_1x$
Swallow Tail	$x^5+a_3x^3+a_2x^2+a_1x$
Elliptic Umblic	$x_1^3-3x_1x_2+a_1\left[x_1^2+x_2^2\right]+a_2x_1+a_3x_2$
Hyperbolic Umblic	$x_1^3+x_2^3+a_1x_1x_2+a_2x_1+a_3x_2$
Butterfly	$x^6+a_1x^4+a_2x^3+a_3x^2+a_4x$
Mushroom	$x_2^4+x_1^2x_2+a_1x_1^2+a_2x_2^2+a_3x_1+a_4x_2$

Table 1

To understand the meaning of the term "essentially" in this theorem we observe that if $V(\mathbf{x},\mathbf{a})$ has a catastrophe set K and $U(\mathbf{y},\mathbf{b})$ has an empty catastrophe set then

$$W(\mathbf{x},\mathbf{y},\mathbf{a},\mathbf{b}) = V(\mathbf{x},\mathbf{a}) + U(\mathbf{y},\mathbf{b}) \tag{2.2}$$

has the same catastrophe set as V. Thus the addition of the variables y and the parameters $\mathbf{b}$ is not essential in classifying the catastrophe set of W.

Example 1: If

$$W(\mathbf{x},\mathbf{y},\mathbf{a}) = V(\mathbf{x},\mathbf{a}) + \sum_{i=1}^{k} \pm y_i^2 \tag{2.3}$$

then an easy computation shows that V and W have the same catastrophe set.

Example 2: Classify the catastrophe set of

$$V(x,y,a,b) = x^3 - 4xy + ax^2 + bx + 4y^2 \qquad (2.4)$$

Solution: The critical points of V must satisfy

$$\frac{\partial V}{\partial x} = 3x^2 - 4y + 2ax + b = 0 \qquad (2.5)$$

$$\frac{\partial V}{\partial y} = -4x + 8y = 0 \qquad (2.6)$$

Furthermore the determinant of the Hessian of V is zero if

$$\begin{vmatrix} 6x+2a & -4 \\ -4 & 8 \end{vmatrix} = 48x + 16a - 16 = 0 \quad . \qquad (2.7)$$

Eliminating x,y using (2.5)-(2.7) we obtain that a point (a,b) is in the catastrophe set of V if and only if

$$b = (1-a)^{2/3} \qquad (2.8)$$

i.e. K is a cusp in the (a,b) - plane. To classify K we recast V in terms of

$$\begin{aligned} u &= x + \frac{1}{3}(a-1) \\ v &= x - 2y \\ c_1 &= a - 1 \\ c_2 &= b - \frac{1}{3}(a-1)^2 \end{aligned} \qquad (2.9)$$

which yields

$$V(u,v,c_1,c_2) = u^3 + c_2 u + v^2 - \frac{1}{3} c_1\left[c_2 + \frac{1}{9} c_1^2\right] \qquad (2.10)$$

To analyze this expression we observe that the addition of a constant (which might depend on **c**) to the potential function does not change the equations of the dynamical system and hence cannot have any effect on the catastrophe set. We deduce, therefore, that the last term in (2.10) can be ignored. Furthermore, we see from (2.10) that the variable v is not "essential" to the classification of K which is obviously a fold catastrophe set.

To gain an insight into the nature of the seven canonical catastrophe sets we analyze further two of them viz. the cusp and the fold.

The Fold Catastrophe

The canonical form of the potential function up to a normalization factor is

$$V = \frac{1}{3} x^3 - ax \tag{2.11}$$

and the corresponding gradient system is

$$\frac{dx}{dt} = -\text{grad } V = -(x^2 - a) \quad . \tag{2.12}$$

Hence the critical points of this system are at $x_0 = \pm \sqrt{a}$, $a > 0$ and the Hessian is zero if

$$\frac{\partial^2 V}{\partial x^2}(x_0) = 2x_0 = 0 \quad . \tag{2.13}$$

We deduce, therefore, that the catastrophe set of this dynamical system consists of one point viz.

$$K = \{a = 0\}. \tag{2.14}$$

To see why this point represents a catastrophe in the system behavior we depict in Figure 4 the form of V for $a > 0$, $a < 0$ and $a = 0$. We see from this figure that when $a > 0$ V has two extrema one stable and the other unstable. However, when $a = 0$ the two points coalesces and disappear. Furthermore when $a < 0$ the "equilibrium state" of the system (which corresponds to the minimum in V) shifts to $x = -\infty$.

Finally we observe that the potential (2.11) is the only one in Table 1 which contains one parameter. Hence we infer that for any gradient system with one parameter the catastrophe set is either empty or is a union of isolated points (in parameter space) each of which is a fold catastrophe.

The Cusp Catastrophe

The canonical form of the potential function in this case is

$$V(x,a,b) = \frac{1}{4}x^4 - ax - \frac{1}{2}bx^2 \qquad (2.15)$$

and the corresponding gradient system is

$$\frac{dx}{dt} = -\frac{\partial V}{\partial x} = -(x^3 - a - bx) \quad . \qquad (2.16)$$

We infer, therefore, that a point (a,b) is in the catastrophe set K of this system if

$$x^3 - a - bx = 0 \qquad (2.17)$$

and

$$3x^2 - b = 0 \quad . \qquad (2.18)$$

Eliminating x between these equations we conclude that K is a cusp curve in parameter space, i.e.

$$K = \left\{(a,b);\ a^2 = \frac{4}{27}b^3\right\} \quad .$$

To clarify the meaning of this catastrophe set in terms of the system behavior we depict in figure 5 the graph of the potential function for various regions of the parameter space. We see from this figure that when a system shifts from region I to III the equilibrium state A disappear and the system moves to equilibrium state B (and vise versa). This fact has some important applications for biological systems which oscillate between two equilibrium points.

For further discussion of this topic the reader is referred to E. C. Zeeman - Catastrophe Theory, Addison Wesley, 1977. T. Potson and I. N. Stewart - Catastrophe Theory and its Applications, Pitman, 1978.

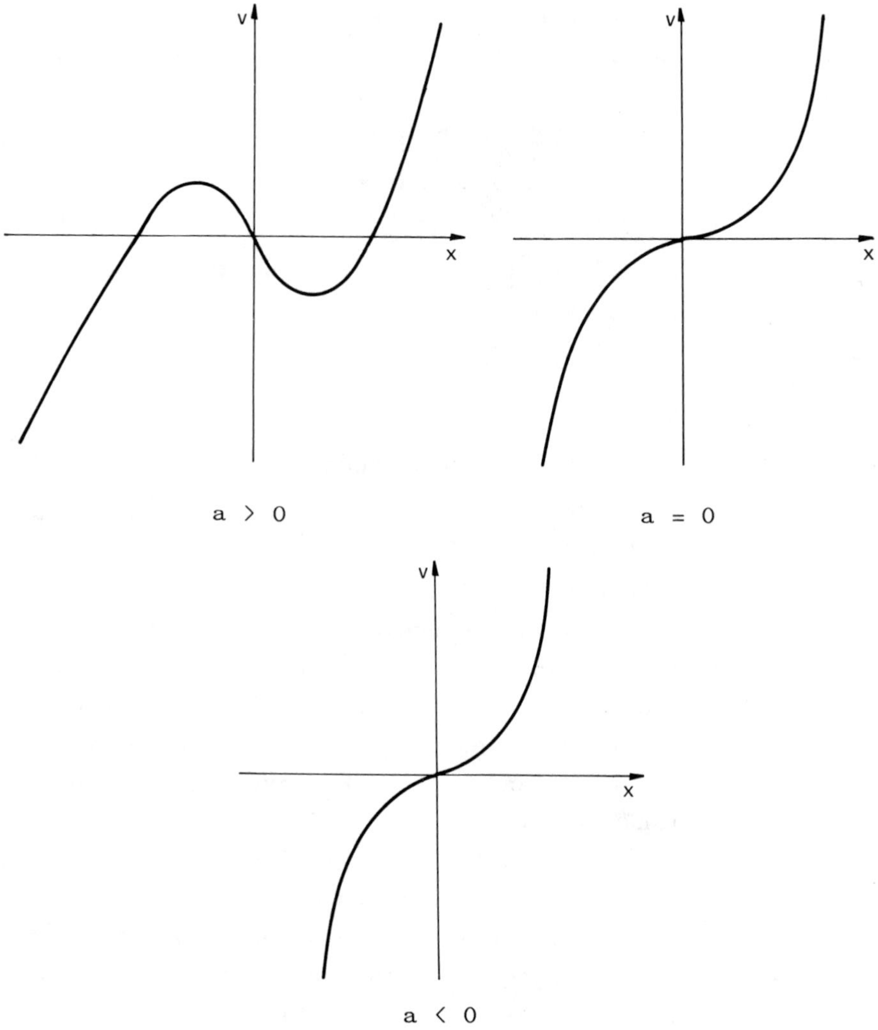

Figure 4

A sketch of the fold catastrophe potential function for various values of the parameter a.

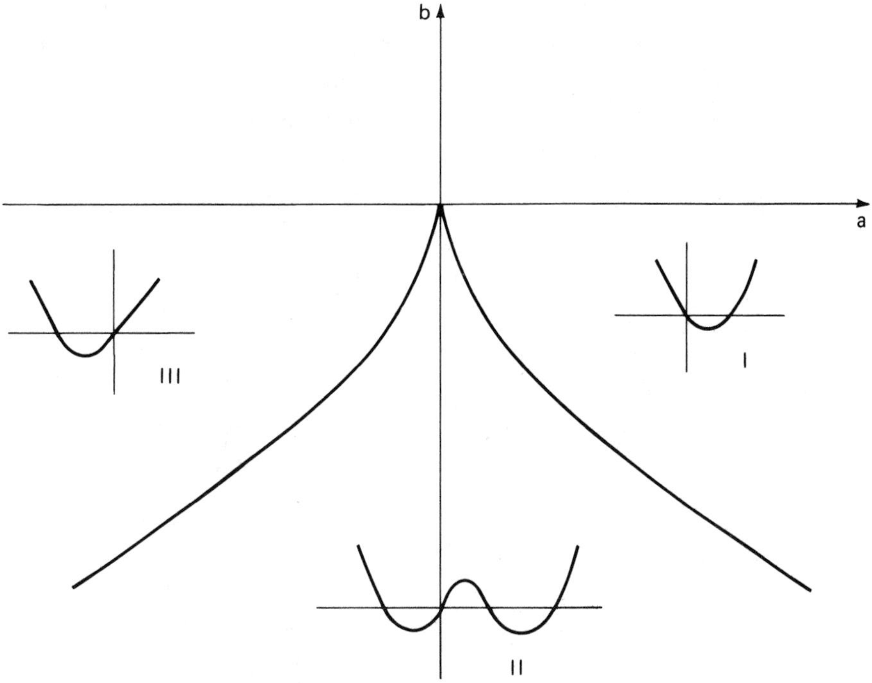

Figure 5
The graph of the cusp potential function for various values of the parameters (a,b).

EXERCISE 2

1. Carry out the computation necessary to verify the statement of Example 1.

3. SOME EXAMPLES OF BIFURCATIONS

Although the terms catastrophes and bifurcations are interchangeable from a strict mathematical point of view they are usually used today in different contexts. Thus, while the term catastrophe is used mostly in relation to R. Thom classification as described in the previous section, the term bifurcation is used in more general contexts. It describes the phenomena whereby at certain values of the system parameters the equilibrium state of the system changes its stability and new equilibrium states of the system appear

We start with some classical examples.

Example 1: The buckling of a rod.

Consider a straight homogeneous metal rod with a circular cross section which is compressed by a force F directed along the rod and applied at one of its extremities (see Figure 6). As we increase the compression the rod at a certain point will flex and its equilibrium shape will be a sine curve.

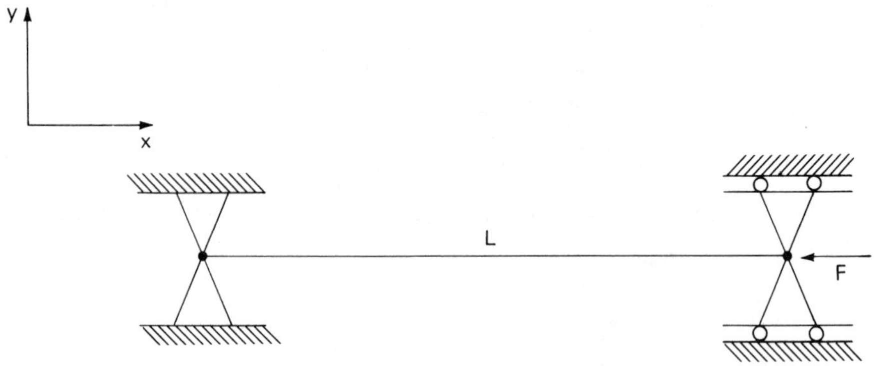

Figure 6

A thin rod of length L subject to a load F.

To give a mathematical description for this phenomena we assume that in its unloaded state the rod coincides with the interval $[0,L]$ on the x-axis. A point $(x,0)$ before the buckling will move thereafter to the position $(s(x),u(x))$ where $s(x)$ is the arc length of the rod and $u(x)$ is the vertical displacement. If we further assume that the rod is inextensible then $s(x) = x$ and the following geometrical and physical relations (see Figure 7), hold

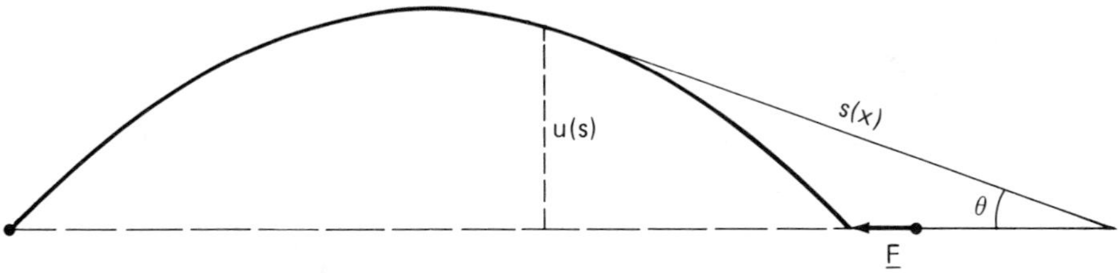

Figure 7
The buckled state of the rod

1. $\frac{du}{ds} = \sin\theta$ (3.1)

where θ is the angle between the tangent to the buckled rod at $(s,u(s))$ and the x-axis.

2. The moments equilibrium condition;

$$M = Fu \tag{3.2}$$

where M is the elastic moment exerted by the rod.

3. Euler-Bernoulli law which relates the elastic moment to the curvature $\frac{d\theta}{ds}$ of the rod

$$M = -EI\frac{d\theta}{ds} \tag{3.3}$$

where E, I are positive physical constants which depend on the material of the rod.

From these equations we easily infer that

$$\frac{d^2\theta}{ds^2} + \lambda\sin\theta = 0 \quad . \tag{3.4}$$

where $\lambda = F/EI$. Furthermore, since $u(0) = u(L) = 0$ we obtain from (3.2),(3.3) the boundary conditions

$$\theta'(0) = \theta'(L) = 0 \quad . \tag{3.5}$$

Since equation (3.4) is nonlinear we try first to see if a linear approximation can yield the desired physical insight. To do so we assume that $|\theta| << 1$ and approximate equation (3.1) by

$$\frac{du}{ds} = \theta \quad . \tag{3.6}$$

Eliminating θ between (3.6),(3.2),(3.3) we obtain then

$$\frac{d^2u}{ds^2} + \lambda u = 0, \quad u(0) = u(L) = 0 \quad . \tag{3.7}$$

Hence we infer that nontrivial solutions for u exist only when $\lambda_n = \frac{\pi^2 n^2}{L^2}$, $n = 1,2,\ldots$ and then

$$u_n(s) = c_n \sin \frac{n\pi s}{L} \tag{3.8}$$

where c_n remains as an arbitrary constant.

From a physical point of view this result means that when $F < \frac{\pi^2 EI}{L^2}$ the rod remains straight (the only solution of (3.7) is $u = 0$). However, when $F = \frac{\pi^2 EI}{L^2}$ the rod buckles (with indeterminate amplitude c) but when $\frac{\pi^2 EI}{L^2} < F < \frac{4\pi^2 EI}{L^2}$ The rod must return to its original state and this scenario repeats itself for other eigenvalues of equation (3.7). Obviously, the picture that emerges from the "linearized theory" is not satisfying since in reality the rod does not return to its original state when $\frac{\pi^2 EI}{L^2} < F < \frac{4\pi^2 EI}{L^2}$ and furthermore, the amplitude of the buckling cannot remain as arbitrary constant.

To gain better insight we must, therefore, reconsider the nonlinear equation (3.4)-(3.5). These can be solved (in terms of elliptic integrals) as follows. Multiplying equation (3.4) by $\frac{d\theta}{ds}$ yields

$$\frac{1}{2}\frac{d}{ds}\left[\left[\frac{d\theta}{ds}\right]^2\right] = -\lambda\sin\theta\,\frac{d\theta}{ds} \quad . \tag{3.9}$$

Hence assuming $\theta(0) = \alpha$ (where $\alpha > 0$ is arbitrary) and using the boundary condition $\theta'(0) = 0$ we infer that

$$\left[\frac{d\theta}{ds}\right]^2 = 2\lambda(\cos\theta - \cos\alpha) \quad . \tag{3.10}$$

Using the identity $1 - \cos\varphi = 2\sin^2\frac{\varphi}{2}$ and the substitution $\sin\psi = \frac{\sin\theta/2}{\sin\alpha/2}$ we obtain from (3.10)

$$A + \sqrt{\lambda}\,s = \int_0^{\psi(s)} \frac{d\varphi}{\sqrt{1-k^2\sin^2\varphi}} = J(\psi,k) \tag{3.11}$$

where A is an integration constant, $k = \sin\alpha/2$ and the integral on the right hand side of (3.11) is an elliptic integral of the first kind. To determine A we now note that $\theta(0) = \alpha$ and, therefore, $\psi(0) = \pi/2$. Hence,

$$A = \int_0^{\pi/2} \frac{d\varphi}{\sqrt{1-k^2\sin^2\varphi}} = J(\pi/2,k) \quad . \tag{3.12}$$

Finally, we must take into account the boundary condition $\theta'(L) = 0$. From (3.10) we obtain $\theta(L) = \alpha + 2\pi n,\ 2\pi n - \alpha$ and, therefore, $\psi = \pi/2 + n\pi,\ n = 0,1,\ldots$ .Hence, from (3.11) we infer that

$$J(\pi/2,k) + \sqrt{\lambda}L = J(\pi/2 + n\pi,k) \tag{3.13}$$

but $J(\pi/2 + n\pi,k) = (2n + 1)J(\pi/2,k)$, i.e.

$$J(\pi/2,k) = \frac{\sqrt{\lambda}L}{2n} \quad . \tag{3.14}$$

This last equation has different number of solutions for different values of λ viz. it has n solutions $(n = 0,1,\ldots)$ when

$$\frac{n^2\pi^2}{L^2} < \lambda < \frac{(n+1)^2\pi^2}{L^2} \quad . \tag{3.15}$$

It follows then that equations (3.4)-(3.5) have only the trivial solution $\theta = 0$ when $0 < \lambda < \frac{\pi^2}{L^2}$, one nontrivial solution on $\left[\frac{\pi^2}{L^2}, \frac{4\pi^2}{L^2}\right]$ etc. Observe, however, that we considered in our treatment only $\theta(0) = \alpha > 0$. If we let $\alpha < 0$ the number of possible solutions for each λ will double (see Figure 8).

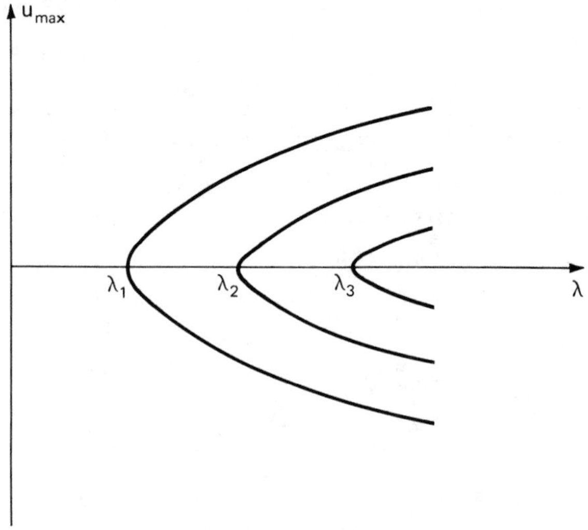

Figure 8
Maximum amplitude of buckling as a function of λ

Although we do not consider here the stability of these solutions for $\theta(s)$ (and hence $u(s)$) we wish to point out that in general whenever a physical system has several possible states only the one with minimum energy will be stable.

Finally, to conclude this example we observe that although the linear approximation to the model equations suffered from several shortcomings it still predicted correctly the bifurcation points of the system under consideration. We shall see later that this is true in many cases.

Example 2: Rotating incompressible fluid in gravitating self interaction.

Over the years several theories were suggested for the origin of the solar system (or the earth-moon system). One such suggestion was that due to its rotation the sun loses its spherical shape and emits from time to time a certain amount of matter which then cools down to form a planet.

To investigate this possibility we consider here a rotating mass of incompressible fluid in gravitating self interaction. This problem was already considered by Newton who showed that the equilibrium figure of such a mass (when the angular velocity is not large) is an oblate axially symmetric ellipsoid through the rotation axis (also referred to as "Maclaurin spheroids"). He also applied this result to the shape of the earth and obtained a reasonable approximation for it.

What happens to such a fixed mass when the angular speed increases (or when the density increases) was extensively investigated by several mathematicians and physicists including Riemann, Dedekind and Poincare, and others.

Their investigations showed that as the angular momentum of the mass increases the system passes through several bifurcations. At first a stable equilibrium state in the shape of three unequal axis ellipsoid (Jacobi ellipsoids) appear. Then as the angular momentum increases further a pear shaped unstable equilibrium state, which is rather suggestive from astrophysical point of view becomes possible (see Figure 9).

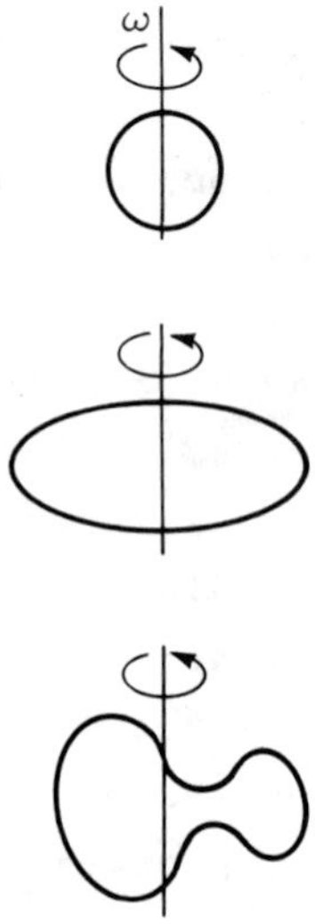

Figure 9
Stages in the evolution of a rotating star. Starting from a spherical shape it turns into an ellipisoid (as ω increases) and then into a pear shaped star

In the following we present a simple mathematical model for this problem from which we derive the first bifurcation point of this system viz. from Maclaurin spheroids to Jacobi ellipsoids.

To begin with we assume that the mass is of ellipsoidal shape with uniform density ρ and semi-axes a, b, c and that it is rotating around the z-axis with angular velocity ω. Hence, the total mass is

$$M = \frac{4}{3} \pi abc\rho \tag{3.16}$$

and the total angular momentum J is given by

$$J = \frac{1}{5} M (a^2+b^2)\omega = I\omega \tag{3.17}$$

where I is the moment of inertia.

The kinetic and potential energy of such a system are then given respectively by

$$T = \frac{J^2}{2I} , \quad E = \frac{1}{2} \int_V \varphi\rho dV \tag{3.18}$$

where φ is the gravitational potential inside the body which satisfies Poisson equation

$$\nabla^2\varphi = 4\pi G\rho \tag{3.19}$$

(G is the gravitational constant). We now note that the expression for φ inside a homogeneous ellipsoid whose external surface is

$$\frac{x^2}{a^2} + \frac{y^2}{b^2} + \frac{z^2}{c^2} = 1 \tag{3.20}$$

is well known to be

$$\varphi(x,y,z) = \pi\rho Gabc \int_0^\infty \left[\frac{x^2}{a^2+s} + \frac{y^2}{b^2+s} + \frac{z^2}{c^2+s} - 1\right] \frac{ds}{\Delta} \tag{3.21}$$

where

$$\Delta^2 = \left[a^2+s\right]\left[b^2+s\right]\left[c^2+s\right] \quad . \tag{3.22}$$

Hence we infer that the total energy of the body is given by

$$H = T + E = \frac{J^2}{2I} - \frac{8}{15} \pi^2\rho^2 Ga^2b^2c^2 \int_0^\infty \frac{ds}{\Delta} \quad . \tag{3.23}$$

Equilibrium states of the system correspond then to the local extrema of the energy functional H. To compute these extrema we first introduce the variables

$$\xi = \left[\frac{a}{c}\right]^2 , \quad \zeta = \left[\frac{b}{c}\right]^2 \tag{3.24}$$

and the parameter

$$\lambda = 25 \left[\frac{4\pi\rho}{3M}\right]^{1/3} \frac{J^2}{3GM^3} \quad . \tag{3.25}$$

H then takes the form

$$H(\xi,\zeta,\lambda) = \frac{3}{10} GM^2 \left[\frac{4\pi\rho}{3M}\right]^{1/3} \left\{\lambda \frac{(\xi\zeta)^{1/3}}{\xi+\zeta} - (\xi\zeta)^{1/6} \int_0^\infty \frac{dt}{[(1+t)(\xi+t)(\zeta+t)]^{1/2}}\right\} \quad (3.26)$$

At a critical point of H

$$\frac{\partial H}{\partial \xi} = \frac{\partial H}{\partial \zeta} = 0$$

which have two sets of solutions. The first are the Maclaurin spheroids and the second are the Jacobi ellipsoids. The latter bifurcate from the former at $\xi = \zeta \cong 3$, $\lambda = 0.769$. It was shown in the literature that Jacobi ellipsoids are always stable (under present assumptions) while the Maclaurin spheroid are stable only when $\lambda < 0.769$. When departures from ellipsoidal shapes are allowed, however, Jacobi ellipsoids become unstable as λ increases and a second bifurcation point appear where pear shaped forms are also possible.

Example 3: Taylor vortices

Another classical example of bifurcating solutions is the Taylor problem which is related to viscous fluid flow in between two concentric cylinders where the inner is rotating at a fixed angular velocity ω while the outer cylinder is at rest.

When ω is small the motion of the fluid is laminar and the fluid velocity has only a θ component (we use cylindrical coordinates r,θ,z) which is a function of r only. This flow is called Couette flow. However, if we increase the angular velocity of the inner cylinder we shall reach a bifurcation point $\omega = \omega_1$, where the laminar flow becomes unstable and there will be superimposed on it uniformly spaced ring vortices (in each ring, fluid will flow both vertically and horizontally). If we increase ω beyond ω_1 a second bifurcation point will be reached where the ring vortices become also unstable and are replaced by "wavey vortices". These waves will rotate around the z-axis and at each point in space the fluid velocity will be periodic in time.

We shall not enter here into the mathematical description of these phenomena since they require partial differential equations.

4. BIFURCATION OF EQUILIBRIUM STATES IN ONE DIMENSION

1. General Setting.

We now proceed to study bifurcations of equilibrium states in autonomous systems of differential equations which depend on a set of parameters $\mu = (\mu_1 \ldots \mu_m)$

$$\frac{d\mathbf{x}}{dt} = \mathbf{f}(\mathbf{x},\mu) \quad , \quad x \in R^n \quad . \tag{4.1}$$

Our objective is to investigate for these systems the dependence of the equilibrium states on μ and identify those values for which new equilibrium states appear (bifurcation points). Furthermore, we will study the properties, viz. stability, of these solutions near the bifurcation points.

Since equilibrium states of (4.1) are characterized by the equation $\frac{d\mathbf{x}}{dt} = 0$ it follows that the problem at hand is equivalent to finding the bifurcation points of the algebraic equation $\mathbf{f}(\mathbf{x},\mu) = 0$. Given such an $\mathbf{f}$ with bifurcation points $\mu^{(1)}, \ldots, \mu^{(k)}$ it is a standard practice to reduce the system (4.1) into a "local form" near each bifurcation point. To do so assume that $\mathbf{x} = \mathbf{x}_0(\mu)$ is a (stable) equilibrium state near $\mu^{(i)}$. We then define

$$\mathbf{u} = \mathbf{x} - \mathbf{x}_0 \quad , \quad \lambda = \mu - \mu^{(i)} \tag{4.2}$$

to obtain

$$\frac{d\mathbf{u}}{dt} = \frac{d(\mathbf{x}-\mathbf{x}_0)}{dt} = \mathbf{f}(\mathbf{u}+\mathbf{x}_0, \ \lambda+\mu^{(i)}) \equiv F(\mathbf{u},\lambda) \ . \tag{4.3}$$

Obviously, for the new system the original equilibrium state corresponds to $\mathbf{u} = \mathbf{0}$ and the bifurcation point under consideration is at $\lambda = 0$.

While the discussion above provides a general setting for the bifurcation problem we shall find it advantageous to discuss first bifurcations in one and two dimensions. The rest of this section as well as the next will be devoted to these special cases.

2. Examples in one-dimension.

Example 1: For the equation

$$\dot{u} = \lambda - u^2 \tag{4.4}$$

$F(u,\lambda) = \lambda - u^2$ and $F(u,\lambda) = 0$ has a real solution only if $\lambda \geq 0$. In this latter case the system has two equilibrium states for each λ, $u = \pm\sqrt{\lambda}$ i.e. $\lambda = 0$ is a bifurcation point of equation (4.4). Furthermore, it is easy to see (using phase diagrams) that $u = \sqrt{\lambda}$ is stable and $u = -\sqrt{\lambda}$ is unstable (see Figure 10).

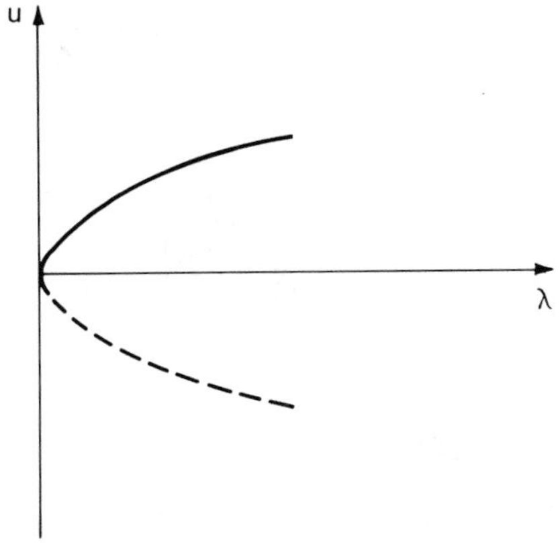

Figure 10
Bifurcation at $\lambda = 0$ for equation (4.4). Unstable solutions are always plotted using a broken line.

Example 2: The equation

$$\dot{u} = u(\lambda - u) \tag{4.5}$$

has two equilibrium solutions $u_1 = 0$ and $u_2 = \lambda$ for all values of λ (see Figure 11). Hence $\lambda = 0$ is a bifurcation point of this equation. Moreover a simple stability analysis show that when $\lambda < 0$ u_1, u_2 are stable and unstable respectively while for $\lambda > 0$, u_2 is stable and u_1 is unstable. From a physical point of view this means that the system will "jump" from one state to another as λ crosses the origin.

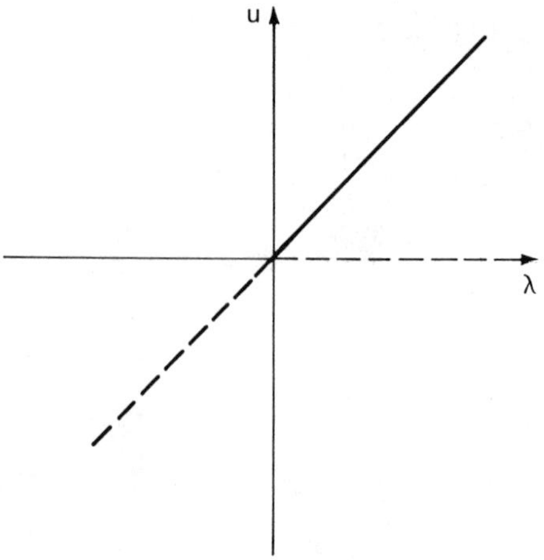

Figure 11
Exchange of stability at $\lambda = 0$ for equation (4.5)

Example 3:

$$\dot{u} = u(\lambda - u^2) \quad . \tag{4.6}$$

In this case $u = 0$ is a solution for all λ. However, when $\lambda > 0$ two nonzero solutions $u_2 = +\sqrt{\lambda}$, $u_3 = -\sqrt{\lambda}$ are also possible. Thus, $\lambda = 0$ is a bifurcation point and the bifurcation itself is called (appropriately) pitch-fork bifurcation (see Figure 12). Once again it is easy to show that $u = 0$ is stable for $\lambda < 0$ and unstable when $\lambda > 0$. The other two nonzero solutions are always stable.

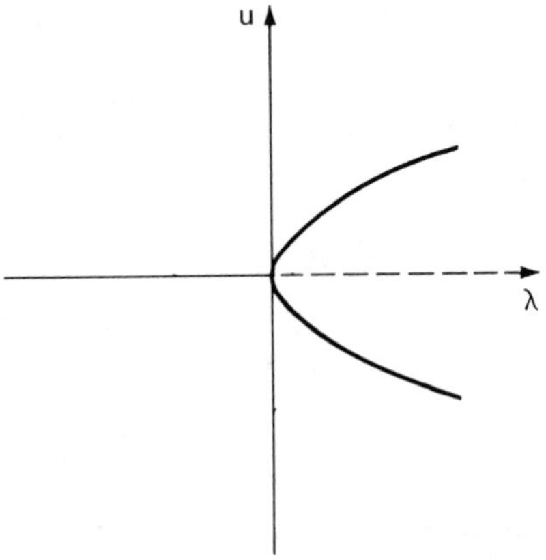

Figure 12
Pitchfork Bifurcation

3. General treatment of bifurcations in one-dimension.

To discuss bifurcations in one dimension in a general setting we need the implicit function theorem which states that under appropriate conditions of differentiability and continuity on F the implicit equation $F(u,\lambda) = 0$ can be solved uniquely for u in terms of λ (or vice versa) in the vicinity of (u_0,λ_0) if

$$F_\lambda(u_0,\lambda_0) = \frac{\partial F}{\partial \lambda}(u_0,\lambda_0) \neq 0 \tag{4.7}$$

or

$$F_u(u_0,\lambda_0) = \frac{\partial F}{\partial u}(u_0,\lambda_0) \neq 0 \quad . \tag{4.8}$$

Accordingly, if we consider F in a region $D \subset R^2$ then a point (u_0,λ_0) in this region which satisfies the equation $F(u_0,\lambda_0) = 0$ is classified as regular if the condition (4.7)-(4.8) is true and as singular if it is false (i.e., $F_\lambda(u_0,\lambda_0) = F_u(u_0,\lambda_0) = 0$). Singular points, however, are classified further as follows;

Definition 1: Let (u_0,λ_0) be a singular point of $F(u,\lambda)$ where not all second order derivatives of F vanish at (u_0,λ_0). We say that (u_0,λ_0) is;

1. A conjugate point if it is an isolated solution of $F(u,\lambda) = 0$,
2. a double point if there are only two solutions of $F(u,\lambda) = 0$ with distinct tangents which pass through (u_0,λ_0),
3. a cusp point if there are only two solutions of $F(u,\lambda) = 0$ which pass through (u_0,λ_0) and have the same tangent at this point.

Finally (u_0,λ_0) is called a higher order singular point of $F(u,\lambda)$ if it is a singular point and all second order derivatives of F also vanish at this point.

In the following we refer to a conjugate, double or cusp point as regular singular points of F.

Example 4: The point $(0,0)$ is obviously a singular point of the equation

$$F(u,\lambda) = au^2 + bu\lambda + c\lambda^2 = 0 \tag{4.9}$$

Furthermore, if $\Delta = b^2 - 4ac < 0$ then $(0,0)$ is conjugate point. On the otherhand if $\Delta > 0$ then this point is a double point since there are two solutions passing through $(0,0)$ with different tangents viz.

$$u_1 = \lambda\left[\frac{-b+\Delta}{2\Delta}\right], \qquad u_2 = -\lambda\left[\frac{b+\Delta}{2\Delta}\right] \quad . \tag{4.10}$$

Finally, when $\Delta = 0$ the origin can be considered as a "degenerate" cusp point, i.e. the two solutions coincide with each other and hence have the same tangent.

Example 5: The equation

$$F(u,\lambda) = (u - 2\lambda^2)(u-\lambda^2) = 0 \tag{4.11}$$

has a cusp point at $(0,0)$ since the two solutions

$$u = 2\lambda^2, \; u = \lambda^2 \tag{4.12}$$

which pass through this point have the same tangent.

We now observe that a regular point of F can still be a bifurcation point if $F(u,\lambda) = 0$ has no nonzero solutions for $\lambda < \lambda_0$ as in example 1 above. However, in general bifurcations occur at singular points of F. Out of these. double point bifurcations are the most common since bifurcations at cusp points and higher order singular point require some special relationship between the higher order derivatives of F (see below). Motivated by this fact we now treat double point bifurcations in greater detail.

4. Bifurcation and stability at a double point.

Near a regular singular point the function F can be approximated by its Taylor expansion to the second order

$$F(u_0+\Delta u,\lambda_0+\Delta\lambda) = \frac{1}{2} F_{uu}(u_0,\lambda_0)(\Delta u)^2 + F_{u\lambda}(u_0,\lambda_0)\Delta u\Delta\lambda + \frac{1}{2} F_{\lambda\lambda}(u_0,\lambda_0)(\Delta\lambda)^2 + O[(|\Delta u| + |\Delta\lambda|)^3]. \tag{4.13}$$

Hence, the solution curves of $F(u,\lambda) = 0$ must satisfy as $\Delta u \to 0, \Delta\lambda \to 0$ the equation

$$\frac{1}{2} F_{uu}(u_0,\lambda_0)(du)^2 + F_{u\lambda}(u_0,\lambda_0)du\, d\lambda$$
$$\frac{1}{2} F_{\lambda\lambda}(u_0,\lambda_0)(d\lambda)^2 = 0. \tag{4.14}$$

From equation (4.14) we deduce that if

$$\Delta(u_0,\lambda_0) = F_{u\lambda}^2(u_0,\lambda_0) - F_{uu}(u_0,\lambda_0)F_{\lambda\lambda}(u_0,\lambda_0)$$

is negative then (u_0,λ_0) is a conjugate point. We now show, however, that if $\Delta(u_0,\lambda_0) > 0$ then the point is a double point. We distinguish two cases;

1. $F_{uu}(u_0,\lambda_0) \neq 0$.

We infer from equation (4.14) that the tangents to the solution curves at (u_0,λ_0) are given by

$$\frac{du_{1,2}}{d\lambda}(\lambda_0) = \frac{-F_{u\lambda}(u_0,\lambda_0) \pm \sqrt{\Delta(u_0,\lambda_0)}}{F_{uu}(u_0,\lambda_0)} \tag{4.15}$$

2. $F_{uu}(u_0,\lambda_0) = 0$.

In this case equation (4.14) reduces to

$$d\lambda[F_{u\lambda}(u_0,\lambda_0)du + \frac{1}{2} F_{\lambda\lambda}(u_0,\lambda_0)d\lambda] = 0 \tag{4.16}$$

and it follows that the tangents to the solution curves at (u_0,λ_0) are given by

$$\frac{d\lambda}{du}(u_0) = 0 \quad , \quad \frac{du}{d\lambda} = -\frac{F_{\lambda\lambda}(u_0,\lambda_0)}{F_{u\lambda}(u_0,\lambda_0)} \tag{4.17}$$

Finally if $\Delta(u_0,\lambda_0) = 0$ then it is evident from (4.14) that the tangent to the two solution curves at (u_0,λ_0) is the same and hence (u_0,λ_0) is a cusp point.

To discuss the stability of the solutions at a double point we first introduce the following definition and then state without proof the appropriate theorem.

Definition 2: A point (u_0,λ_0) which satisfies $F(u_0,\lambda_0) = 0$ is called

1. A regular turning point of F if it is a regular point of F and $\frac{d\lambda}{du}$ changes it sign at this point and $F_\lambda(u_0,\lambda_0) \neq 0$.
2. A singular turning point if it is a double point of $F(u,\lambda) = 0$ at which the sign of $\frac{d\lambda}{du}$ on one of the solution curves changes its sign.

Theorem 1: If all singular points of $F(u,\lambda) = 0$ are double points then the stability of the solutions to this equation must change at

1. Regular turning points.
2. Singular points which are not turning points

and only at these points.

Examples 1,3 above demonstrate the validity of this theorem.

Example 6: Discuss the stability of the equilibrium solutions for the equation

$$\dot{u} = (u^2 + \lambda^2)^2 - 2a^2(u^2 - \lambda^2) \tag{4.18}$$

Remark: The equation of the lemniscate in polar coordinates is

$$r^2 = 2a^2 \cos 2\theta \tag{4.19}$$

hence in Cartesian coordinates it is represented by the equation

$$(x^2 + y^2)^2 - 2a^2(x^2 - y^2) = 0 \quad . \tag{4.20}$$

Thus, the equilibrium states of equation (4.18)

$$F(u,\lambda) = (u^2+\lambda^2)^2 - 2a^2(u^2-\lambda^2) = 0 \tag{4.21}$$

are on a lemniscate curve in the $\lambda - u$ plane (see Figure 14).

Solution: Since

$$F_u = 4u(u^2+\lambda^2) - 4a^2u, \quad F_\lambda = 4\lambda(u^2+\lambda^2) + 4a^2\lambda$$

$$F_{uu} = 4(u^2+\lambda^2) + 8u^2-4a^2,$$

$$F_{\lambda\lambda} = 4(u^2+\lambda^2) + 8\lambda^2 + 4a^2, \quad F_{u\lambda} = 8u\lambda \quad . \tag{4.22}$$

We infer that $(0,0)$ is a double singular point ($F_u = F_\lambda = 0$ and $\Delta > 0$). Moreover, this is the only singular point on $F(u,\lambda) = 0$. As to the regular turning points we observe that

$$\frac{d\lambda}{du} = -\frac{F_u}{F_\lambda} = -\frac{u(u^2+\lambda^2-a^2)}{\lambda(u^2+\lambda^2+a^2)} \quad . \tag{4.23}$$

Hence, it follows (using (4.21)) that the curve has

four regular turning points viz. $\left[\frac{a\sqrt{3}}{2}, \pm \frac{a}{2}\right]$, $\left[\frac{-a\sqrt{3}}{2}, \pm \frac{a}{2}\right]$. To see the change in stability at the singular and turning points we choose a fixed $\lambda \left[\text{e.g., } \lambda = \pm \frac{a}{4}\right]$ and plot the phase diagram of equation (4.18). We obtain that (for both values of λ)

$$u_2 = \frac{a}{4}\sqrt{15-8\sqrt{3}}, \quad u_4 = -\frac{a}{4}\sqrt{15+8\sqrt{3}}$$

are stable while

$$u_1 = \frac{a}{4}\sqrt{15+8\sqrt{3}}, \quad u_3 = -\frac{a}{4}\sqrt{15-8\sqrt{3}}$$

are unstable (see Figure 13). When these points are plotted on the lemniscate curve they clearly corroborate the result stated in theorem 1 (see Figure 14).

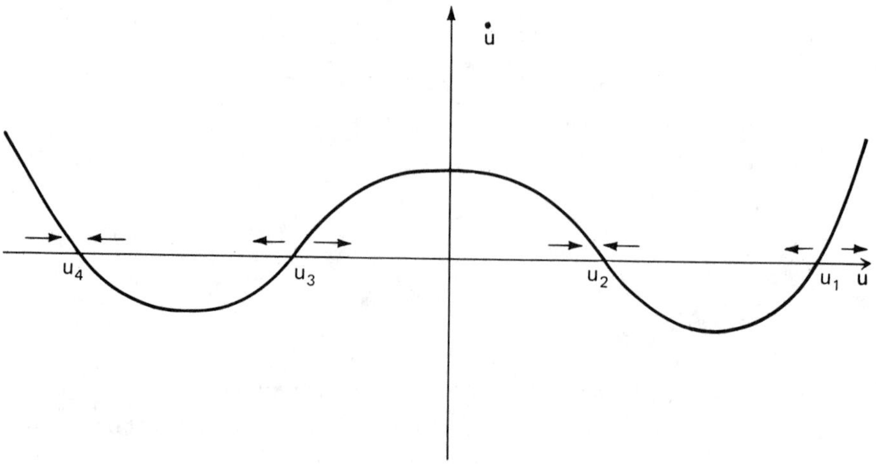

Figure 13

Phase diagram illustrating the stability of the equilibrium solutions of equation (4.18) at $\lambda = \pm a/y$.

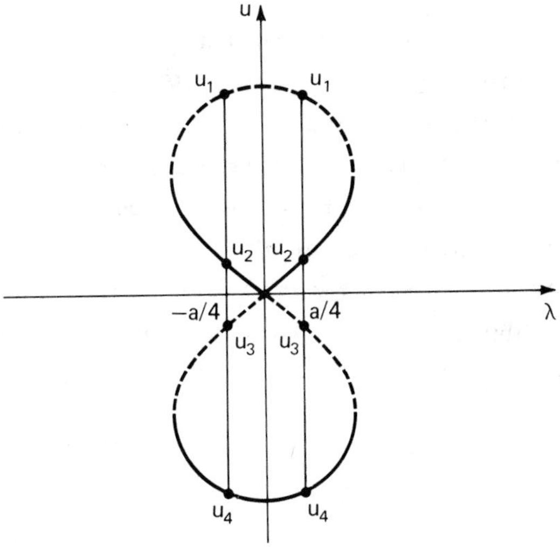

Figure 14

Exchange of stability at (0,0) and the regular turning points in equation (4.18).

EXERCISE 3

For the following curves discuss the singular points, turning point and the stability of the equilibrium solutions for the equation $\dot{u} = F(u,\lambda)$.

1. $F(u,\lambda) = (\lambda^2+u^2)^2 - 2a^2(\lambda^2-u^2)$
2. $F(u,\lambda) = \lambda^3 + u^3 - 3a\lambda u$
3. $F(u,\lambda) = u^2(2a-\lambda) - \lambda^3 = 0$
4. $F(u,\lambda) = (u^2+\lambda^2)^3 - (\lambda^2-u^2)^2$
5. $F(u,\lambda) = u^2 - au + \lambda^2$
6. $F(u,\lambda) = u^2 + \lambda^2 - a\lambda$

(in each case plot the curve $F(u,\lambda) = 0$).

7. Verify the statements made in examples 1,2,3 of this section.

5. HOPF BIFURCATION

As one can easily guess the structure and variety of bifurcation phenomena in two dimensions is much richer than in one-dimension and the corresponding mathematical treatment is more complicated. We shall, therefore, limit ourself to the consideration of only one such bifurcation - the Hopf bifurcation - which is a prototype for a whole class of bifurcations in higher dimensions and is useful in many applications.

We introduce this bifurcation using a classical example.

Example 1: Consider the system

$$\frac{du}{dt} = -v - u(u^2+v^2) + \lambda u \tag{5.1}$$

$$\frac{dv}{dt} = u - v(u^2+v^2) + \lambda v \quad . \tag{5.2}$$

To find the equilibrium states of this system we must solve simultaneously

$$- v - u(u^2+v^2) + \lambda u = 0 \tag{5.3}$$

$$u - v(u^2+v^2)^2 = \lambda(u^2+v^2) \tag{5.4}$$

Multiplying (5.3) by u and (5.4) by v and adding we obtain;

$$(u^2+v^2)^2 = \lambda(u^2+v^2) \quad . \tag{5.5}$$

Hence $u = v = 0$ is an equilibrium state of this system for all λ. However, when $\lambda > 0$ the system has also a continuous set of equilibrium points

$$u^2 + v^2 = \lambda \quad . \tag{5.6}$$

To treat the equilibrium state $u = v = 0$ we note that the linear approximation to the system (5.1)-(5.2) near this point is

$$\frac{d}{dt}\begin{bmatrix} u \\ v \end{bmatrix} = \begin{bmatrix} \lambda & -1 \\ 1 & \lambda \end{bmatrix}\begin{bmatrix} u \\ v \end{bmatrix} = A\begin{bmatrix} u \\ v \end{bmatrix} \quad . \tag{5.7}$$

since the eigenvalues of A are

$$\mu = \lambda \pm i \tag{5.8}$$

we infer that when $\lambda < 0$ the state $(0,0)$ is a

stable spiral point which becomes unstable when λ is positive.

To see the significance of the other set of equilibrium states that exists for $\lambda > 0$ we introduce polar coordinates

$$u = r \cos \theta \quad , \quad v = r \sin \theta \tag{5.9}$$

and obtain (see previous chapter)

$$r \frac{dr}{dt} = u \frac{du}{dt} + v \frac{dv}{dt} = r^2(\lambda - r^2) \tag{5.10}$$

$$r^2 \frac{d\theta}{dt} = v \frac{du}{dt} - u \frac{dv}{dt} = -r^2 \quad . \tag{5.11}$$

It follows then that the system has a stable limit cycle at $r^2 = \lambda$. To summarize: The system (5.1)-(5.2) has a stable spiral point at $(0,0)$ for $\lambda < 0$ which becomes unstable as λ crosses the origin. Moreover, for $\lambda > 0$ the system has a stable limit cycle. Hence, $\lambda = 0$ is a bifurcation point of this system viz. the stable spiral bifurcates into a limit cycle. Such a bifurcation is called Hopf bifurcation.

In 1942 Hopf proved that such a behavior is rather general whenever the linear approximation to the system near an equilibrium state has a pair of conjugate eigenvalues which move from the left half of the complex plane to the right one.

The exact formulation of this theorem (in R^n) is as follows:

Theorem 1 (Hopf): Suppose that the system

$$\dot{\mathbf{x}} = F(\mathbf{x},\lambda) = \begin{bmatrix} f_1(\mathbf{x},\lambda) \\ \vdots \\ f_n(\dot{\mathbf{x}},\lambda) \end{bmatrix}, \quad \mathbf{x} \in R^n, \ \lambda \in R \tag{5.12}$$

has an equilibrium state $(\mathbf{x}(\lambda_0),\lambda_0)$ so that

1. The Jacobian

$$J(x(\lambda_0),\lambda_0) = \left[\frac{\partial f_i}{\partial x_j}\right] (\mathbf{x}(\lambda_0),\lambda_0) \tag{5.13}$$

has a simple pair of pure imaginary eigenvalues $\pm ia$ with all other eigenvalues having non-zero real part.

2. The eigenvalues $\mu(\lambda)$, $\mu(\lambda)$ of $J(\mathbf{x}(\lambda),\lambda)$ which are imaginary at $\lambda = \lambda_0$ are smooth functions of λ and

$$\frac{d}{d\lambda}\,(\mathrm{Re}\mu(\lambda)) \neq 0 \quad . \tag{5.14}$$

Then $(\mathbf{x}(\lambda_0),\lambda_0)$ is a bifurcation point of the equilibrium state $\mathbf{x}(\lambda_0)$ to a limit cycle.

We now describe a physical system which exhibits a Hopf bifurcation.

Example 2: Brusselator Reaction.

It is well known that many chemical systems exhibit oscillatory behavior viz. the concentration of some of the chemicals participating in the chemical reactions of the system vary periodically with time. The Brusselator reaction is a model for such a system.

The model considers the following set of reaction

$$\begin{aligned} A &\to x \\ B + x &\to y + C \\ 2x + y &\to 3x \\ x &\to D \quad . \end{aligned} \tag{5.15}$$

Here A,B,C,D are the initial and final products of these reactions whose concentrations are assumed to be constants. Using the same letter to denote the chemical and its concentration we now derive equations for the rate of change of x,y. To this end we observe that the production rate of x in the first and third reactions are (proportional to) A and x^2y while its rate of loss in the second and fourth are Bx and x respectively.

Hence

$$\frac{dx}{dt} = A - (B + 1)x + x^2y \quad . \tag{5.16}$$

Similarly we obtain for y

$$\frac{dy}{dt} = Bx - x^2y \quad . \tag{5.17}$$

Assuming $A = 1$ we deduce that the system (5.16)-(5.17) has an equilibrium state $x = 1$, $y = B$. Reducing (5.16)-(5.17) to a local form by the transformation

$$x = 1 + u \quad , \quad y = B + v \tag{5.18}$$

we obtain

$$\dot{u} = (B-1)u + v + u^2(B+v) + 2uv \tag{5.19}$$

$$\dot{v} = -Bu - v - u^2(B+v) - 2uv \quad . \tag{5.20}$$

(Hence B plays the role of λ). Hence at the equilibrium state $(0,0)$ we have

$$J(0,0,B) = \begin{bmatrix} b - 1 \, , & 1 \\ -B \, , & -1 \end{bmatrix} \tag{5.21}$$

whose eigenvalues are

$$\mu = \frac{B-2\pm\sqrt{(B-2)^2-4}}{2} \tag{5.22}$$

we see that when $B < 2$, $(0,0)$ is a stable spiral point. At $B = 2$ the Jacobian has two pure imaginary eigenvalues and all other conditions of Hopf theorem hold. Hence $B = 2$ is a Hopf bifurcation point for the system, i.e. for $B > 2$ the system has a limit cycle and the concentrations of x,y will vary periodically with time.

EXERCISE 4

1. Discuss the equilibrium states and Hopf bifurcations for different values of λ,μ in the system

$$\dot{x} = y + (\lambda x + \mu y)(x^2 + y^2)$$
$$\dot{y} = -x + (\mu x - \lambda y)(x^2 + y^2)$$

2. Show that the equation
$$\ddot{u} + (u^2 + \lambda)\dot{u} + (u^2 + \mu)u = 0$$
undergoes Hopf bifurcation when
a. $\lambda = 0 \ , \ \mu > 0$ b. $\lambda = \mu \ , \ \lambda,\mu < 0$.

6. BIFURCATIONS IN R^n.

As we saw in the previous sections the bifurcation problem for the equilibrium states is equivalent to solving the algebraic equation $F(\mathbf{u},\lambda) = 0$ and identifying those values of λ for which new solutions appear. While this problem can be solved directly in low dimensions it becomes rapidly very complicated in higher dimensions. In this section we present the solution to this problem when

$$F(u,\lambda) = T(u) - \lambda u \tag{6.1}$$

where T is a general mapping of a vector space into itself.

We start with some general definitions;

Definition 1: Let T be an operator (i.e., a mapping, nonlinear in general) mapping the vector space (more precisely Banach space) V into itself. We say that T is linearizable at $\mathbf{v}$ if there exist a linear operator $L_{\mathbf{v}}$ so that

$$T(\mathbf{v}+\mathbf{h}) - T(\mathbf{v}) = L_{\mathbf{v}}\mathbf{h} + R_{\mathbf{v}}(\mathbf{h}) \tag{6.2}$$

where

$$\lim_{|\mathbf{h}|\to 0} \frac{|R_{\mathbf{v}}(\mathbf{h})|}{|\mathbf{h}|} = 0 \quad . \tag{6.3}$$

The operator $L_{\mathbf{v}}$ is called the Frechet derivatives of T at $\mathbf{v}$ and is also denoted by $T'(\mathbf{v})$.

Example 1: If T is a linear operator $R^n \to R^n$ then obviously

$$T(\mathbf{v}+\mathbf{h}) - T(\mathbf{v}) = T(\mathbf{h}) \quad . \tag{6.4}$$

Hence $T'(\mathbf{v})(h) = T(h)$ and $R_{\mathbf{v}}(\mathbf{h}) = 0$.

Example 2: Let T be a nonlinear mapping $R^n \to R^n$. For $\mathbf{x} = (x_1, \ldots, x_n)$ we then have

$$T(\mathbf{x}) = \begin{bmatrix} f_1(x_1, \ldots, x_n) \\ \\ f_n(x_1, \ldots, x_n) \end{bmatrix} \quad . \tag{6.5}$$

If f_i, $i = 1, \ldots, n$ are smooth functions then it is easy to see using Taylor expansion that $T'(\mathbf{x})$ is the Jacobian of these functions viz.

$$T\ (\mathbf{x}) = \begin{bmatrix} \frac{\partial f_1}{\partial x_1} & \cdots & \frac{\partial f_1}{\partial x_n} \\ & \cdots & \\ \frac{\partial f_n}{\partial x_1} & \cdots & \frac{\partial f_n}{\partial x_1} \end{bmatrix} \tag{6.6}$$

Example 3: Consider the differential operator

$$D(u) = u'' + u^2u' \tag{6.7}$$

where primes denote differentiation with respect to x. To compute $D'(u)$ we note that

$$\begin{aligned} D(u+h) - D(u) &= [(u+h)'' + (u+h)^2(u+h)'] \\ &- [u'' + u^2u'] = h'' + u^2h' + (2uu')h \\ &+ (\text{nonlinear terms in } h) \quad . \end{aligned} \tag{6.8}$$

Hence

$$D'(u)(h) = h'' + u^2h' + (2uu')h \tag{6.9}$$

Example 4: Consider the integral operator

$$T(u) = \int_a^b k(x,t)f(u(t))dt \tag{6.10}$$

where k is bounded and f is smooth. Then

$$\begin{aligned} T(u+h) - T(u) &= \int_b^a k(x,t)[f(u+h) - f(u)]dt = \\ &= \int_a^b k(x,t)f'(u(t))h(t)dt + R_u(h) \end{aligned} \tag{6.11}$$

i.e.

$$T'(u)(h) = \int_a^b k(x,t)f'(u(t))h(t)dt \tag{6.12}$$

Consider now the equation

$$T(u) - \lambda u = 0 \tag{6.13}$$

where T is a mapping $V \to V$ and $T(0) = 0$. In this equation $u = 0$ is always a solution and is referred to as the "basic solution". We now state two theorems which characterize the bifurcations points of this equation from the basic solution, i.e. those values of λ for which a nontrivial solution becomes available.

Definition 2: λ_0 is said to be in the spectrum of the linear operator $L: V \to V$ if there exists $u \neq 0$, $u \in V$ so that $Lu = \lambda_0 u$. Obviously, if $V = R^n$ then λ_0 is in the spectrum of L if and only if λ_0 is an eigenvalue of L.

Theorem 1: λ_0 is a bifurcation point of equation (6.13) only if λ_0 is in the spectrum of $T'(0)$.

Thus, theorem 1 is a necessary but not sufficient condition for λ_0 to be a bifurcation point. The next theorem is a protype for such sufficient condition.

Definition 3: An operator $T: V \to V$ is said to be completely continuous if it is continuous and compact.

Remark: T is compact if for every bounded set S in V, $T(S)$ is a compact set in $T(V)$(= range of T).

Theorem 2: (Leray-Schauder) if T is completely continuous and $\lambda^0 \neq 0$ is an eigenvalue of odd multiplicity of $T'(0)$ then λ_0 is a bifurcation point of the basic solution in equation (6.13).

We now illustrate these theorems through various examples in R^n.

Example 5: Let T be a linear operator $R^n \to R^n$ then obviously the equation $T\mathbf{v} = \lambda\mathbf{v}$ has nontrivial solutions only when λ is an eigenvalue of T which is equal in this case to $T'(0)$(see example 2).

Example 6: Let $T:R\to R$

$$T(x) = ax + bx^3 \quad , \quad b > 0 \quad . \tag{6.14}$$

Obviously, $T'(0) = ax$ and hence $\lambda = a$ is an eigenvalue of odd multiplicity for $T'(0)$. Accordingly, it must be a bifurcation point of T. In fact equation (6.13)

$$ax + bx^3 = \lambda x \tag{6.15}$$

has two nontrivial solutions

$$x = \pm\sqrt{\frac{\lambda - a}{b}} \tag{6.16}$$

when $\lambda > a$.

Example 7: Let $T:R^2\to R^2$ where

$$T\begin{bmatrix} x \\ y \end{bmatrix} = \begin{bmatrix} ax + bx(x^2+y^2) \\ ay + cy(x^2+y^2) \end{bmatrix}, \quad b \neq c, b, c > 0 \ . \tag{6.17}$$

In this case

$$T'(0) = \begin{bmatrix} a & 0 \\ 0 & a \end{bmatrix} \tag{6.18}$$

which has only one eigenvalue $\lambda = a$ of multiplicity 2. To see whether this is a bifurcation point of T (note that theorem 2 does not apply) we have to solve

$$ax + bx(x^2+y^2) = \lambda x \tag{6.19}$$

$$ay + cy(x^2+y^2) = \lambda y \quad . \tag{6.20}$$

We see that if $\lambda > a$ then we have the following nontrivial solutions;

$$1. \quad y = 0, \ x = \pm\sqrt{\frac{\lambda - a}{b}} \qquad 2. \quad x = 0, y = \pm\sqrt{\frac{\lambda - a}{c}} \tag{6.21}$$

Thus, $\lambda = a$ is a "double" bifurcation point.

Example 8: Let $T:R^2\to R^2$

$$T\begin{bmatrix} x \\ y \end{bmatrix} = \begin{bmatrix} ax - by^3 \\ ay + bx^3 \end{bmatrix} \cdot \ b \neq 0 \tag{6.22}$$

Then $T'(0)$ is the same as in the previous example and $\lambda = a$ is an eigenvalue of multiplicity 2. However, in this case the nonlinear problem has no bifurcation at this point (nor at any other value of λ in view of theorem 1). In fact from

$$ax - by^3 = \lambda x$$
$$ay + bx^3 = \lambda y$$

we obtain by multiplying by y and x respectively and subtracting that

$$x^4 + y^4 = 0$$

i.e., $x = y = 0$.

EXERCISE 5

1. Compute the Frechet derivative of the following operators.

   a. $D(u) = (uu')' + xu^2$

   b. $T\begin{bmatrix} x \\ y \end{bmatrix} = \begin{bmatrix} \sin(x+y) \\ \cos(x+y) \end{bmatrix}$

   c. $T(u(x)) = u'(x) + u^2$

   d. $T(u(x)) = \int_0^1 k(x,t,u(t))dt$

   where k is analytic.

2. Discuss the bifurcation points and the nontrivial solutions of

   $$T\begin{bmatrix} x \\ y \end{bmatrix} = \begin{bmatrix} a_1 x + bx(x^2+y^2) \\ a_2 y + cy(x^2+y^2) \end{bmatrix}$$

   where $a_1 \neq a_2$, $b \neq c$ and all constant are

positive.

3. Repeat exercise 2 if

   $$T\begin{bmatrix} x \\ y \end{bmatrix} = \begin{bmatrix} x(1+y) \\ y+x^2+2y^2 \end{bmatrix}$$

4. Do the same for

$$T\begin{bmatrix} x \\ y \end{bmatrix} = \begin{bmatrix} x+xy \\ -x+x^2+2y^2 \end{bmatrix}$$

BIBLIOGRAPHY

1. J. Guckenheimer and P. Holmes - Nonlinear Oscillations, Dynamical Systems and Bifurcations of Vector Fields, Springer-Verlag, 1983.
2. G. Ioos and D. D. Joseph - Elementary Stability and Bifurcation Theory, Springer-Verlag, 1980.
3. J. M. T. Thompson - Instabilities and Catastrophes in Science and Engineering, J. Wiley, 1982.
4. B. Hassard, N. Kazarinoff and Y. H. Wan - Theory and Applications of Hopf Bifurcation, Cambridge University Press, 1981.
5. R. Gilmore - Catastrophe Theory for Scientists and Engineers, J. Wiley, 1980.
6. M. Golubitsky and D.G. Schaeffer - Singularities and Groups in Bifurcation Theory, Vol. I, Springer-Verlag, 1985.
7. J. K. Hale and S. N. Chow - Methods of Bifurcation Theory, Springer-Verlag, 1982.

CHAPTER 10. STURMIAN THEORY

1.1 SOME MATHEMATICAL PRELIMINARIES.

In many applications we deal with differential equations or systems which cannot be solved in closed form. Previously we showed how to cope with this situation when one considers only the equilibrium solutions of such equations. However, in many other instances when the time dependent solution is needed it is important to be able to derive upper and lower bounds as well as the properties (e.g. number of zeros) of the solution by solving proper approximations to the original equations. In this chapter we present an introduction to the theory developed by Sturm and others in an attempt to answer such questions.

In this section we discuss some unrelated topics which are needed as a background for our subsequent discussion. Some of these results are elementary yet others are of a more specialized nature. We begin with two well known theorems from elementary calculus, Rolle's theorem and the Bolzano-Weierstrass theorem.

Definition 1: A *zero* (or a root) of a function $f(x)$ is a number $x = c$ such that $f(c) = 0$. If $f(c) = f'(c) = \ldots = f^{(n-1)}(c) = 0$, then f has a zero of order n at $x = c$.

Example 1: The function $f(x) = x(x-1)^3$ has a zero of order 1 (sometimes called a simple zero) at $x = 0$ and a zero of order 3 at $x = 1$.

Theorem 1 (Rolle's): Let $f(x)$ be continuous on $[a,b]$ and differentiable on (a,b). If $f(a) = f(b) = 0$ then there exists a point $c \in (a,b)$ so that $f'(c) = 0$.

Thus between two roots of $f(x)$ there exist at least one root of $f'(x)$.

Definition 2: A point $\mathbf{x}_0$ is said to be an *accumulation point* of a set $S \subset R^n$ if every neighborhood $|\mathbf{x}-\mathbf{x}_0| < \epsilon$ of $\mathbf{x}_0$ contains at least one point of the set distinct from $\mathbf{x}_0$.

Example 2: Let $S = \left\{\frac{1}{n}\right\}$, $n = 1,2,\ldots,$ then $x_0 = 0$ is an accumulation point of S.

Theorem 2 (Bolzano-Weierstrass): Let $S = \{\mathbf{x}_n\}$, $n = 1,2,\ldots$ be a bounded sequence of points in R^n (i.e, there exists an $M > 0$ so that $|x_n| < M$ for all n), then S has at least one accumulation point.

1.2 SYSTEMS OF FIRST ORDER EQUATIONS.

An nth order differential equation or a system of such equations is equivalent to a system of first order equations. Accordingly many theorems and results about differential equations are formulated in terms of such systems.

Example 3: The system

$$y^{iv} + x^2 y''y' + z'' = 0$$

$$z''' + z' + xyz = x$$

is equivalent to the following system with seven equations in seven unknowns:

$$y' = y_1,\ y_1' = y_2,\ y_2' = y_3,\ z' = z_1,\ z_1' = z_2$$
$$y_3' + x^2 y_2 y_1 + z_2 = 0$$
$$z_2' + z_1 + xyz = x \quad .$$

1.3 EXISTENCE THEOREM FOR DIFFERENTIAL EQUATIONS.

Definition 3: A function $f(x,\mathbf{y})$ is said to satisfy a *Lipschitz condition* on a domain $D \subset R^{n+1}$ if there exists an $M > 0$ such that for all $(x,\mathbf{y}_1)$, $(x,\mathbf{y}_2) \in D$

$$|f(x,\mathbf{y}_1) - f(x,\mathbf{y}_2)| \leq M|\mathbf{y}_1-\mathbf{y}_2| \ .$$

Example 4: The function $f(x,y) = xy$ satisfies the Lipschitz condition on the rectangle $|x| < 1$, $|y| < 1$ since

$$|f(x,y_1) - f(x,y_2)| = |x||y_1-y_2| \leq |y_1-y_2|.$$

The Lipschitz condition is utilized in many theorems to ensure the local existence of a solution on an initial value problem. The following is an example.

Theorem 3: Let $f(x,\mathbf{y})$ be continuous on the closed domain

$$|\mathbf{y}-\mathbf{c}| < K, \qquad |x-a| < L$$

and satisfy a Lipschitz condition there. Then the initial value problem

$$\mathbf{y}' = f(x,\mathbf{y}), \quad \mathbf{y}(a) = \mathbf{c}$$

has an unique solution on some interval $|x-a| \leq b$.

1.3 THE WRONSKIAN.

Definition 4: The *Wronskian* of any n functions $f_1(x),\ldots,f_n(x)$ where $f_i \in C^{n-1}$ on $[a,b]$ is defined as

$$W(f_1,\dots,f_n,x) = \begin{vmatrix} f_1(x) & \dots & f_n(x) \\ f_1'(x) & \dots & f_n'(x) \\ & \dots & \\ f_1^{(n-1)}(x) & \dots & f_n^{(n-1)}(x) \end{vmatrix}.$$

The main application of this function is to characterize the independent solutions of a differential equation.

Theorem 4: The Wronskian of any two solutions f and g of the differential equation

$$y'' + p(x)y' + q(x)y = 0 \tag{1.1}$$

satisfies the differential equation

$$W' + p(x)W = 0$$

and hence

$$W(f,g,x) = W(f,g,a)e^{-\int_a^x p(t)dt}.$$

Thus, the Wronskian either vanishes identically or is never zero.

Theorem 5: If f,g are two linearly independent solutions of equation (1.1), then $W(f,g,x)$ never vanishes.

As to the general solution of a linear differential equation we have the following:

Theorem 6: An nth order linear homogeneous equation

$$a_0(x)y^{(n)} + a_1(x)y^{(n-1)} + \dots + a_n(x)y = 0 \tag{1.2}$$

has n independent solutions. The general solution of this equation is a linear combination of these solutions.

b. The general solution of the inhomogeneous equation

$$a_0(x)y^{(n)} + a_1(x)y^{(n-1)}+\dots+a_n(x)y = r(x) \tag{1.3}$$

is a sum of a particular solution of equation (1.3) and the general solution of equation (1.2).

1.5 STANDARD FORMS OF SECOND ORDER DIFFERENTIAL EQUATIONS

While discussing the properties of the solution of a second order differential equation

$$a_0(x)y'' + a_1(x)y' + a_2(x)y = 0 \tag{1.4}$$

where $a_0(x)$, $a_1(x)$, $a_2(x)$ are C^2, C^1 and C^0 functions respectively and $a_0(x) \neq 0$ over some interval, it is useful in many instances to transform it into some standard form. We discuss here only two such transformations.

a. Normal form.

If $a_0(x) \neq 0$ on the interval $[a,b]$ then equation can be transformed on this interval into the normal form

$$u'' + p(x)u = 0$$

through the substitution

$$y(x) = u(x)\exp\left[-\frac{1}{2}\int_a^x \frac{a_1(t)}{a_0(t)}\,dt\right] \quad . \tag{1.5}$$

The explicit relationship between $p(x)$ and the coefficients $a_i(x)$ is given by

$$p(x) = \frac{1}{a_0^2}\left[a_2a_0 - \frac{1}{4}\,a_1^2 - \frac{1}{2}\,(a_1'a_0 - a_1a_0')\right] \quad .$$

Note that the transformation (1.5) does not change the location or number of roots in the solution of equation (1.4), since the exponential factor is never zero.

b. Self-adjoint form

(A comprehensive discussion and motivation of this concept will be presented in the next chapter).

Definition 5: The *adjoint of the differential operator*

$$L(u) = a_0(x)u^{(n)} + a_1(x)u^{(n-1)} + \dots + a_n(x)u$$

is defined as

$$L^*(v) = (-1)^n(a_0(x)v)^{(n)} + (-1)^{n-1}(a_1(x)v)^{(n-1)}$$
$$+ \dots - (a_{n-1}(x)v)' + a_n v \quad .$$

Example 5: The adjoint of the second order differential operator

$$L(u) = a_0(x)u'' + a_1(x)u' + a_2(x)u \tag{1.6}$$

is

$$L^*(v) = a_0(x)v'' + (2a_0'(x) - a_1(x))v' +$$
$$+ (a_0''(x) - a_1'(x) + a_2(x))v \quad .$$

Definition 6: An operator L is said to be *self-adjoint* if $L^* = L$. If $L(u) = L^*(u) = 0$, then $L(u) = 0$ and $L^*(u) = 0$ are called *self-adjoint equations*.

Thus, a necessary and sufficient condition for the operator (1.6) to be self-adjoint is that

$$a_0' = a_1$$

Theorem 7: If $a_0(x) \neq 0$, then the second order differential operator (1.6) is equivalent to a self-adjoint one.

Proof: If we multiply (1.6) by

$$h(x) = \frac{1}{a_0(x)} \int_a^x \frac{a_1(t)}{a_0(t)} dt$$

we obtain the self-adjoint operator

$$\ell(u) = (p(x)u')' + q(x)u$$

where

$$p(x) = a_0 h(x), \; q(x) = a_2 h(x) \quad .$$

Example 6: The self-adjoint form of Bessel's equation

$$x^2 y'' + xy' + \left[x^2 - \nu^2\right]y = 0$$

is

$$(xy')' + \left[x - \frac{\nu^2}{x}\right]y = 0.$$

Example 7: The self-adjoint form of Legendre equation

$$\left[1-x^2\right]y'' - 2xy' + n(n+1)y = 0$$

is

$$\left[(1-x^2)y'\right]' + n(n+1)y = 0 \quad .$$

EXERCISE 1

1. Rewrite the following differential equations in normal and self-adjoint forms.
   a. $x^2y'' + 2xy' + y = 0, \quad x > 0$
   b. $x(x-1)y'' + y' + k^2x^2y = 0, \quad x > 1$
   c. $(1-x^2)y'' + 4xy' + n(n+3)y = 0, \quad x > 1$
   d. $y'' + ay' + by = 0, \quad a,b$ constants
2. Find the adjoint of the following equations
   a. $y''' + xy'' + x^2y' - 2y = 0.$
   b. $x^2y'' - 5x^3y' + 2y = 0 \quad .$
   c. $x^3y''' - x^2y' + 2xy = 0.$

2. STURMIAN THEORY FOR FIRST ORDER EQUATIONS.

Our objective in this section is to compare the behavior and properties of the solutions of first order differential equations.

To begin with we investigate how the solutions of an equation depend on the initial conditions.

Theorem 1: Let $f(x,y) \in C^1$ and satisfy a Lipschitz condition on the rectangle

$$D: \quad |x-a| < \alpha, \ (y-c_1) < \beta$$

and let $y(x,c)$ be a solution of

$$y' = f(x,y) \quad . \tag{2.1}$$

with the initial condition

$$y(a) = c \quad . \tag{2.2}$$

For every $\epsilon > 0$ there exist a $\delta > 0$ so that if

$$|c_1-c_2| < \delta$$

then

$$|y(x,c_1) - y(x,c_2)| < \epsilon$$

for all $|x-a| < \alpha$.

Less formally the theorem states that if the initial conditions on the solutions are close then the solutions themselves are close to each other on the whole domain, i.e. $y(x,c)$ is continuous in c.

Proof: From equation (2.1) we infer that

$$y(x,c) = c + \int_a^x f(t,y(t,c))dt \tag{2.3}$$

and therefore

$$y(x,c_1) - y(x,c_2) = (c_1-c_2) + \int_a^x [f(t,y(t,c_1)) - f(t,y(t,c\ ))]dt.$$

Hence

$$\begin{aligned}\left|y(x,c_1) - y(x,c_2)\right| &\le \left|c_1-c_2\right| \cdot \\ &+ \left|\int_a^x f[t,y(t,c_1)]-f[t,u(t,c\ )]dt\right| \\ &\le \left|c_1-c_2\right|+\int_a^x \left|f(t,y(t,c\ ))-f(t,y(t,c\ ))\right|dt\end{aligned} \tag{2.4}$$

But f satisfies Lipschitz conditions and therefore there exists $m > 0$ so that

$$|f(x,y_1) - f(x,y_2)| < M|y_1-y_2| \ .$$

We conclude, therefore, from equation (2.4) that

$$|y(x,c_1) - y(x,c_2)| \le |c_1-c_2| + M\int_a^x |y(t,c_1) - y(t,c_2)dt \ . \tag{2.5}$$

Introducing

$$z(x) = \int_a^x \left|y(t,c_1) - y(t,c_2)\right|dt$$

we rewrite equation (2.5) as

$$z'(x) - Mz(x) \le |c_1-c_2| \le \delta \quad . \tag{2.6}$$

Multiplying both sides by $e^{-M(x-a)}$ we have

$$\frac{d}{dx}\left[z(x)e^{-M(x-a)}\right] \le \delta e^{-M(x-a)} \quad .$$

Integrating this differential inequality over $[a,x]$ we obtain

$$z(x)e^{-M(x-a)} \le \frac{\delta}{M}\left[1-e^{-M(x-a)}\right]$$

or

$$z(x) \le \frac{\delta}{M}\left[e^{M(x-a)}-1\right] \quad .$$

Substituting this result in equation (2.6) yields

$$\left|y(x,c_1)-y(x,c_2)\right| \le \delta e^{M(x-a)} \quad . \tag{2.7}$$

If we choose $\delta = \epsilon e^{-M\alpha}$ then for all $|x-a| \le \alpha$

$$|y(x,c_1) - y(x,c_2)| < \epsilon$$

Corollary 1: When $f(x,y) \in C^1$ and satisfies a Lipschitz condition, the solution of (2.1)-(2.2) is unique.

Proof: Suppose initial value problem (2.1)-(2.2) has two solutions $y_1(x,c)$ and $y_2(x,c)$. Then since $\delta = 0$ we infer from (2.7) that

$$|y_1(x,c) - y_2(x,c)| = 0$$

for all $|x-a| < \alpha$, from which it follows that $y_1(x,c) = y_2(x,c)$.

The results of Theorem 1 can be generalized to systems of equations and also strengthened by showing that the dependence on the initial conditions is not only continuous but also differentiable. We state this result without proof.

Theorem 2: The solution $\mathbf{x}(t,\mathbf{c})$ of

$$\frac{d\mathbf{x}}{dt} = \mathbf{F}(\mathbf{x},t)$$

where $F \in C^1$ and satisfies a Lipschitz condition, is continuously differentiable with respect to $\mathbf{c}$.

Example 1: For a linear system of equations

$$\frac{d\mathbf{x}}{dt} = A\mathbf{x} + \mathbf{f}(t) \;, \quad \mathbf{x}(t_0) = \mathbf{c}$$

where A is a matrix with constant coefficients the solution $\mathbf{x}(t,\mathbf{c})$ is given by

$$x(t,c) = e^{A(t-t_0)}\mathbf{c} + e^{At}\int_{t_0}^{t} e^{-As}\mathbf{f}(s)ds \quad .$$

From this expression it is obvious that the dependence of $\mathbf{x}$ on $\mathbf{c}$ is differentiable.

Example 2: When f in equation (2.1) is not C^1 or does not satisfy a Lipschitz condition, equations (2.1)-(2.2) might have more than one solution, e.g.

$$y' = \frac{1}{y} \;, \quad y(0) = 0$$

has two solutions $y(x) = \pm\sqrt{2x}$.

On the other hand these conditions are sufficient but not necessary for the uniqueness of the solutions. Thus, even though

$$f(x,y) = \begin{cases} y(1-3x^2) & x > 0 \\ y(3x^2-1) & x < 0 \end{cases}$$

is discontinuous at $x = 0$. The solution of equation (2.1) with the initial condition $y(1) = 1$ which is given by

$$y(x) = \begin{cases} e^{x-x^3} & x \geq 0 \\ e^{x^3-x} & x < 0 \end{cases}$$

is continuous and unique for all x.

To avoid such complications we shall always assume in what follows that we consider only well-behaved differential equations, in the sense of Theorem 1 and

Corollary 1, whose solutions subject to proper initial conditions are unique.

Example 3: If equation (2.1) models a physical system then equations (2.3) provides an estimate on the possible variation in the system behavior when the initial condition of the system is subject to a maximum error δ.

If for example

$$y' = xy \quad , \quad |y(0)-1| \leq \delta, \quad |x| < 1 \quad ,$$

then since

$$\left|f(x,y_1) - f(x,y_2)\right| \leq \left|y_1-y_2\right|$$

it follows from equation (2.7) that

$$|\Delta y| = |y(x,1+\delta) - y(x,1)| \leq \delta e^x < \delta e \quad .$$

Therefore all solutions with initial conditions $1-\delta < y(0) < 1+\delta$ must satisfy

$$e^{x^2/2} - \delta e \leq y(x) \leq e^{x^2/2} + \delta e \quad .$$

We now proceed to compare solutions of two first order equations.

Lemma 1: Let $F(x,y)$ satisfy a Lipschitz condition for $x > a$. If $g(x)$ satisfies the differential inequality $g'(x) \leq F(x,g(x))$, then the solution $f(x)$ of

$$y'(x) = F(x,y)$$

$$f(a) = g(a) = c$$

satisfies the inequality $g(x) \leq f(x)$ for $a \leq x$.

Proof: Assume to the contrary that there exists x_1 so that $f(x_1) < g(x_1)$ Consider the set

$$S = \left\{x \middle| f(x)-g(x) \geq 0, \ a \leq x < x_1\right\} \quad .$$

S is a non-empty $(a \in S)$, bounded, and closed set $(f-g)$ is continuous). It follows then that there exists a point $b = \max\{x \in S\}$. Thus on the interval $(b,x_1]$

$$r(x) = g(x) - f(x) > 0 \text{ and } r(b) = g(b) - f(b) = 0.$$

However, $r(x)$ satisfies the following differential inequality

$$r'(x) = g'(x) - f'(x) \leq F(x,g(x)) - F(x,g(x))$$
$$\leq M(f(x) - g(x)) = Mr(x).$$

Hence, we infer that on $[b,x_1]$

$$0 \leq r(x) \leq r(b)e^{M(x-a)} = 0$$

i.e. $r(x)$ vanishes indentically on $\left[b,x_1\right]$ which contradicts the assumption that $f(x) < g(x)$ on this interval. This shows that $g(x) \leq f(x)$ for all $a \leq x$.

Theorem 3: Let f,g be solutions of

$$y' = F(x,y) \quad , \quad w' = G(x,w)$$

respectively satisfying $f(a) = g(a) = c$. If F and G satisfy a Lipschitz condition and $G(x,y) \leq F(x,y)$ on some domain D then $g(x) \leq f(x)$ on D.

Proof: Since g is a solution of $w' = G(x,w)$ we infer that

$$g'(x) = G(x,g) \leq F(x,g) \quad .$$

Hence, from the preceding lemma it follows that $g(x) \leq f(x)$.

This result can be sharpened as follows:

Theorem 4: Let f,g be as in Theorem 3. If $g(a) < f(a)$ then $g(x) < f(x)$ for $a < x$.

Example 4: Consider the solutions f,g to the differential equations

$$y' = x^2y \quad , \quad w' = xw$$
$$f(1) = g(1) = 1 \quad .$$

Since for $1 < x$, $0 < y$ we have $xy \leq x^2y$, we infer from Theorem 3 that on this domain these solutions must satisfy the inequality $g(x) \leq f(x)$. As a matter of fact

$$f(x) = e^{[x^3-1]/3} \quad , \quad g(x) = e^{[x^2-1]/2}$$

and it is easy to show that these functions do satisfy the desired inequality for $1 < x$.

EXERCISE 2

1. For the differential equation
$$\frac{dy}{dx} = k\sqrt{|y|}$$
   a. Show that $f(x,y) = k\sqrt{|y|}$ does not satisfy a Lipschitz condition on any region which includes $y = 0$ .Hint: $y_2 = y_1 + (y_2-y_1)$.
   b. Show that this equation with the initial condition $y(0) = 0$ has a nontrivial solution (beside $y(x) = 0$).
2. Solve
$$\frac{dy}{dx} = k\frac{y}{x} \qquad y(x_0) = y_0 \quad , \quad x_0 \neq 0 \; .$$
   What happens at the origin? Explain.
3. Let $F(x,y,s) \in C^1$ on $|x-a| \leq \alpha$, $|y-c| \leq \beta$, $|s-s_0| \leq \gamma$ and satisfy the Lipschitz condition
$$|F(x,y_1,s) - F(x,y_2,s)| < M|Y_1-y_2| \; .$$
   Show that the solution $y(x,s)$ of
$$\frac{dy}{dx} = F(x,y,s) \quad , \quad y(a,s) = c$$
   is continuous in s.

3. STURMIAN THEORY FOR SECOND ORDER EQUATIONS.

In this section we relate, at first, the roots of two solutions of a given second order differential equation and then proceed to compare solutions of two such equations. We shall assume that
$$L(y) = a_0(x)y'' + a_1(x)y' + q_2(x)\; y = 0$$
where $a_0(x), a_1(x), a_2(x)$ are C^2, C^1, C^0 respectively and $a_0(x) \neq 0$ for some interval.

We begin with some elementary observations.

Lemma 1: All roots of a nontrivial solution to the equation

$$L(y) = a_0(x)y'' + a_1(x)y' + a_2(x)y = 0 \qquad (3.1)$$

are simple.

Proof: Suppose a nontrivial solution $y(x)$ of $L(y) = 0$ has a zero at $x = x_0$ which is not simple. Then $y(x_0) = y'(x_0) = 0$ and $y(x)$ is the trivial solution by the uniqueness theorem in contradiction to the assumptions.

Theorem 1: A nontrivial solution $y(x)$ of $L(y) = 0$ can have at most a finite number of zeros on a finite interval $[a,b]$.

Proof: Suppose to the contrary that $y(x)$ has a infinite number of zeros on $[a,b]$. Then by Bolzano-Weirstrass theorem these zeros must have an accumulation point c on this interval. Furthermore, by Rolle's theorem $y'(x)$ must also vanish at an infinite number of points of $[a,b]$ and c is an accumulation point of this set. Hence $y(c) = y'(c) = 0$. But by the previous lemma this implies that $y(x)$ is the trivial solution in contradiction with the assumptions. This contradiction shows that $y(x)$ must have a finite number of zeros on $[a,b]$.

We now proceed to compare the number of zeros that two solutions of a second order homogeneous equation might have.

Theorem 2 (Separation Theorem): Let y_1, y_2 be two linearly independent solutions of the differential equation (3.1) and let x_1, x_2 be two consecutive zeros of y_1, then y_2 vanishes exactly once on (x_1, x_2).

Proof: First note that y_2 cannot vanish at x_1 or x_2 since otherwise the Wronskian of y_1, y_2 will vanish contrary to the assumption that y_1, y_2 are linearly independent. Now consider the function $\varphi(x) = y_1(x)/y_2(x)$. If y_2 does not vanish on (x_1, x_2), then $\varphi(x)$ is continuous and differentiable on $(x_1 x_2)$ and furthermore $\varphi(x_1) = \varphi(x_2) = 0$. By Rolle's theorem this result implies that at some point $\xi \in (x_1, x_2)$, $\varphi'(\xi) = 0$. But since

$$\varphi'(\xi)) = \frac{y_2(\xi)y_1'(\xi)-y_1(\xi)y_2'(\xi)}{y_2^2(\xi)} = 0 ,$$

the Wronskian of y_1, y_2 vanishes at ξ. This contradiction proves that y_2 must have a zero on (x_1, x_2). Moreover y_2 cannot have more than one zero in (x_1, x_2) since if it has two then y_1 would have a zero in between them and x_1, x_2 would not be consecutive zeros of y_1.

Example 1: The functions

$$y_1(x) = A \cos kx + B \sin kx$$

$$y_2(x) = C \cos kx + D \sin kx$$

are independent solutions of

$$y'' + k^2 y = 0$$

whenever $AD-BC \neq 0$. From Theorem 2 we infer that the roots of these two functions must separate each other.

Example 2: We observe that Theorem 2 does not imply that each solution of a differential equation must have a root. For example

$$y'' - k^2 y = 0$$

has a solution $y(x) = A \cosh kx$ which does not vanish for any x.

We now compare solutions of two second order equations.

Theorem 3 (Sturm Fundamental Theorem): If $q(x) \leq p(x)$ on the interval $[a,b]$ and if y, w are nontrivial solutions of

$$y'' + p(x)y = 0 \tag{3.2}$$

$$w'' + q(x)w = 0 \tag{3.3}$$

respectively, then $y(x)$ vanishes at least once between any two consecutive zeros of $w(x)$ unless $p = q$ and y is a multiple of w.

Remark: We showed in Subsection 1.5 that a second order differential equation

$$a_0(x)y'' + a_1(x)y' + a_2(x)y = 0, \quad a_0(x) \neq 0 \tag{3.4}$$

can always be transformed to the form given by equations (3.2)-(3.3) without altering the number of its roots. Hence, the theorem can be applied essentially to any pair of equations on a proper interval.

Proof: Let x_1, x_2 be two consecutive zeros of $w(x)$ and suppose that $y(x)$ does not vanish on (x_1, x_2). Since equations (3.2)-(3.3) are homogeneous we can assume without loss of generality that $y, w > 0$ (since $-y$, $-w$ are also solutions of these equations) on (x_1, x_2). However, on (x_1, x_2) the Wronskian function of y, w will satisfy

$$W(y,w,x_1) = y(x_1)w'(x_1) > 0,$$

$$W(y,w,x_2) = y(x_2)w'(x_2) < 0 \tag{3.5}$$

and

$$W'(y,w,x) = yw'' - y''w = yw(p-q) \geq 0 \quad .$$

Thus, $W(y,w,x)$ is a nondecreasing function on (x_1, x_2) which contradicts equation (3.5). This result shows that $y(x)$ must have a zero on (x_1, x_2).

Especially note that if y, w have a common zero at x_1, the theorem implies that y will vanish again before w does so that $y(x)$ will oscillate more rapidly than $w(x)$.

Example 3: Show that if $q(x) \leq 0$ on $[a,b]$ then any nontrivial solution of

$$w'' + q(x)w = 0 \tag{3.6}$$

has at most one zero on $[a,b]$.

Solution: If a solution $w(x)$ of (3.6) has two zeros on $[a,b]$, then by Theorem 3, the solution $y(x) = 1$ of $y'' = 0$ $(p(x) = 0 \geq q(x))$ should have at least one root on this interval. This contradiction shows that $w(x)$ can have at most one zero on (a,b).

Although as noted above any second order differential equation of the form (3.4) with $a_0(x) \neq 0$ can be transformed to the form (3.2) it is still useful for many applications to have a direct statement of Sturm fundamental theorem for self-adjoint second order equations.

Theorem 4: Let $w(x), y(x)$ be solutions of

$$\frac{d}{dx}\left[k_1 \frac{dw}{dx}\right] - G_1 w = 0$$

$$\frac{d}{dx}\left[k_2 \frac{dy}{dx}\right] - G_2 y = 0$$

respectively where $K_1 \geq K_2 > 0$, $G_1 \geq G_2$. Then in between any two consecutive zeros of $w(x)$ there exists at least one root of $y(x)$.

Thus the solutions of a self-adjoint second order equation oscillate more rapidly when we decrease the functions G, K.

Example 4: To determine the oscillatory behavior of the solutions of

$$\frac{d}{dx}\left[(1+x)\frac{dy}{dx}\right] - x^2y = 0 \tag{3.7}$$

on $[0,1]$ we compare this equation with

$$y'' = 0 \quad . \tag{3.8}$$

Since $K_1 = 1 + x \geq K_2 = 1 > 0$, and $G_1 = x^2 \geq G_2 = 0$ and $g(x) = 1$ is a solution of (3.8), we infer from Theorem 3 that no solution of equation (3.7) can have more than one root on this interval.

Example 5: Bessel Functions

Bessel functions are solutions of

$$x^2y'' + xy' + (x^2-\nu^2)y = 0 \quad , \quad \nu \text{ real} \quad . \tag{3.9}$$

This equation can be transformed for $x > 0$ to the form (3.2) by the substitution

$$y = \frac{w}{\sqrt{x}} \tag{3.10}$$

which yields

$$w'' + \left[1 + \frac{1-4\nu^2}{4x^2}\right]w = 0 \quad .$$

Using Theorem 4 and comparing the solutions of (3.9) with those of

$$y'' + y = 0$$

we infer that each interval $[k\pi,(k+1)\pi]$ of the positive x-axis contains:

a. At least one root of any solution to Bessel's equation of order 0, i.e. $\nu = 0$.

b. At most one root of any nontrivial solution to Bessel equation of order $\nu > \frac{1}{2}$.

Remark: Note that the substitution (3.10) does not change the number or location of the zeros of a solution to Bessel equation for $x > 0$.

EXERCISE 3

1. Suppose on the interval $[a,b]$

   $$0 < m \leq K(x) \leq M$$

   and

   $$n \leq G(x) \leq N$$

   where m,n,M,N are constants.

   a. Show that a solution of

   $$\frac{d}{dx}(K(x)y') - Gy = 0 \qquad (3.11)$$

   has at most one root on $[a,b]$ if

   $$\frac{n}{m} > -\frac{\pi^2}{(b-a)^2} .$$

   Hint: Compare (3.11) with

   $$\frac{d}{dx}(mw') - nw = 0 .$$

   b. Show that a solution of (3.11) has at least s roots on $[a,b]$ if

   $$\frac{N}{M} \leq -\frac{s^2\pi^2}{(b-a)^2} .$$

   Hint: Compare (3.11) with

   $$\frac{d}{dx}(Mw') - Nw = 0 .$$

2. If $0 < m < p(x) < M$ on $[a,b]$, show that any two successive roots x_1,x_2 of a solution to

   $$y'' + p(x)y = 0$$

   must satisfy

   $$\frac{\pi^2}{M} < \left[x_2-x_1\right]^2 < \frac{\pi^2}{m} .$$

3. Prove Theorem 3 Section 2 without assuming that F or G satisfy a Lipschitz condition if $G(x,w) < F(x,y)$ for all x,y and w in D.

   Hint: $r'(x) = f'(x) - g'(x)$
   $= F(x,f(x)) - G(x,g(x)) > 0$

4. Prove the following generalization to Theorem 1: A nontrivial solution of a homogeneous nth order

differential equation has at most a finite number of zeros on a finite interval $[a,b]$.

4. PRÜFER TRANSFORMATIONS

In an article entitled "New derivations of the Sturm-Liouville series development for continuous functions", Heinz Prüfer introduced another approach for attacking Sturmian theory.

Basically his idea is similar to that used when converting rectangular coordinates to polar coordinates.

We shall apply this transformation to the second-order initial-value problem.

$$\begin{aligned} &[r(x)y'(x)]' + q(x)y(x) = 0 \quad , \\ &y(x_0) = c \\ &r(x_0)y''(x_0) = d \end{aligned} \tag{4.1}$$

where r, q are continuous and $r > 0$ on the interval $a \leq x \leq b$ and where x_0 is chosen so that $a \leq x_0 \leq b$. The quantities c and d are constants.

Prüfer found that many properties concerning the solutions to the differential equation could be developed in a straightforward manner if the following transformations were used:

$$y(x) = \rho(x) \sin \theta(x) \tag{4.2}$$

$$r(x)y'(x) = \rho(x) \cos \theta(x) \ . \tag{4.3}$$

These equations are called *Prüfer transformations*. Using (4.2) and (4.3) we find that

$$\rho(x) = \pm \sqrt{y^2(x)+[r(x)y'(x)]^2}$$

and

$$\theta(x) = \tan^{-1} \frac{y(x)}{r(x)y'(x)} \quad .$$

Obviously if $y(x)$ and $r(x)y'(x)$ are differentiable

in $[a,b]$ so is $\rho(x)$. Furthermore, if we are considering nontrivial solutions only, then $y(x_0)$ and $r(x_0)y'(x_0)$ cannot be zero simultaneously. Therefore, $\rho(x) \neq 0$.

Solving (4.2) for $\sin\theta(x)$, we find

$$\sin\theta(x) = \frac{y(x)}{\rho(x)}$$

and we see that $\sin\theta(x)$ must be differentiable, and this result implies $\theta(x)$ must be differentiable in $[a,b]$.

Applying the transformations (4.2) and (4.3) to the initial value problem (4.1) we are lead to the first order system

$$\rho' = \left\{\left[\frac{1}{r(x)} - q(x)\right] \sin\theta \cos\theta\right\}\rho(x) \tag{4.4}$$

$$= \frac{1}{2}\left\{\left[\frac{1}{r(x)} - q(x)\right]\sin 2\theta(x)\right\}\rho(x)$$

$$\theta'(x) = \frac{\cos^2\theta}{r(x)} + q(x)\sin^2\theta \quad , \tag{4.5}$$

$$\rho\left[x_0\right] = \pm \sqrt{c^2+d^2}$$

$$\theta\left[x_0\right] = \tan^{-1} \frac{c}{d} \quad d \neq 0 \quad \text{or} \tag{4.6a}$$

$$\theta\left[x_0\right] = \cot^{-1} \frac{d}{c} \quad c \neq 0 \quad . \tag{4.6b}$$

In equations (4.6a) and (4.6b) $\tan^{-1}$ and $\cot^{-1}$ are relations rather than functions, i.e. $\theta(x_0)$ can be any suitable value. This means there are an infinite number of choices for $\theta(x_0)$. In what follows we shall always pick $\theta(x_0) = 0$ if $y(x_0) = 0$ and $0 < \theta(x_0) < \pi$ if $y(x_0) > 0$.

It can be shown that each of the equations (4.4) and (4.5) can be transformed into the original second order equation (4.1) through the use of the Prüfer transformations.

Our purpose in this section is to investigate the zeros of solutions of the second order differential equation. Since $\rho(x) \neq 0$, it follows from (4.2) that a zero of a solution y can only occur when θ is a multiple of π.

If we return to the first order differential equation (4.5), we notice it is nonlinear and contains only the function $\theta(x)$. From the theory of first order equations the initial value problem

$$\begin{aligned} \theta' &= \frac{\cos^2\theta}{r(x)} + q(x)\sin^2\theta \\ \theta(x_0) &= K \quad \text{(a constant)} \end{aligned} \tag{4.7}$$

has a unique solution over an interval I containing x_0.

If we could solve the problem (4.7) for $\theta(x)$, it might be possible for us to locate the values of x which make $\theta(x)$ a multiple of π. These values of x would be the zeros of a solution y of the initial value problem (4.1).

But even if we cannot solve the problem (4.7) we can use inequalities, graphs, etc. to locate approximately the zeros of $y(x)$.

Example 1: For the differential equation $y'' + y = 0$ equation (4.5) becomes

$$\theta' = \cos^2\theta + \sin^2\theta = 1.$$

Therefore, $\theta = x + C$. As $x \to +\infty$ so does $x + C$ and therefore $\theta(x)$ equals a multiple of π an infinite number of times. Since $y(x) = \rho(x)\sin\theta(x)$, $y(x)$ has an infinite number of zeros in $[a, +\infty]$ where a is any number.

It is easy to solve $y'' + y = 0$ for $y = C_1\sin x + C_2\cos x$ and verify our result since y is a sinusoidal wave.

Example 2: Examine the behavior of solutions to the differential equation

$$x^2y'' + xy' + \left[x^2 - \frac{1}{9}\right] y = 0 \tag{4.8}$$

on the interval $[1, +\infty]$.

We notice the differential equation is Bessel's equation of order $\frac{1}{3}$; however, it is not in self-adjoint form. It can be made so by multiplying both sides by $\frac{1}{x}$. The equation (4.8) can be written as

$$(xy')' + \left[x - \frac{1}{9x}\right]y = 0 \quad .$$

Now $r(x) = x$ and $p(x) = x - \frac{1}{9x}$ and thus (4.5) becomes

$$\theta' = \frac{\cos^2\theta}{x} + \left[x - \frac{1}{9x}\right] \sin^2\theta \tag{4.9}$$

Instead of solving (4.9) we observe that if we choose $x \geq 2$,

$$\theta' = \frac{\cos^2\theta}{x} + \left[x - \frac{9}{x}\right] \sin^2\theta = \frac{1}{x} + \left[x - \frac{10}{9x}\right]\sin^2\theta$$

$$\geq \frac{1}{x} \quad \text{for all} \quad x,\ 2 \leq x.$$

Integrating both sides from 2 to x we find

$$\theta(x) - \theta(2) \geq \ln x - \ln 2 \quad .$$

Therefore as $x \to +\infty$, $\ln x \to +\infty$ and $\theta(x)$ takes on the value $n\pi$ an infinite number of times. It follows that $y(x)$ has an infinite number of zeros in $[2,+\infty) \subset [1,+\infty)$. See Figure 1.

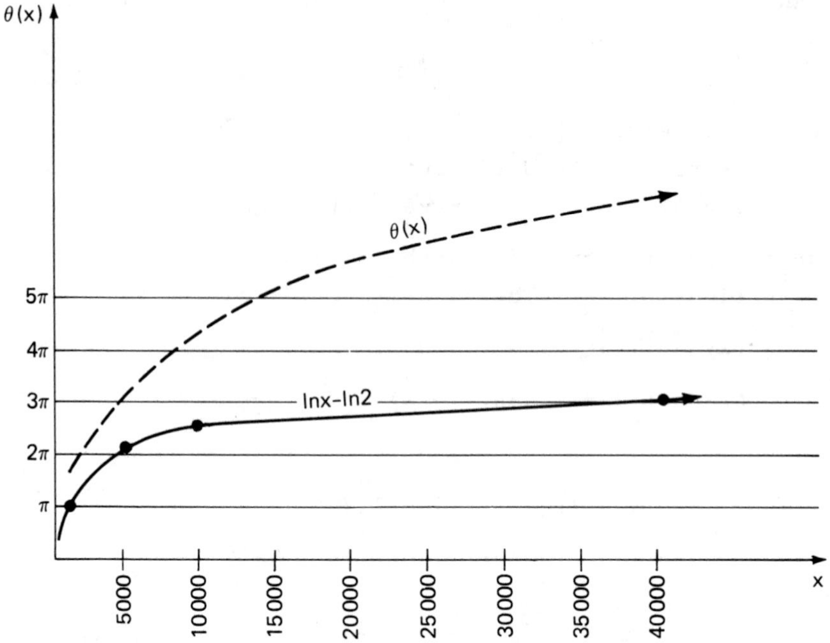

Figure 1

Theorem 1: Given the differential equation

$$\theta' = \frac{1}{r}\cos^2\theta + q\sin^2\theta \quad ,$$

a particular Prüfer component $\theta(x)$ can take on a multiple of π at most once.

Proof: Let x_0 be a value of x for which $\theta(x_0) = n\pi (n=0 , \pm 1, \pm 2, \ldots)$. At this point the differential equation leads to the inequality

$$\theta'(x_0) = \frac{1}{r(x)} > 0 \quad . \tag{4.10}$$

Therefore, it is impossible for the solution $\theta(x)$ to return and take on the same multiple of π again at $x = x_1$, $x_0 < x_1$ because in this situation $\theta'(x_1) < 0$ which violates (4.10). See Figure 2.

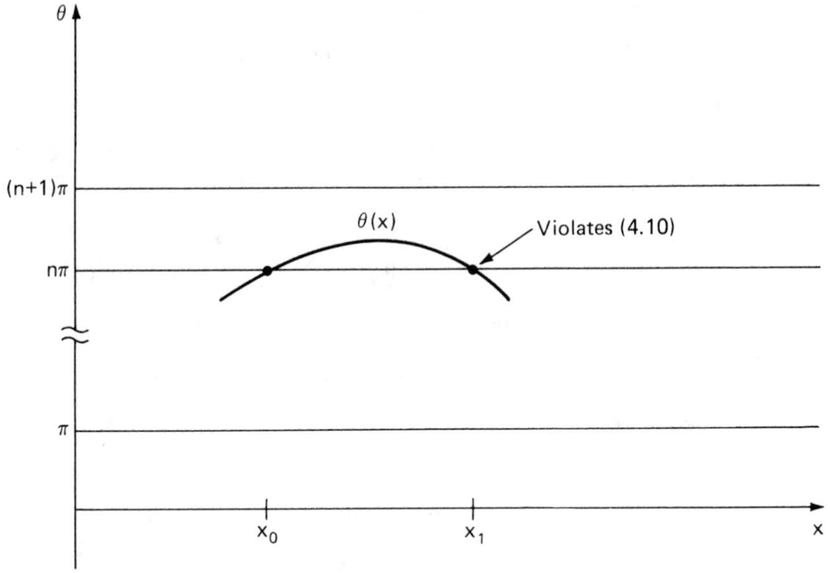

Figure 2

The following theorem was proved earlier. (See Theorem 2, Section 3.) But it is helpful in our current study to see it done by Prüfer transformation methods.

Theorem 2 [Theorem 2, Section 3]: If y_1 and y_2 are linear independent solutions of $(ry')' + py = 0$ where r, p are continuous and $r > 0$ on $[a,b]$ and x_0, x_1, $a \le x_0 < x_1 \le b$, are consecutive zeros of y_1, then there exists a point x_2, $x_2, x_0 < x_2 < x_1$, such that $y_2(x_2) = 0$.

Proof: First notice that y_1, y_2 cannot be zero simultaneously, otherwise they would be linearly dependent.

Choose the Prüfer component θ_1, θ_2 corresponding to y_1, y_2 respectively such that

$$0 = \theta_1(x_0) < \theta_2(x_0) < \pi \tag{4.11}$$

Since θ_1, θ_2 satisfy the same first order equation, they cannot intersect for if they did it follows from the uniqueness theorem of first order differential equations that $\theta_1 \equiv \theta_2$ in $[a,b]$. Thus, for all

$$x,\ x_0 \leq x \leq x_1\ \theta_1(x) < \theta_2(x) \quad . \tag{4.12}$$

Since x_1 is the next zero of y_1 to the right of x_0, then $\theta_1(x_1) = \pi$. Using results from (4.11) and (4.12), we see that

$$\theta_2(x_0) < \pi < \theta_2(x_1) \quad .$$

Since $\theta_2(x)$ is a continuous function there exists, by the intermediate theorem, a point $x_2, x_0 < x_2 < x_1$, such that $\theta_2(x_2) = 0$. Thus, $y_2(x_2) = 0$, and the theorem is proved. See Figure 3.

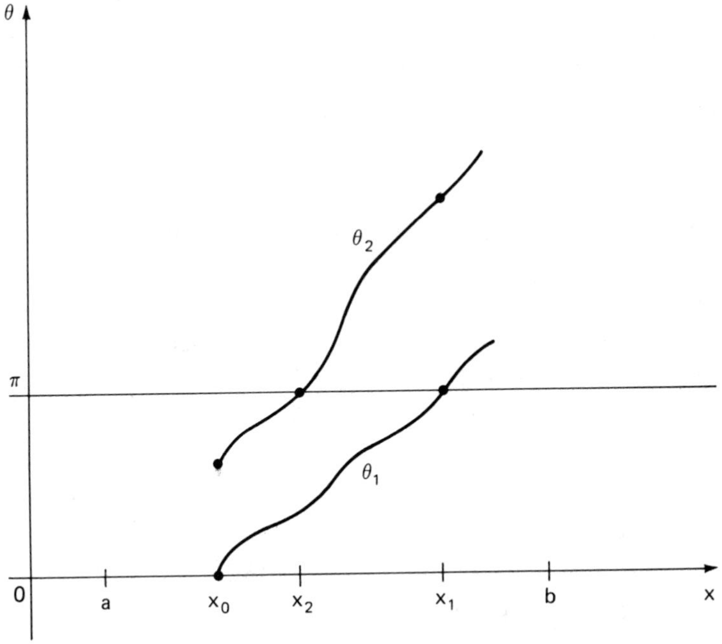

Figure 3

The justification of the previous theorem is certainly more involved than our earlier proof. But there are other theorems from Sturmian theory which can be handled very elegantly through the use of Prüfer transformations.

5. OSCILLATION THEOREMS

So far we have studied the location of zeros of two solutions of the same differential equation which leads to *separation* theorems.

This idea was followed by *comparison* theorems which compare the location of zeros of two different differential equations.

In this section we shall examine if the number of zeros of a solution of a differential equation in an interval is finite or infinite. These ideas are contained in *oscillation* theorems.

Definition: If the number of zeros of a solution of a differential equation in the interval $[a, +\infty)$ is infinite, then the solution is said to be *oscillatory* in $[a, +\infty)$. Otherwise the solution is *nonoscillatory* in $[a, +\infty)$.

Example 1: (a) A solution of $y'' + 4y = 0$ is $y = \sin 2x$ which has an infinite number of zeros on $[a, +\infty)$. This solution is oscillatory in $[a, +\infty)$.

(b) A solution of $y'' - 4y = 0$ is $y = \sinh 2x$ which has one zero at $x = 0$. Therefore, the solution is nonoscillatory in $[a,+\infty)$.

Given the differential equation

$$(r(x)y')' + p(x)y = 0 \tag{5.1}$$

where $r(x)$ and $p(x)$ are continuous and $r(x) > 0$ on $0 < x < +\infty$, we are able under certain circumstances to determine whether the solutions are oscillatory or nonoscillatory in $[a, +\infty)$ without actually solving differential equation. The following theorem due to Leighton is typical.

Theorem 1: Any solution of the differential equation (5.1) is oscillatory in $[a, +\infty)$ if

$$\int_a^{+\infty} \frac{dx}{r} = +\infty \quad \text{and} \quad \int_a^{+\infty} p dx = +\infty$$

Walter Leighton, Ordinary Differential Equations, Wadsworth Publishing Co., Belmont, CA (1970).

Proof: We recall from the Sturmian separation theorem that if one solution of equation (5.1) has an infinite number of zeros in $[a, +\infty)$, all solutions of equation (5.1) are oscillatory in $[a, +\infty)$.

Suppose there is a solution y of (5.1) which is nonoscillatory on $[a, +\infty)$. Then there is a point $x = x_1$, $a < x_1$, such that $y(x) > 0$ for $x_1 \leq x$.

Since y does not vanish in $[x_1, +\infty)$, we can write (5.1) in the form

$$\frac{(ry')'}{y} + p = 0 \quad .$$

Integrating this equation by parts from x_1 to x we have

$$\frac{r(x)y'(x)}{y(x)} - \frac{r(x_1)y'(x_1)}{y\,(x_1)} + \int_{x_1}^{x} \frac{ry'^2}{y}\,dx = -\int_{x_1}^{x} p\,dx.$$

Since the integral on the left-hand side is nonnegative

$$\frac{r(x)y'(x)}{y(x)} \leq \frac{r(x_1)y'(x_1)}{y(x_1)} - \int_{x_1}^{x} p\,dx \quad .$$

From this inequality and the second integral in the hypothesis the

$$\lim_{x\to+\infty} \frac{r(x)y'(x)}{y(x)} = -\infty \quad ,$$

and for sufficiently large x, say $x_2 > x_1$, y and y' have opposite signs which imply y is monotonically decreasing and bounded on $[x_2, +\infty)$.

Therefore,

$$y(x_2) > y(x) \quad \text{on} \quad [x_2, +\infty) \quad . \tag{5.2}$$

Let z be another linear independent solution of (5.1) which is also positive for $x_1 < x$ such that the Wronskian $r(yz' - zy') = \kappa > 0$. Using the same argument as before z satisfies the inequality $z(x_2) > z(x)$ on $[x_2, +\infty)$.

Define v as follows:

$$v = \text{Tan}^{-1}\frac{z}{y} \quad .$$

Then

$$v' = \frac{r(yz'-zy')}{r(y^2+z^2)} = \frac{k}{r(y^2+z^2)} > 0 \quad \text{for} \quad x_1 < x \tag{5.4}$$

Integrating (5.4) from x_2 to $+\infty$, then using the fact that $y^2(x_2) + z^2(x_2) > y^2(x) + z^2(x)$ and finally referring to the first integral in the hypothesis we find

$$v = \int_{x_2}^{+\infty} \frac{k}{r(y^2+z^2)}\,dx > \frac{k}{y^2(x_2)+z^2(x_2)}\int_{x_2}^{+\infty}\frac{dx}{r} = +\infty \tag{5.5}$$

However, we see from (5.3) that $v < \frac{\pi}{2}$ which contradicts the statement (5.5) above, and the theorem is proved.

This is quite a powerful theorem because there is no restriction on the sign of p, only that $\int_a^{+\infty} p\,dx = +\infty$.

Example 2: Show that the solutions of Bessel's equation $x^2y'' + xy' + \left[x^2-\nu^2\right]y = 0$ are oscillatory on $[a, +\infty)$, $0 < a$.

First we must write this equation in self-adjoint form which is

$$(xy')' + \left[x - \frac{\nu^2}{x}\right]y = 0 \quad .$$

Therefore, $r(x) = x$ and $p(x) = x - \frac{\nu^2}{x}$. We see that

$$\int_a^{+\infty}\frac{dx}{r} = \lim_{b\to+\infty}\int_a^b \frac{dx}{x} = \lim_{b\to+\infty}(\ell nb - \ell na) = +\infty$$

and

$$\int_a^{+\infty} p\,dx = \lim_{b\to+\infty}\int_a^b \left[x - \frac{\nu^2}{x}\right]dx \quad .$$

Now if $\nu < x$ then $x - \nu \leq x - \frac{\nu^2}{x}$, and

$$\int_a^{+\infty} p dx = \lim_{b\to\infty} \int_a^b \left[x - \frac{\nu^2}{x}\right] dx = \int_a^{\nu} \left[x - \frac{\nu^2}{x}\right] dx$$

$$+ \lim_{b\to+\infty} \int_{\nu}^b \left[x - \frac{\nu^2}{x}\right] dx$$

$$\geq \int_a^{\nu} \left[x - \frac{\nu^2}{x}\right] dx + \lim_{b\to+\infty} \int_{\nu}^b (x-\nu) dx$$

$$= \int_a^{\nu} \left[x - \frac{\nu^2}{x}\right] dx + \frac{1}{2}\left\{(b - \nu)^2 - (a - \nu)^2\right\} = +\infty$$

Bessel's functions are oscillatory on $[a, +\infty)$.

Let us now turn to solutions which are nonoscillatory. Although there are a number of theorems which are related to the nonoscillatory behavior of solutions, we shall look at just one.

Theorem 2: Any solution of (5.1) is nonoscillatory in $[a, +\infty)$ if $p(x) < 0$ for $a \leq x$.

Proof: We recall that if one solution is nonoscillatory in $[a, +\infty)$, all solutions are nonoscillatory on that interval.

Let $y(x)$ be the solution satisfying the initial conditions $y(a) > 0$, $y'(a) > 0$.

Suppose y crosses the x-axis at a point x_1, $a < x_1$ and let x_1 be the first such point after a. From (5.1) we have

$$(ry')' = - py > 0 \quad \text{on} \quad [a, x_1).$$

Integrating both sides of the inequality from a to x_1 yields

$$r(x_1)y'(x_1) - r_1(a)y'(a) > 0$$

or

$$r(x_1)y'(x_1) > r_1(a)y'(a) > 0$$

since $y'(a) > 0$.

Therefore, the slope of the curve y is positive at $x = x_1$ but in order for a curve to cross from above x-axis to below x-axis its slope must be negative. We have arrived at a contradiction. Thus, $y(x)$ never crosses the x-axis in $[a, +\infty)$.

If we use this theorem in conjunction with Sturm's separation theorem, Theorem 2, Section 3, we arrive at the following corollary.

Corollary 1: If $p(x) < 0$ for $a \leq x$ a solution of (5.1) can have at most one zero on $[a, +\infty)$.

EXERCISE 5

In problems 1-7 show that the solutions are oscillatory on $(a,+\infty)$

1. $y'' + 81y = 0$
2. $y'' + y' + y = 0$
3. $x^2y'' + 3xy' + 3y = 0$
4. $y'' + xy = 0$
5. $y'' + \sqrt{x}y = 0$
6. $x^2y'' + xy' + x^2y = 0$
7. $x^2y'' + xy' + \left[x^2-9\right]y = 0$
8. Solve the differential equation $y''-4y = 0$.
   (a) Find the initial conditions for a solution which has no zeros in $[10, +\infty)$.
   (b) Find the initial conditions for a solution which has one zero in $[10, +\infty)$.
9. Find a value for $x = a$ such that all solutions of $y'' + (4-x)y = 0$ are nonoscillatory on $[a, +\infty)$.

SUBJECT INDEX